D1564361

NEURONAL MECHANISMS OF THE ORIENTING REFLEX

NEURONAL MECHANISMS OF THE ORIENTING REFLEX

Edited by

E. N. SOKOLOV
O. S. VINOGRADOVA

WITH THE ASSISTANCE OF
A. Pakula and A. L. Yarmizina

TRANSLATION UNDER THE SUPERVISION OF
E. N. SOKOLOV

EDITOR FOR THE ENGLISH EDITION
NORMAN M. WEINBERGER
Department of Psychobiology
School of Biological Sciences
University of California, Irvine

LAWRENCE ERLBAUM ASSOCIATES, PUBLISHERS
1975 Hillsdale, New Jersey

DISTRIBUTED BY THE HALSTED PRESS DIVISION OF
JOHN WILEY & SONS
New York Toronto London Sydney

Lawrence Erlbaum Associates, Inc., Publishers
62 Maria Drive
Hillsdale, New Jersey 07642

Distributed solely by Halsted Press Division
John Wiley & Sons, Inc., New York

Library of Congress Cataloging in Publication Data

Main entry under title:
Neuronal mechanisms of the orienting reflex.

Translation of the Soviet papers delivered at an international symposium held in Moscow. The complete papers were originally published under title: Neĭronnye mekhanizmy orientirovochnogo refleksa.
Includes bibliographical references and indexes.
1. Orienting reflex-- Congresses. 2. Neurons-- Congresses. I. Sokolov, Evgeniĭ Nikolaevich, 1920- II. Vinogradova, O. S.
QP372.N38133 1975 599'.01'88 75-23135
ISBN 0-470-92562-0

Printed in the United States of America

Contents

Part B: Neuronal Behavior in Selected Brain Loci

Part C: Overview and Hypothetical Schema

SECTION III: EVOKED POTENTIAL AND ELECTROENCEPHALOGRAPHIC STUDIES

Preface to the Translated Edition

There is no need here to trace the very long and productive tradition of neuroscience research in the Soviet Union. Unfortunately, most of the Soviet findings are inaccessible to Western scientists because of the language barrier and a paucity of translated material. The resultant information void is particularly critical with regard to the orienting reflex, for we owe both its discovery and most detailed subsequent study to investigators working in the Soviet Union.

Perhaps the names most closely linked to the orienting reflex are I. P. Pavlov, its discoverer, and E. N. Sokolov, a major protagonist in the analysis of this reflex. In 1960 Professor Sokolov himself reintroduced the orienting reflex to the general community of Western scientists in the third and final interdisciplinary conference, "The Central Nervous System and Behavior," which was sponsored jointly by the Josiah Macy, Jr. Foundation and the National Science Foundation.[1] Although this publication stimulated research on the orienting reflex in the West, a natural consequence was to heighten the need for further information on Soviet investigations. Major publications since that time have been few, the most widely known being Professor Sokolov's article in the *Annual Review of Physiology* (1963),[2] a translation of his book *Perception and the Conditioned Reflex* (1963),[3] and a translation of *Orienting Reflex and Exploratory Behavior* (1965).[4] Publication of the present volume will alleviate

[1] Sokolov, E. N. Neuronal models of the orienting reflex. In M.A.B. Brazier (Ed.), *The central nervous system and behavior.* New York: Josiah Macy, Jr. Foundation, 1960. pp. 187–276.

[2] Sokolov, E. N. Higher nervous functions. The orienting reflex. *Annual Review of Physiology,* 1963, **26**, 545–580.

[3] Sokolov, E. N. *Preception and the conditioned reflex.* Oxford: Pergamon Press, 1963.

[4] Voronin, L. G., Leontiev, A. N., Luria, A. R., Sokolov, E. N., & Vinogradova, O. S. (Eds.) *Orienting reflex and exploratory behavior.* Washington, D.C.: American Institute of Biological Sciences, 1965.

the problem, but not to any great extent. Clearly, there remains a critical need for information about advances in Soviet works on higher nervous activity.

"Neuronal Mechanisms of the Orienting Reflex" was originally published in the Soviet Union in 1970. It contains a collection of papers that were presented at an international symposium previously held in Moscow. This translation was kindly provided by Professor Sokolov. In order to best serve the need for obtaining information on research in the Soviet Union, this edition was restricted to Soviet papers. The central findings of most of the chapters that could not be included have appeared elsewhere in English.

A second type of alteration from the original Russian volume consists of organizing the chapters into three major sections, according to the type of electrophysiological data obtained. This was felt to be preferable to publishing the chapters in the original order, to provide a frame of reference for each contribution. Brief commentaries have been prepared and are located at the beginning of each section.

The dominant concern in preparing this edition has been to enhance communication of ideas and data from the original authors to the reader, and to accomplish this without permitting personal viewpoints and biases to surface. In this regard, there has been no systematic attempt to Anglicize the translation. Thus, phrases such as "under conditions of the action of the stimulus" have not been altered to "during stimulus presentation"; the Russian emphasizes the active properties of the stimulus, while the English phrase places the stimulus in a more passive role. Although the two phrases may be operationally equivalent, they do have a different flavor. In other cases, the Russian phrase may have no direct English equivalent, and so "clarification" could not have been achieved by a change to more common phraseology. Thus, the term "functional state" would seem to encompass more than "arousal level," or "degree of excitability."

It is with great pleasure that I acknowledge the invaluable assistance to Dr. Michael Cole in preparing this volume for publication.

NORMAN M. WEINBERGER

NEURONAL MECHANISMS OF THE ORIENTING REFLEX

SECTION I

INTRACELLULAR MECHANISMS

Part I contains three papers by Sokolov and his colleages, all concerned with the behavior of single giant neurons in the mollusk *Limnaea stagnalis.* At first glance, it would appear that only the last paper of this trilogy (Sokolov and Yarmizina) is concerned with the orienting reflex (OR), as it involves repeated electrical stimulation of a neuron. However the first two papers represent important links toward an understanding of the effects of repeated stimulation. The first paper by Sokolov and Pakula investigates simply the general effects of electrical stimulation of the cells and reveals, among other phenomena, modulation of pacemaking. In the second paper by Pakula, Arakelov, and Sokolov, the authors are concerned with the effects of impalement with a microelectrode upon cellular behavior. This is a topic that has been insufficiently investigated and brings to mind the Heisenberg Principle, derived from physics, referring to the unavoidable effect of the measuring device upon the subject's behavior. Certainly, most investigators recognize that impalement alters a neuron's discharges; but few if any have considered impalement to be an external stimulus to which the neuron must "adapt" (the term being used in a behavioral sense, not to receptor adaptation). This paper is of particular interest because it concerns itself with data that constitute "error variance" for many investigators, i.e., are ignored. The third chapter, by Sokolov and Yarmizina, involves a more traditional approach to the OR, except that the repeated stimulation is applied directly rather than synaptically. The response decrement observed fulfills "dishabituation" criterion of extinction of the OR, suggesting that the process under investigation in these giant neurons may serve as a model for the OR which is, of course, a phenomenon characteristic of the integrated behavior of an entire organism. The authors conclude that the habituation effect is due to the weakening of activation effects upon the pacemaker mechanism.

1

RESPONSES OF PACEMAKER NEURONS IN THE VISCERAL GANGLION OF *Limnaea stagnalis* TO INTRACELLULAR ELECTRIC SHOCKS

E. N. Sokolov
A. Pakula

Moscow State University

In classifying the responses of the neuronal membrane, Bullock and Horridge (1965) singled out a specific group of endogenous pacemaker potentials, which to a considerable degree determine spontaneous neuronal rhythmic activity. The functional role of the pacemaker neurons is the maintenance of this kind of activity. Bennet *et al.* (1967a, b, c, d) have found that in the medulla oblongata of fishes, the pacemaker neurons possessing an electrical synapse determine the genetically fixed frequency of the discharges of the electrical organ. Each spike of a pacemaker neuron is accompanied by a corresponding discharge of the electrical organ. The pacemaker neurons also participate in the organization of the rhythmic work of the muscles which set the wings of the locust in motion during flight, and ensure the rhythm of sound generation in the cicada (Svidersky & Karlov, 1967). In these cases the afferent stimulation triggers the genetically fixed rhythm of the pacemaker potential. The possibility of such triggering was confirmed by the work of Tasaki (1959).

Apparently, pacemaker neurons are encountered in those organisms for which the rhythmicity of certain processes is of vital importance. The assumption of Bullock and Horridge concerning the endogenous origin of pacemaker potentials might mean that information about the frequency of pacemaker potentials is transmitted by a genetic mechanism.

Alving (1968) demonstrated that isolation–by means of ligature–of the soma of a pacemaker cell in *Aplysia* from other cells and from the region where synaptic transmission takes place, does not prevent the emergence and persistence of a pacemaker potential. At the same time, all the synaptic signals on their way to the soma are blocked by ligature.

These facts support the hypothesis that pacemaker potentials originate in the soma and make more comprehensible the role of the soma in the electrophysiological activity of the giant mollusk neurons.

A more detailed analysis (Grundfest, 1965) showed that in fishes there exist both areactive neurons (pacemakers whose rhythm does not change under the action of afferent stimuli), and another type of pacemaker neuron which is affected by afferent stimulation.

Sokolov and co-workers (Sokolov, Arakelov, & Pakula, 1969) have described a representative neuron in the visceral ganglion of the snail *Limnaea stagnalis*, with firing rate determined by pacemaker potentials. Their frequency of discharge proved to be high owing to the insertion of the microelectrode into the neuron. With the development of adaptation to the inserted microelectrode, the pacemaker potentials gradually diminished and, as a result of this, the firing rate also declined. This decline took place at constant levels of membrane potential and of firing threshold.

The emergence of intensification of pacemaker potentials due to the insertion of the microelectrode into the neuron allowed us to infer indirectly that a pacemaker neuron might be triggered by intracellular electrical shocks. The goal of the present work was to discover some relationships between the polarity and amplitude of intracellular electrical shock, and the generation of pacemaker potentials.

METHOD

Whole-animal preparations of *Limnaea stagnalis* were used. During preparation the ring of ganglia was impaled on a plastic pivot fixed in a paraffin chamber. A micromanipulator (MM-01) was used to impale a neuron. The electrical activity was recorded with glass micropipettes, of 1 micron (μ) O.D. at the tip, filled with 2.75 *M* $FeCl_3$. They also served for intracellular stimulation through a bridge circuit. Stimuli were electrical shocks of 100 msec duration, differing in their sign and amplitude.

A set of Nihon Kohden (Japan) amplifying and recording devices was used for recording. Additionally, the slow oscillations of membrane potential were recorded with an EPP-09 M2 penwriter. The frequency of spikes, level of transmembrane potential, firing threshold, and the amplitude of the action potentials (AP) were all measured. The record was divided in periods of 10 sec, and the number of spikes for each period was counted. For the first spikes of periods, we measured the resting potential (as a point of inflection during the transition of an afterhyperpolarization to a pacemaker potential), the firing threshold (as the level of the resting potential at which generation of the spike begins), as well as the amplitude of the spike. The duration of the interspike interval preceding the spike was also measured. The stimulus artifact served as the reference point for the beginning of the period, if a stimulation was used.

RESULTS

Dynamics of the Modifications of the Firing Rate During the Experiment

A calculation of the number of spikes generated over every 10 sec enabled us to better evaluate activity of the neuron after the insertion of the microelectrode.

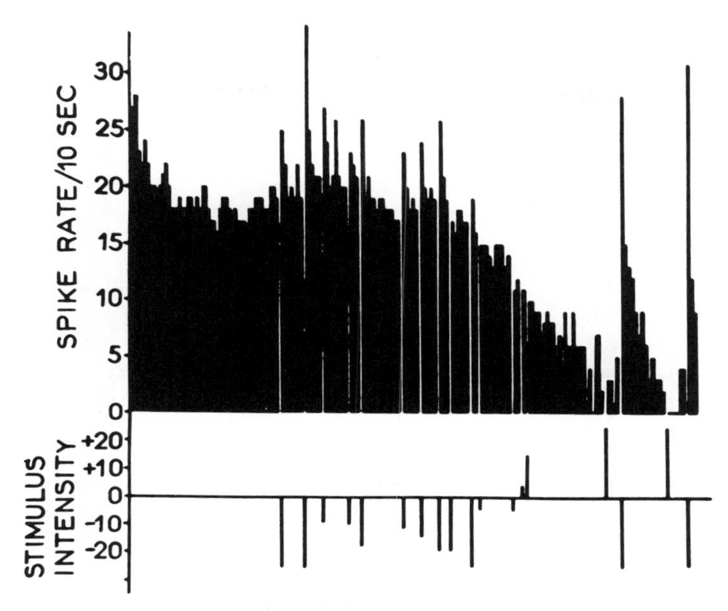

FIG. 1 Dynamics of spike activity of a pacemaker neuron. Abscissa–period (10-sec) number; ordinate–number of spikes in the period (upper graph) and amplitude of the electrical stimuli (lower graph), in 10^{-9} amp units.

Adaptation to the microelectrode is seen initially. An increase in activity was produced by the negative stimuli; suppression of spike activity was caused by positive electrical shocks.

The experiment was divided into four stages: (*a*) modifications evoked by impalement; (*b*) responses elicited by negative electrical shocks; (*c*) responses elicited by positive electrical shocks; (*d*) responses to the alternative application of electrical shocks of opposite signs.

An analysis of the record of the background activity (400 sec) showed that the frequency of discharges, which was 2.7 spikes per sec after the insertion of the microelectrode, declined at the end of this period to 1.7–2.0 spikes per sec. This decrease of frequency occurred following a waxing and waning course (Fig. 1).

The application of single negative electrical shocks of differing intensities restored the initial spike frequency for some 20–30 sec. The background activity remained within the limits of 1.8–2.0 spikes per sec. Threshold stimuli (-3.5×10^{-9} amp) did not increase the frequency of discharges (Fig. 1). In these cases a general tendency consisting of a decrease of the firing rate began to manifest itself.

Alternative applications of electrical shocks of similar intensities but of differing polarities showed that while a negative shock (-25×10^{-9} a) evokes a 30-sec increase of the spike frequency, a positive shock completely suppresses spike activity for a similar period of time.

TABLE I
Dependence of the Response to Electrical Stimulation on the Frequency of Spikes in the Background

Intensity of the stimulus (100 msec), 10^{-9} amp	Average number of AP's for 10 sec in response to stimulation	Difference "Response-background"
− 3.5	18	+ 1.5
− 9.5	25	+ 4.5
−14.5	24	+ 5.0
−17.5	26	+ 5.0
−19.5	27	+ 8.5
−25.0	30	+17.0
+ 3.5	13	− 1.0
+14.5	10	− 1.0
+25.0	1.5	− 0.5

The modifications of discharge frequency taking place during the experiment developed as follows: immediately after the impalement the frequency increased, followed a little later by a gradual decline. Responses to the electrical stimuli became "superimposed" on this gradually developing reduction of spike activity. Whereas the negative electrical shocks augmented the spike activity, the positive ones either did not change it at all, or reduced it.

The general dependence of the rate of action potentials (AP) on the intensity of stimulus is presented in Table 1. (The response was evaluated by the number of spikes during the first 10 sec after the stimulus presentation.) It can be seen that with the intensification of the negative electrical shock, the response increased from 1.8 to 3.0 spikes per sec on the average. However, when evaluating the neuronal response, it is necessary to take into account the preceding background rate. The difference between the number of spikes per 10 sec generated by the neuron before and after the stimulus presentation showed a direct relationship between negative stimulus intensity and the amount of increase in discharge rate. Responses to positive electrical shocks showed a feeble relationship with stimulus intensity, but always were characterized by a small decline of the spike frequency in comparison with that of the background.

It is reasonable to ask how the neuronal response is associated with the background activity. A comparison of the action of a negative electrical shock (-25×10^{-9} amp) at different stages of the experiment led to the conclusion that whatever the background firing rate, the response reached a frequency of 2.5–3.0 spikes per sec. As to the difference between the response and the background frequency, it gradually grew with the decline of the background firing rate in the course of the experiment.

Initially the positive electrical shocks did not evoke any phasic inhibitory reaction. However, when the background firing rate decreased, the same stimulus evoked a lasting (up to 30 sec) suppression of the spike activity.

Correlation between the Interspike Intervals, Resting Potential, Firing Threshold, and AP Amplitude during the Adaptation of the Neuron to the Inserted Microelectrode

To reveal the factors responsible for the decline of the firing rate after the insertion of the microelectrode into the neuron, the values of the resting potential (RP), critical firing threshold, action potential (AP) amplitude, and duration of the interspike interval preceding the spike were determined every 10 sec. It was characteristic of this experiment that the threshold "followed" the RP during both the hyperpolarization and the depolarization of the cell. Thus, the interspike intervals were largely independent of the absolute level of the RP, and the critical firing threshold value (Fig. 2A). At the same time, the amplitude of spikes changed over a range of 55–45 mV.

An analysis of the records showed that spike generation was related to the development of a smoothly growing pacemaker potential which, as a rule, followed immediately after the wave of the afterhyperpolarization. Thus, the decline in the frequency of spikes during adaptation to the inserted microelectrode was conditioned by a decline in the frequency of the pacemaker potentials.

Mechanisms of the Effects of Negative Electrical Stimulation

An increase of spike frequency depended a great deal on both the intensity of the stimulus and the level of the spike activity before stimulus presentation. Measurement of the RP level, critical firing threshold, and interspike intervals showed that negative electrical shocks at first suppressed spike generation for a short time and thus one interspike interval increased. On the other hand, negative shocks evoked a lasting after-activation of the neuron, which resulted in a decrease of the interspike intervals (Fig. 2B). This decrease, most considerable during the first period following the inhibition of spike activity, was accompanied by an increase of the critical firing threshold. The RP remained at the previous level, so that an increase of a difference between the RP and the threshold resulted. Therefore, the shift of RP necessary to initiate an AP was increased.

Afterward the RP drifted towards the firing threshold and so facilitated spike generation, but the interspike interval increased at the same time. It is noteworthy that the maximum decrease of interspike intervals (increase of frequency) was in the period in which the difference between the RP and the threshold was greatest, that is, just when the greatest *decline* of frequency could be expected. An analysis of the records showed that this paradoxical effect was due to an intensification and acceleration of the pacemaker potentials, which

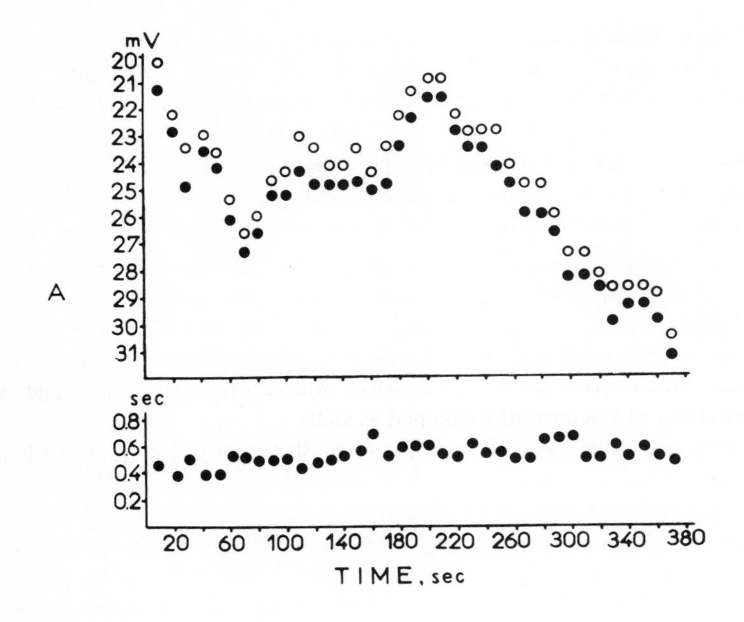

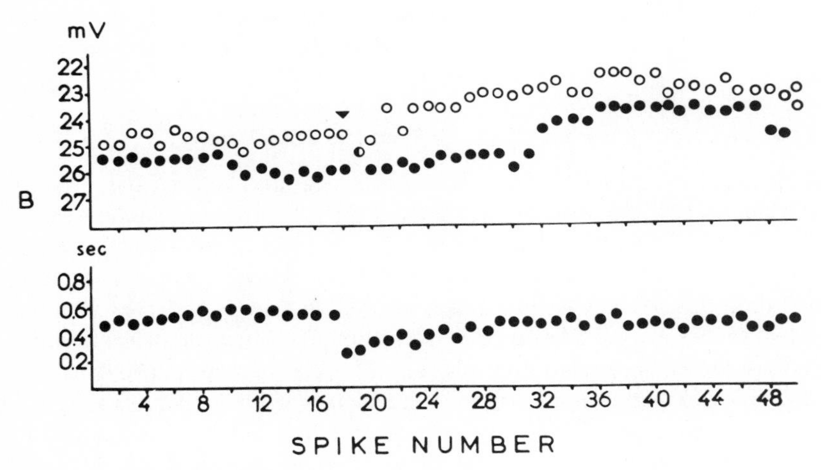

FIG. 2 (A) Dynamics of interspike intervals, membrane potential and firing threshold. Measurements were performed every 10 sec. Abscissa–time in sec; ordinate–membrane potential (solid circles) and threshold (open circles), in the upper graph; interspike intervals (solid circles), in the lower one. Joint modification of the firing threshold and the membrane potential. Constancy of interspike intervals in the course of the experiment.

(B) Dynamics of interspike intervals, membrane potential and firing threshold after a single stimulation by a negative electrical impulse. Abscissa–spike number; ordinate–membrane potential (solid circles) and firing threshold (open circles), in the upper graph, and interspike intervals (solid circles), in the lower graph. Abrupt changes of the interspike intervals in different directions are accompanied by variations of the membrane potential within the range ±1 mV.

proved to be sufficient for the increase in spike generation rate in spite of the increased threshold.

The frequency of both the pacemaker potentials and the spikes gradually returned to the initial level after shock. At the same time the RP and the amplitude of spikes did not change (Fig. 2B). The effect of negative electrical shocks depended on their parameters; the stronger the electrical shock, the longer was the first interspike interval and the more pronounced was the subsequent decrease of interspike intervals. The hyperpolarizing impulse and the afterhyperpolarization of the AP were similar in a certain sense: in both cases the hyperpolarization was followed by the development of the pacemaker potential.

When a weaker stimulus was applied, the effect of the electrical current diminished and the corresponding interspike intervals increased little. The after-activation effect of the current weakened as well.

The action of negative electrical shocks of similar amplitudes depended a great deal on the initial level of activity. When the background frequency of pacemaker potentials was high, a negative shock modified the frequency of the discharges negligibly (Fig. 3A). When the frequency of the pacemaker potential was low, an identical impulse increased the frequency considerably. However, the reaction rapidly dissipated, since a part of the pacemaker potentials, whose amplitude started to decline, failed to reach the firing threshold (Fig. 3B).

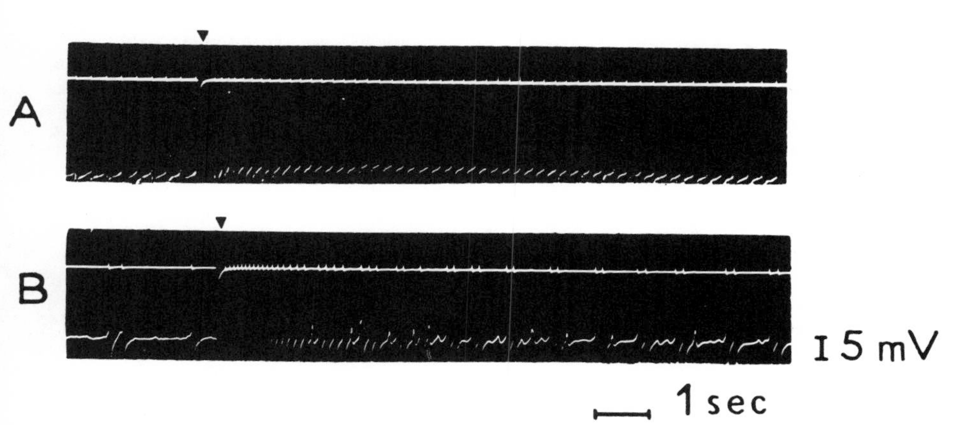

FIG. 3 Responses of a pacemaker neuron to a negative electrical impulse (-25×10^{-9} amp) at the beginning (A) and at the end (B) of the experiment. Upper trace–low gain, lower trace–high gain. At high frequency of the spike discharges in the background activity, after a short delay, the stimulus evokes a long-lasting but relatively small increase in the firing rate, without variations of the resting potential and threshold (A). In the case of low firing frequency, the identical stimulus evokes a greater increase in the frequency which, however, reaches the same value as at a high-frequency background activity (B).

It is the pacemaker potential which is responsible for the increase of frequency. Grouped discharges supercede repetitive activity because certain pacemaker potentials are of a low amplitude and do not reach the firing threshold. The intervals between the spikes are divisible by the pace of the pacemaker.

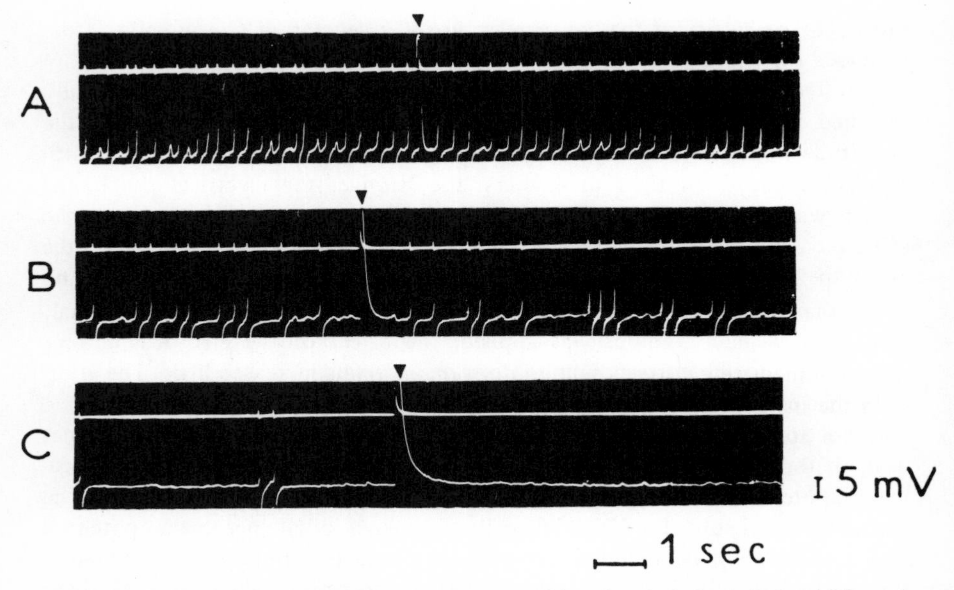

FIG. 4 Response of a pacemaker neuron to a positive electrical shock (3.5×10^{-9} amp) at the beginning (A), and to the same impulse (25×10^{-9} amp) at both, the middle (B) and the end (C) of the experiment. A weak stimulus (3.5×10^{-9} amp) at the high background frequency does not evoke any response. Not is it evoked in the middle of the experiment although a stronger stimulus (25×10^{-9} amp) is applied, since the pacemaker potential is not yet distinct. Only after a reduction of activity to the sparse spikes does the same stimulus (25×10^{-9} amp) produce a long-lasting (30-sec) inactivation of the pacemaker potential, without any modification of the membrane potential.

Mechanism of the Effects of Positive Electrical Stimulation

At the beginning, no direct nor delayed responses to positive electrical shocks were observed if compared with the background spike activity; however, positive shocks contributed to a gradual decline of the frequency of the latter. Later, when the spike activity diminished, positive stimulation started to suppress the firing completely for some 20–30 sec. Then the spike activity recovered. All these alterations of the neuron's activity occurred without variations of firing threshold or membrane potential.

The instances of a long-lasting inhibitory action of the positive electrical shocks at the various levels of pacemaker activity are presented in Fig. 4A, B, and C. Whereas the positive electrical shocks did not modify the neuronal discharges during a high background frequency of pacemaker potentials (Fig. 4A, B), it suppressed the latter completely in conditions of decreased pacemaker activity (Fig. 4C).

DISCUSSION

According to Bullock and Horridge (1965), on the membrane of some neurons there are, along with the subsynaptic patches, special loci which generate pacemaker potentials. When making models of a neuron of this type, the membrane is believed to be a system of loci, generating postsynaptic and pacemaker potentials, and connected with the locus of spike generation in the axon (Lewis, 1964).

Usually, in the case of intracellular stimulation by a negative electrical shock, a neuron stops spike activity for a period of stimulation, as a result of developing hyperpolarization. On the contrary, positive impulses produce spike generation which immediately ceases with the offset of a stimulus.

In the instance of the pacemaker neuron described above, a negative electrical impulse at first acted in the usual manner, that is, inhibiting the neuron; this manifested itself in a delay of the generation of the next spike. However, with a termination of an evoked short-term hyperpolarization, an activation of the pacemaker potentials started, with a simultaneous increase in the firing threshold but without changes, or even with a small hyperpolarizing shift, of the membrane potential.

A similar activation of spike discharges was evoked by an application of repetitive negative electrical shocks. In this case the spike generation started at the end of the hyperpolarizing impulse which initiated the pacemaker potential. The spike activity persisted even after the stimulation had been stopped (Fig. 5A). When the same stimulus was repeated at intervals of 5 sec, the pattern of the response remained unchanged as well. No habituation phenomena could be observed (Fig. 5B).

It is of interest to compare our results with those obtained by Luco (1964) on cockroaches. Using a tetanic stimulation (100 per sec) of the abdominal chord of the cockroach, he recorded, from the efferent fiber of the metathoracic ganglion, a specific neuronal activity which he calls a "natural response." This "natural response" consists of the following: if the tetanic stimulation is applied throughout a definite period which exceeds the "utilization time," the neuron responds, according to the "all-or-nothing" law, by a long-lasting burst of a very stable pattern. On the other hand, the "utilization" period is a function of the frequency of stimuli; the higher the frequency, the shorter is the "utilization" period. Luco applied stimuli of equal intensity and duration, modifying only their frequency in the tetanus series, which is apparently equivalent to a prolongation of the shock duration. In this respect an analogy with our data can be established easily. In our experiments, in conditions of an invariable duration of the impulses, (100 msec), we changed their intensity, and depending on the magnitude of the latter, the "natural response" was or was not evoked, in accordance with the Horweg-Weiss law. The main difference was that in our case the response resulted from an intracellular stimulation as a "rebound" after the

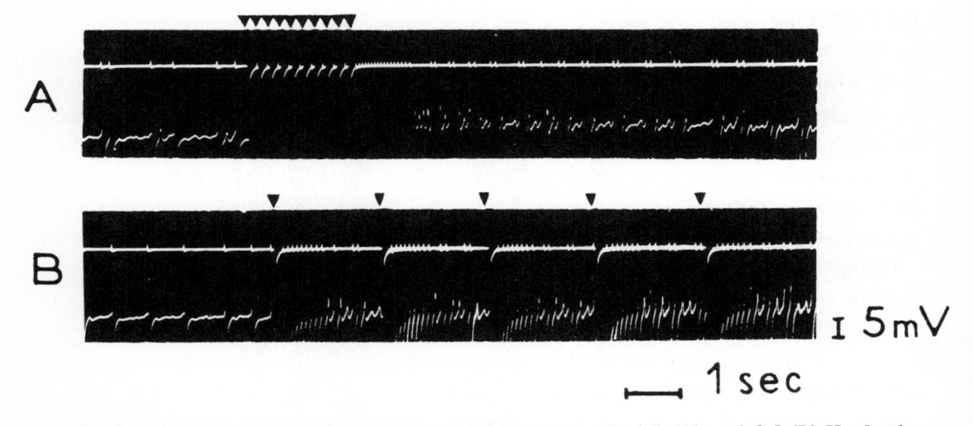

FIG. 5 Response to repetitive negative electrical stimuli of 5 (A) and 0.5 (B) Hz. In the course of the rhythmical stimulation spikes arise after the termination of each impulse. After the series of stimuli have been stopped, a long-lasting activation of the pacemaker potentials is observed. The amplitude of the pacemaker potentials gradually declines and only a part of them reaches the firing threshold. A transition to a rhythmical activity takes place. Paired spikes are due to the fact that after the generation of a spike the amplitude of the subsequent pacemaker potential proves to be somewhat higher and leads to the generation of another one. The interspike intervals are divisible by the pace of the pacemaker. The application of a stimulus at a lower frequency shows that in the case of intracellular stimulation the pattern of the spike train remains invariable and is determined by the dynamics of the decline in the amplitude of the pacemaker potentials.

hyperpolarization. Luco assumed that the pattern of the "natural response" is inherent for certain neurons. Our results, obtained by way of intracellular stimulation and recording, from the neuron under discussion as well as from other ones, showed that the "natural response" was due to modifications in the generation of pacemaker potentials, these potentials being apparently of an endogenous origin.

Positive shocks evoking no generation contributed to the inactivation of the spontaneous pacemaker potentials and to the decline of spike frequency. On the other hand, this inactivation of the spontaneous pacemaker potentials apparently reflects a "habituation" of the neuron to the microelectrode. The evidence that this is not a result of an accumulation of remote effects, produced by the injury of the cell, is provided by the following phenomenon: to stimuli of a negative sign, the neuron responded with a pronounced reaction, the threshold of this reaction remaining invariant over time.

Carpenter (1967) arrived at the conclusion that the generation of pacemaker potentials may be triggered by certain external factors. He likewise observed a protracted recovery of the initial functional state of the cell after the insertion of the microelectrode. The demonstration of the relative independence of the spike discharges of the level of the membrane potential was the most important result of this work. This independence is explained by existence of the autonomous mechanism for pacemaker potentials' activation. The following instances bore witness to this. As a result of an increase in the environmental temperature, the membrane potential of the pacemaker neuron increased. However, at the same time, the frequency of the spikes increased, due to the temperature-dependent activation of the pacemaker potential. A decline of the temperature led to the depolarization of the neuron and to a simultaneous reduction of spike activity, as a result of inactivation of the pacemaker potential.

The possibility of triggering pacemaking by external stimuli is a very important feature of neuronal activity. As shown above, pacemaking can be elicited by single or repetitive intracellular electric shocks. We have found it to be possible to start a delayed pacemaking also by prolonged injection of subthreshold DC current (Sokolov, Pakula, & Arakelov, 1970).

It is common practice to classify neurons into "pacemakers" and "nonpacemakers." The observation of these two sorts of activity in one neuron renders such a classification doubtful. If pacemaking is understood as the mere generation of pacemaker potentials, we believe most, if not all, giant molluscan neurons to be endowed with this faculty which may or may not be evidenced during an experiment. The latter fact apparently depends on the concurrence of the natural and experimental conditions. We have called the neuron described above a "pacemaker" because it displayed spike activity driven by pacemaker potentials for a certain period after impalement. One may argue this since mechanical injury in the giant neurons of *Limnaea stagnalis* often results in pacemaker activity. However, the problem of the criteria for classification of neurons must be dealt with separately. For the moment we may state that a pacemaker, latent or active, is a very important physiological mechanism which greatly widens the integrative possibilities of a neuron.

2

SOME FEATURES OF ADAPTATION TO MECHANICAL INJURY IN THE GIANT "A" NEURON OF *Limnaea stagnalis*

A. Pakula
G. G. Arakelov
E. N. Sokolov

Moscow State University

As a rule, neurons having different properties are investigated when studying electrophysiological phenomena at the level of the single cell. The data obtained from such experiments characterize a population rather than a single neuron. This variation of data is determined by both the fluctuations of electrophysiological parameters from experiment to experiment in neurons of the same type, and by the sampling of neurons of different types.

The variability of the parameters of neuronal responses, depending on the type of the neuron, can be avoided by recording electrical reactions from a neuron identified from experiment to experiment. A limited number of neurons and their morphological specificity in the nervous center may contribute to such an identification. From this point of view the giant neurons of Mollusca are very convenient for investigation. The five giant neurons in the left parietal ganglion of *Limnaea stagnalis,* and neuron "A" in particular, have previously been described by us (Yarmizina, Sokolov, & Arakelov, 1969–1970), so it was not accidental that this neuron became the object of our attention.

The interest shown by some researchers for molluscan neurons is due to the fact that investigations of mechanisms of various electrophysiological phenomena of higher organisms can be carried out better under conditions of a less complex nervous system that possesses rudiments of the higher nervous activity under investigation. Thus, experiments carried out on mollusks and arthropods demonstrated at the neuronal level such phenomena as selective "habituation," temporal conditioning, and disinhibition, which play an important role in higher nervous activity.

GENERAL DATA ON THE "A" NEURON

The nervous system of the *Limnaea stagnalis* occupies an intermediate position among the mollusks in terms of organization. Here we see a distinct division into a central nervous system—the clusters of neurons called ganglia—and a peripheral

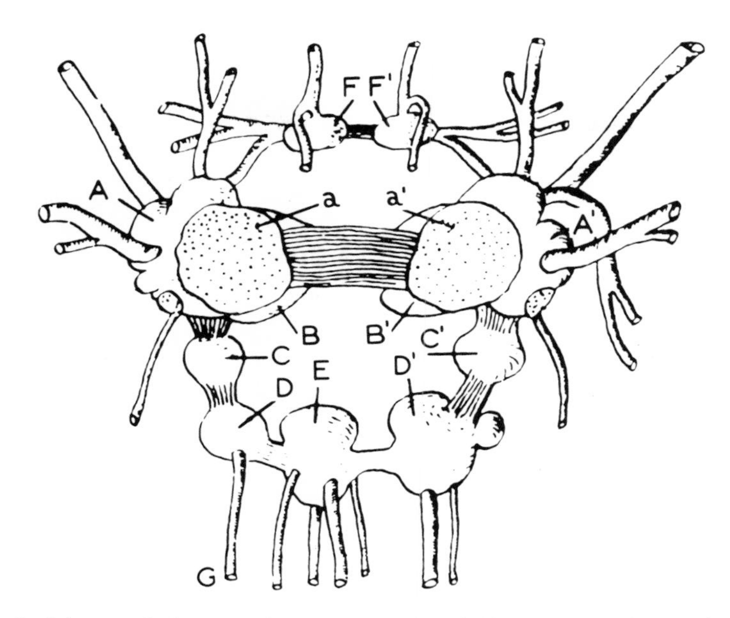

FIG. 1 Scheme of the central nervous system of *Limnaea stagnalis.* (AA′) cerebral ganglia; (BB′) pedal ganglia; (CC′) pleural ganglia; (DD′) parietal ganglia; (E) visceral ganglion; (FF′) buccal ganglia; (G) nervus pallialis sin.

one, which represents a diffuse net of neurons. However, in *Limnaea stagnalis* its 11 ganglia are not merged into a brainlike formation, but form a chain (Fig. 1), contrary to what happens in *Helix pomatia,* in which the nervous system is more developed.

Giant neurons are found in all the ganglia of *Limnaea stagnalis.* The left parietal ganglion is particularly convenient for study since it contains only 4–5 giant neurons (150–300 μ in diameter), easily identified by their size, pigmentation, and location in the ganglion (Fig. 2A). Neuron A, as a rule, is of a larger size and has a more intense pigmentation (the neurons of many Mollusca contain carotinoid substances).

Like all other molluscan giant neurons, the A neuron, in which the body size ranges from 160 to 290 μ, has no dendritic apparatus. The only process of the neuron, coursing from the soma to the interior of the ganglion, branches in two, one of which goes to the parietovisceral connective and the other to the parietopleural connective. The body of the neuron is surrounded by medium-sized and small neurons, as well as by neuroglia.

Absence of a dendritic apparatus, a specific organization of the input on the axon branches, and an existence of several spike triggering zones located on them and eventually also on the soma, all strongly suggest that the giant neuron of Mollusca performs a role of a centre which in higher organisms consists of several neurons (Saharov, 1965; Tauc, 1966; Tauc & Hughes, 1963).

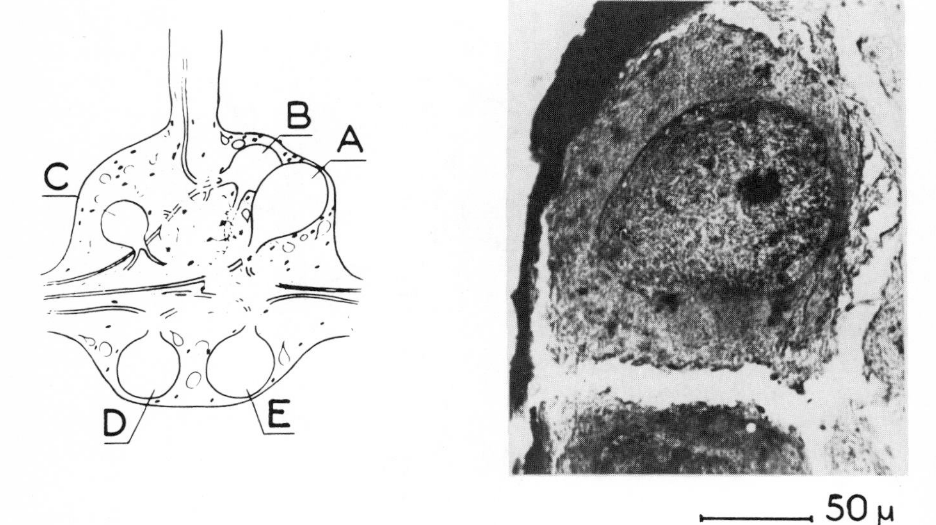

FIG. 2 (Left-hand side) a scheme of the left parietal ganglion. The letters A, B, C, D, and E mark the five giant neurons of the ganglion. (Right-hand side) the A neuron. The "mark" of the microelectrode's tip can be seen in the nucleus as a dark spot.

The present paper deals with the properties of the A neuron observed during its spontaneous activity. However, it should be noted that in such conditions the behaviour of the neuron is also affected by mechanical stimulation exerted by the microelectrode. Thus, the dynamics of modifications taking place in the electrical activity during the experiment reflect mainly the process of development of the neuronal adaptation to the inserted microelectrode.

METHOD

The experiments were carried out on the snails from the pond of the Botanical Gardens of the Moscow State University. Whole-animal preparations were used.

The procedure of preparation was the following. The snail was removed from its shell, and the "foot" was slit; then the mantle was slit along the back, beginning from the oral cavity. The sides of the "back slit" were pinned, the esophagus was cut in the area of the pharynx and its long end was extracted from the ring of ganglia. Such preparation was then put in a special chamber. A plastic pivot erected in the centre of the chamber was fitted to enter the "foot slit," and then through the middle of the ring, to fix the position of ganglia.

As the experiments were usually performed on the A neuron, it was necessary to identify it before the experiment. For this, the whole-animal preparation already being impaled on the pivot, the left nervus pallialis, coursing from the left parietal ganglion to the mantle, was cut off at its peripheral end, picked up and rolled upon the pivot. When the parietal ganglion was located this way, the location of the A neuron proved to be almost invariable in various specimens. As a rule, it was protuberant in the upper right quadrant of the ganglion and was exposed to the observer. The identification was facilitated by some other attributes of the A neuron (see Yarmizina, Sokolov, & Arakelov, 1968). The reliability of such visual identification was confirmed by a histological control carried out after the experiment. In those few instances when the identification of the A neuron before the experiment was impossible because of the irregular location of the giant neurons in the ganglion, we gave up an idea of experiment on such preparation.

During the experiment the whole-animal preparation was immersed in the Ringer's solution for Mollusca (6.5 gm NaCl; 0.14 gm KCl; 0.2 gm $NaCO_3$; 0.12 gm $CaCl_2$; distilled H_2O up to 1 liter).

Insertion of the microelectrode (ME) into the cell was accomplished by means of an MM - 1 micromanipulator under visual control with the aid of a binocular magnifier.

Glass micropipettes filled with a 2.75 *M* solution of $FeCl_3$ served as electrodes. The micropipettes were made from Pyrex glass tubing of 1.2–1.4 mm O.D. using the vertical puller of Marinytchev's system. Micropipettes were selected for filling only if the outer diameter of the tips did not exceed 1 μ. A resistance of such ME in saline ranged from 5 up to 25 MΩ (megohms), the tip potential of the electrode being 10–15 mV.

$FeCl_3$ was chosen as an electrolyte for microelectrodes first of all because of the possibility of marking the position of the tip in the cell. During the histological procedure, the location of the tip in the neuron was detected by a blue spot with a diameter of 4–20 μ. Ions necessary for marking were injected iontophoretically (Sokolov, Arakelov, & Levinson, 1967). The iontophoresis did not affect the precision of recording, since the resistance of the ME's filled with 2.75 *M* $FeCl_3$ exhibited negligible modifications when currents up to 10^{-8} amp were passed through them.

The potential recorded with the microelectrode was connected through an agar–agar bridge (300 mg of agar–agar per 10 ml of 2.5 *M* KCl) to a microelectrode amplifier MZ-3B with a movable cathode follower. The output of the amplifier was led directly to the input amplifier of universal dual-beam VC-7A oscilloscope with a continuous-recording camera (all products of the Nihon Kohden Co. Ltd., Tokyo, Japan). To better study slow variations of the membrane potential (MP), an additional registration with an EPP-09 M2 pen-writer was performed, a pH meter serving as cathode follower.

Electrical current was passed into a cell from a Nihon Kohden electronic MSE-3R stimulator through an isolation unit MSE-JM and Wheatstone bridge circuit (see Frank, 1959). The latter allowed us to avoid the artificial shifts of the potential in recordings when current was injected.

Records were analysed with the aid of a Microphoto magnifying apparatus as described below.

RESULTS

Transmembrane Potential (TMP)

Some investigators (Kopylov, 1966; Nadvodnyuk, 1967; Tauc, 1966) have found the resting potential of neurons in different representatives of Gastropoda (mollusks) to be within a range of 40–60 mV. For *Limnaea stagnalis* the following values were established: 40–55 mV during the summer and 15–35 mV during winter (Kopylov, 1966).

In our experiments only 30% of the A neurons (from 87 investigated) had a transmembrane potential, which at the fifth minute of the experiment exceeded 35 mV. The range of values were 5–60 mV.

The wide range of MP values in the A neurons might be explained mainly by the fact that we took into account all neurons irrespective of their initial resting potential values, while Eccles (1959), describing the mean resting potential values for the motoneurons, pointed out that he had established them without taking into account those neurons in which depolarization effects predominated. Apparently the authors who established the values of resting potential in molluscan neurons also ignored the "inconvenient" cells.

We did not eliminate the neurons with "inconvenient" initial RP values because of the necessity to study the possibilities of the functional recovery of the neuron after the impalement with the microelectrode.

The low values of the RP observed in our experiments might be partially caused also by the following factors:

1. During the procedure of preparation the connective tissue capsule of the ganglion was not removed; thus, the A neuron during the experiment was in the hemolymph, which, however, in various experiments could be more or less diluted by the physiological solution diffusing from the pool as a result of a puncture in the capsule. The data published by other investigators (Kopylov, 1966; Nadvodnuyk, 1967; Tauc and Hughes, 1963) usually refer to experiments carried out on "bared" neurons, i.e., when practically the same artificial medium was used in various experiments.

2. Adrian (see Meshchersky, 1960, p. 50) has pointed out that measurement of the transmembrane potential is impeded greatly by the differing tip potentials of the microelectrode when in the saline and in the neuron, which automatically results in an increase or decrease of the real values. As the connective tissue capsule was not removed, a microelectrode impaled the neuron abruptly and in most of cases penetrated into the nucleus. It is possible that the tip potential in the nucleoplasm might have some specific shift when the microelectrodes are filled with 2.75 M $FeCl_3$. Other factors may also contribute to the wide range of the RP values.

Characteristic dynamics of the potential variations in the A neuron were observed after insertion of the microelectrode. In experiments of different duration (ranging from 20 min to 8 hr), without the application of electrical stimuli, modifications of the TMP as a reaction to impalement developed in all A neurons in the same way. Immediately after the initial sudden shift of the potential, the neuron exhibited a rapid depolarization, after which the TMP returned to the resting potential. Three types of reactions to impalement were delineated, taking modifications of the TMP as a criterion for classification:

(a) A gradual long-lasting increase in TMP (Fig. 3C).
(b) A stabilization of the TMP for 20 to 40 min (Fig. 3B).

The following three instances of such stabilization were distinguished: (1) a rapid stabilization of the TMP (during a period of 1–3 min) which developed at the low level of polarization (5–16 mV); (2) a stabilization which developed later than that in the previous instance and at a higher level of polarization (15–40 mV); (3) a "compromise" stabilization which developed, after a more or less prolonged wave of polarization, at a lower than the maximal level reached during this wave.

3. A modification of the TMP in the shape of a depolarization wave (Fig. 3A). The rate of growth of the depolarization varies: Some cells became depolarized during the first five min, others within 20–40 min, and still others exhibited only a slight tendency to develop depolarization.

In 19 neurons the TMP increased during the experiment (40 min), in 50 neurons it attained a stable level, and in 18 neurons a depolarization developed.

Nadvodnyuk (1967), who studied the giant neurons of *Helix pomatia*, observed the TMP stabilization at once after the insertion of ME or after a depolarization which lasted several seconds. Our data showed that the reactions of the A neurons to the insertion of the microelectrode were of a more complex character. Two basic factors seem to be responsible for the dynamics of the TMP modifications observed: an ionic shunt in the membrane after the impalement, and the action of the ME as a damaging intracellular agent.

One can infer that modifications of the TMP during the experiment should have the following sequence: Due to the pressure exerted by the ME on the membrane, the neuron becomes depolarized to a certain degree before the impalement. This value is recorded at the moment of the impalement, but immediately thereafter numerous positive ions start to penetrate the cell through the shunt, which shifts the TMP still more in the direction of depolarization. The subsequent modifications of the TMP depend on the rate at which the shunt is eliminated by the cell itself, as well as on the effectiveness of the system ensuring the selective permeability of the membrane for different ions. This effectiveness is determined by the degree to which the cytoplasmatic structures of the neuron are injured. In our case, this relates chiefly to the nucleus, since, because of its large size, it was, as a rule, damaged to some degree as a result of impalement. It is a combination of these two factors which apparently determines all the variants of the TMP modifications observed. It is unlikely that A neurons, which form a homogeneous population, have genuine idiosyncratic TMP modifications. Evidently, effects which are related to the ionic shunt

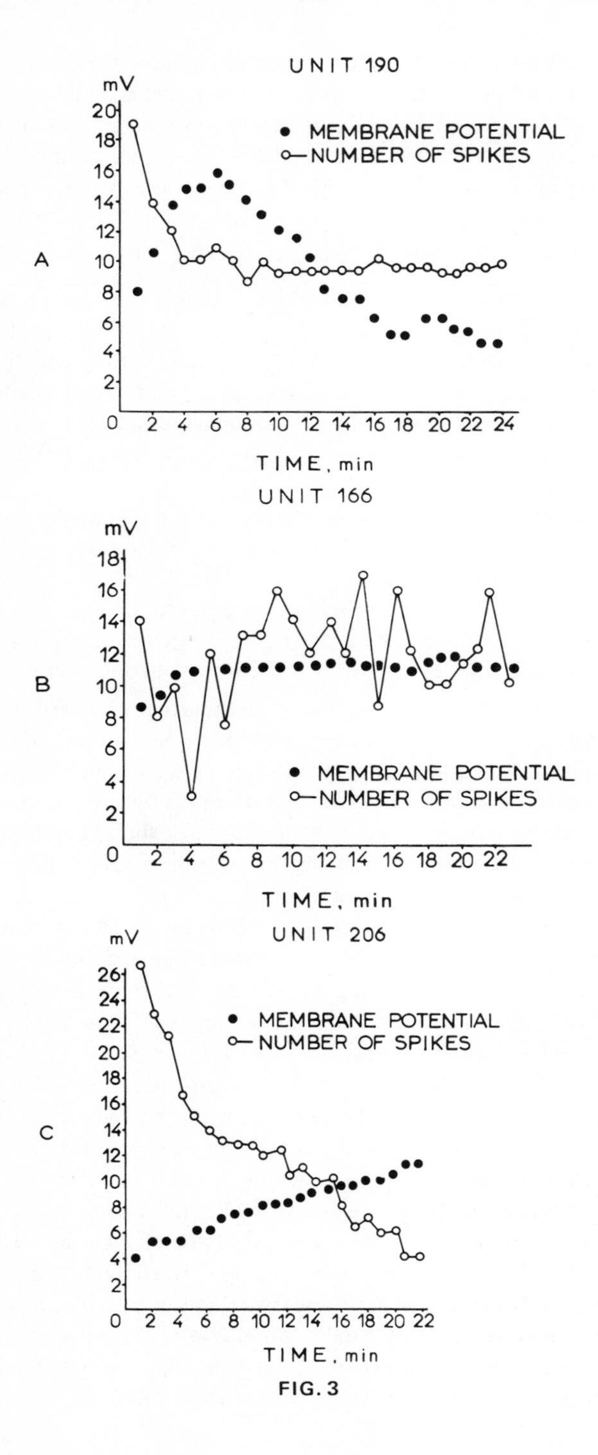
UNIT 190
mV
A
MEMBRANE POTENTIAL
NUMBER OF SPIKES
TIME, min
UNIT 166
mV
B
MEMBRANE POTENTIAL
NUMBER OF SPIKES
TIME, min
UNIT 206
mV
C
MEMBRANE POTENTIAL
NUMBER OF SPIKES
TIME, min

FIG. 3

usually manifest themselves in rapid modifications of the TMP; this means that the elimination of the shunt leads to a more or less rapid repolarization, while the nonelimination produces a rapidly developing depolarization, and even the death of the cell.

The microelectrode injures not only the cellular membrane, but also the intracellular organelles. As is known, the response to different damaging agents in various cells consists of similar unspecific changes. Alterations in the viscosity, in the dye-binding properties of the protoplasm, in the biochemistry of the cell, etc., have been observed (Braun, Levin, & Rosenthal, 1966). These phenomena are believed to be present also due to mechanical injury, after the insertion of the microelectrode into the neuron. As a result of these unspecific changes, the specific function of the neuron—the electrogenesis—also becomes altered.

The phenomenon of the "reparative adaptation," described by Aleksandrov (1966) deserves particular attention. This phenomenon consists of the following: During the long-lasting action of a damaging agent (if not too intensive), the cell adapts to it and recovers its normal functional state. This is explained by the fact that protein structures at the higher levels of organization may recover even without an elimination of the damaging agent, if destruction of the bonds in the protein molecules does not take place. Greater destruction, which destroys the primary or secondary protein structures, is irreversible, and the "reparative adaptation" is impossible in this case.

Thus, the three types of TMP modification during the experiment may be understood as resulting from differing degrees of destruction by insertion of the microelectrode. The gradual development of polarization and the stabilization of the TMP are apparently examples of the development of a "reparative adaptation" to a mechanical damaging agent, whereas the depolarization reflects the results of an irreversible destruction.

Background Spike Activity

In some cases no background spike activity was observed in spite of the fact that after impalement a normal level of the RP was restored. It was quite difficult to draw a definite conclusion concerning the spontaneous spike activity of the A neuron, since experiments were carried out at different times of the day. It is known that many neurons, including some giant neurons of the sea hare *Aplysia*, possess a circadian rhythm of activity, with the quiescent and active periods alternating "spontaneously" (Harker, 1960; Strumwasser, 1962). The possible existence of circadian rhythms may prevent one from obtaining knowledge about the real time course of various electrophysiological reactions,

FIG. 3 Examples of the types of modifications taking place in the TMP level and AP frequency during the experiment. In (C) the TMP increases, in (B) it becomes stable, and in (A) it decreases. At the same time, (C) represents a decline of the firing rate, (B) group activity, and (A) stabilization of the frequency. The open circles are for the number of the APs, the solid ones are for the TMP, in millivolts from the reference line.

because the absence or presence of "spontaneous" spike activity might result from a coincidence of the time of the experiments with the various phases of the circadian rhythm. Unfortunately, from the obtained data nothing could be stated about the laws governing the circadian rhythm, if present, in the A neuron. To unveil them is a task of primary importance.

After the impalement, A neurons, in most cases, exhibited spike activity which manifested the following types of changes: (*a*) the frequency of the AP decreased, and, some time later, the discharges ceased altogether (Fig. 3C); (*b*) the spike frequency declined to a certain level, after which the first became stable (Fig. 3A); (*c*) some time after impalement the cell exhibited group discharges (Fig. 3B). In this case during a prolonged recording (4–8 hr) the groups used to be superseded by single spikes.

An increase in the frequency of spikes during the experiments was observed only as phasic reactions, whereas the general tendency was a decline of spiking.

Considering the rather great variety of modifications of the TMP and firing rate, it was of interest to study the correlation between them. It could be expected that in conditions of hyperpolarization, spike frequency would decrease, during depolarization, would tend to increase, and in the case of a constant RP level, would remain stable or would switch to group discharges.

In order to test these expectations, we plotted for 36 A neurons the average values of the TMP and the number of APs per 10 sec for every minute of the experiments. The generalized qualitative picture of the TMP modifications and of the firing rate for these neurons is presented in Table 1.

TABLE 1
Distribution of A Neurons of Different TMP Types According to the Character of their Spike Activity

Type of the TMP modification / Character of the spike activity	Hyperpolarization	Stabilization	Depolarization	Total
Decrease	3	8	5	16
Stabilization	2	3	2	7
Group activity	5	8	0	13
Total:	10	19	7	36

TABLE 2
Coefficients of Linear Correlation between the TMP Level and Spike Frequency for the Whole Population of the A Neurons and for Separate Types of the TMP Modifications

	Correlation coefficient r	Level of significance of the r P_r	Level of significance of the difference between the correlation coefficients $P(r_x - r_y)$
Overall	+0.11	0.02	
Hyperpolarization	−0.27	0.001	$P(r_h - r_s) = 0.64$
Stabilization	−0.32	0.001	$P(r_s - r_d) = 0.001$
Depolarization	+0.41	0.001	$P(r_h - r_d) = 0.01$

From the table it can be seen that the firing rate changed rather independent of the TMP modifications. For example, during hyperpolarization a decrease of the spike frequency was as probable as stable or group activity, although, according to classical concepts, a decrease of the firing rate would be expected.

To test the idea of relative independence of firing rate and polarization in a quantitative fashion, a correlation analysis was carried out. For 32 A neurons the average values of the TMP and APs were calculated for every minute except the first five (in order to reduce to a minimum the contribution of the transitory processes in neuronal activity due to impalement). For more than 500 pairs of values obtained in this way we calculated the overall coefficient of correlation, as well as the correlation coefficients for separate groups of A neurons, which were divided into groups according to the character of their TMP modifications (Table 2).

On the basis of the relationship between the TMP level and the firing rate in motoneurons (Eccles, 1965b), a negative linear correlation between them would be expected. However, this was the case only during the development of hyperpolarization and stabilization, though in these instances too, the correlation proved to be a small one. During the development of depolarization there was a positive correlation, contrary to popular opinion about the laws governing this relationship. Also the overall correlation coefficient proved to be a positive one.

The results of the correlation analysis lead to the conclusion, which seems a paradox at first glance. It is that the firing frequency in the A neuron is determined by the level and oscillations of the TMP only in a few cases. Next, the correlation coefficients between the TMP and the firing rate were calculated

TABLE 3
Some Coefficients of Correlation between the TMP and the Firing Rate in the A Neurons

Call number	Type of TMP	r	P_r
190	hyperpolarization	−0.95	0.001
194	hyperpolarization	+0.75	0.001
115	stabilization	+0.52	0.06
166	stabilization	−0.41	0.035
182	depolarization	+0.91	0.001
195	depolarization	+0.85	0.01

for single neurons. From Table 3 it can be seen that in some neurons this relationship, being of a quite strong character (neuron No. 190), proved to be negative, while in others it was positive. The number of positive correlations was greater, and this was reflected in the overall correlation coefficient (Table 2).

The positive correlation between the TMP and spike frequency during the development of depolarization was partly due to the fact that for many A

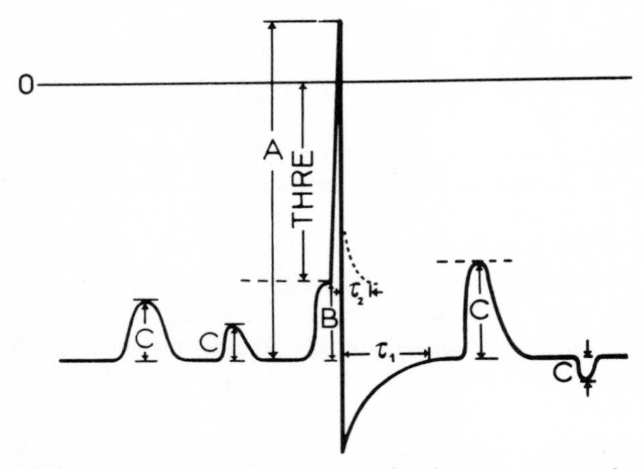

FIG. 4 Scheme illustrating the measurement of various parameters of neuronal activity. (C) zero level of membrane potential; (A) amplitude of the action potential; (B) amplitude of the prepotential (effective input signal); (C) amplitude of the ineffective input signal. Firing threshold "THRE" indicates the level at which the generation of a spike starts. The sum of "Threshold level + B" corresponds to the TMP level at rest, i.e., a resting potential. The value of the spike amplitude is obtained by subtraction of B from A. τ_1 is duration of the afterhyperpolarization; τ_2 is duration of the threshold recovery after spike generation.

neurons, the spontaneous decline of the firing rate during an experiment was inherent, even if no TMP modifications were present. On the other hand, this natural decline resulted in a negative correlation coefficient when it coincided with the development of hyperpolarization.

Theoretically a high negative coefficient of correlation between the TMP and spike frequency is probable if both the firing threshold and the input signals, i.e., the EPSPs and IPSPs, are more or less constant. The absence of a negative correlation in the bulk of the neurons caused us to measure not only the TMP and the firing frequency, but also other electrophysiological parameters of the neuronal responses, which, apparently, often determined the generation of spikes recorded from the soma of the A neuron. These parameters included (see Fig. 4): the input signals to the neuron, the firing threshold, the duration of the after hyperpolarization (τ_1), and the duration required for recovery of threshold (τ_2) after the generation of the regular spike. The last two parameters are of great importance in the case when pacemaking is elicited due to a permanently suprathreshold level of the resting potential. Since in the A neurons such pacemaking conditions had only a random character, we have focused mainly on changes of the threshold and of the input signals. The bases of measurements of various parameters are shown in Fig. 4.

When examining the threshold modifications in the A neurons, one can see that within a definite range of the TMP mean values, the firing threshold "follows" the TMP quite strictly. The possibility of an electronic drift of the recording devices was excluded. In the case of a pronounced hyperpolarization, the threshold begins to lag more and more (see Table 4).

However, in many cases there was no decline of the firing frequency despite the increasing polarization. In the case of development of depolarization, the

TABLE 4
Time Course of the TMP and Firing Threshold in the A Neuron (Cell No. 59) with Hyperpolarization Developing Spontaneously

Minute of the experiment:	1	5	10	15	20	25	30	35	40	45	50
TMP, millivolts	30	26.5	34.75	39.25	41.4	43.25	45	47.5	49.25	50	51.75
Level of the firing threshold, mV	29.8	26.25	34.5	38.75	41.1	42.25	44.5	45	44.75	44.71	45.5
Difference between the levels of the TMP and threshold, mV	0.2	0.25	0.25	0.5	0.3	1.0	0.5	2.5	4.5	5.25	6.25

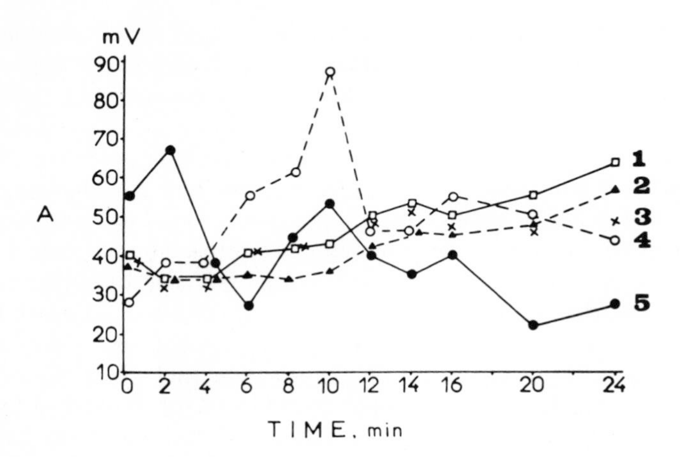
mV
90
80
70
60
50
40
30
20
10
A
0 2 4 6 8 10 12 14 16 20 24
TIME, min
1
2
3
4
5

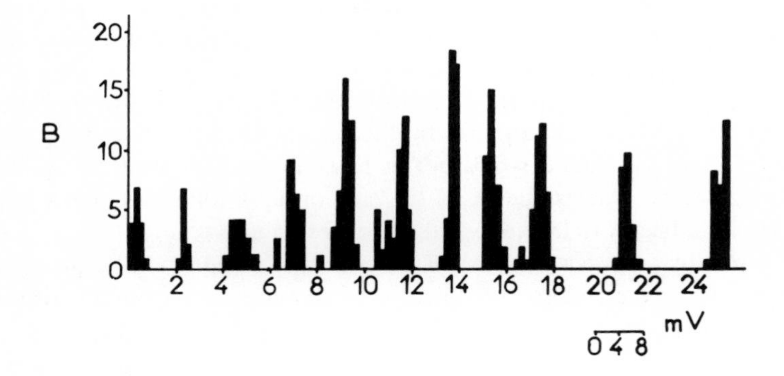
B
20
15
10
5
0
2 4 6 8 10 12 14 16 18 20 22 24
0 4 8 mV

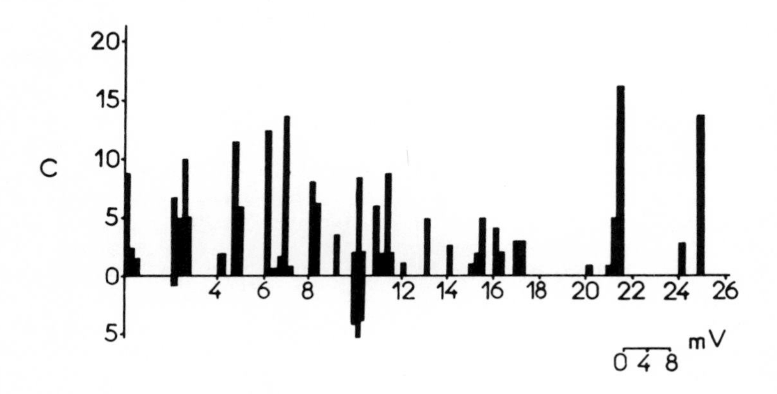
C
20
15
10
5
0
5
4 6 8 12 14 16 18 20 22 24 26
0 4 8 mV

FIG. 5

firing rate either remained invariable or tended to decline. The latter effect may be explained by the accomodation of the firing threshold to the slowly growing depolarization, but the maintenance of a constant frequency in conditions of developing hyperpolarization requires the involvement of some additional mechanisms.

Measurement of the electrophysiological parameters of the A neuron showed that the frequency of the discharge is not merely a linear function of the TMP level; it depends as well on the amplitudinal and temporal distributions of the EPSPs and IPSPs, on the long- and short-term modifications of the threshold, and on the duration of the afterhyperpolarization. A very important role in the maintenance of the spike frequency is that of the pacemaker potentials which are characterized, in contradistinction to the EPSPs, by a slower depolarization. It is of interest that in various neurons, and even at different phases of activity of one neuron, different factors can predominate in influencing and determining spike activity. Such dependence of the firing rate proved to be inherent not only for the A neuron. Some neurons of the visceral ganglion also displayed a complex dependence of spike activity on a host of electrophysiological parameters. The giant neuron of the visceral ganglion is discussed here as being, in a sense, more representative of a multifactor dependence of the spike generation.

Figure 5A depicts the number of APs per 20 sec at every minute of the experiment, the average TMP level, the mean level of the firing threshold, and the mean AP amplitude. Recording the activity at two differing gains allowed us to measure all the necessary electrophysiological parameters, and to increase our precision in evaluating the excitatory and inhibitory messages arriving at the soma. The distributions of the amplitudes of the effective and ineffective input signals are presented in Fig. 5B and C, respectively.

The TMP of this neuron increased over time while its firing rate declined, waxing and waning. It is noteworthy that in this case the dependence of spike frequency on TMP level was small. Phasic changes taking place in the discharge rate testified to it. The firing frequency was determined, however, not only by the character of the input signals, but was inferred from the fact that maximum AP frequency did not coincide with the maximum number of input signals. Although the amplitude of the EPSPs increased along with an increase in their number, the maximum discharge frequency was found earlier in the experiment,

---

FIG. 5 Kinetics of modifications of the electrophysiological parameters in the giant neuron from the visceral ganglion of *Limnaea stagnalis* over the experiment. (A) The mean values of the TMP, threshold, amplitude, and frequency of the APs, as well as of the total number of input signals for the 20-sec intervals. (1) TMP; (2) firing threshold; (3) amplitude of the APs; (4) number of the input signals (all); (5) number of APs.

(B) Time course of the amplitudinal distribution of effective signals. The reference points for the distributions are the time marks. The scale of the distribution is given in the Figure. The width of each column is 0.8 mV.

(C) Analogous distributions for ineffective input signals. The IPSPs are plotted downwards.

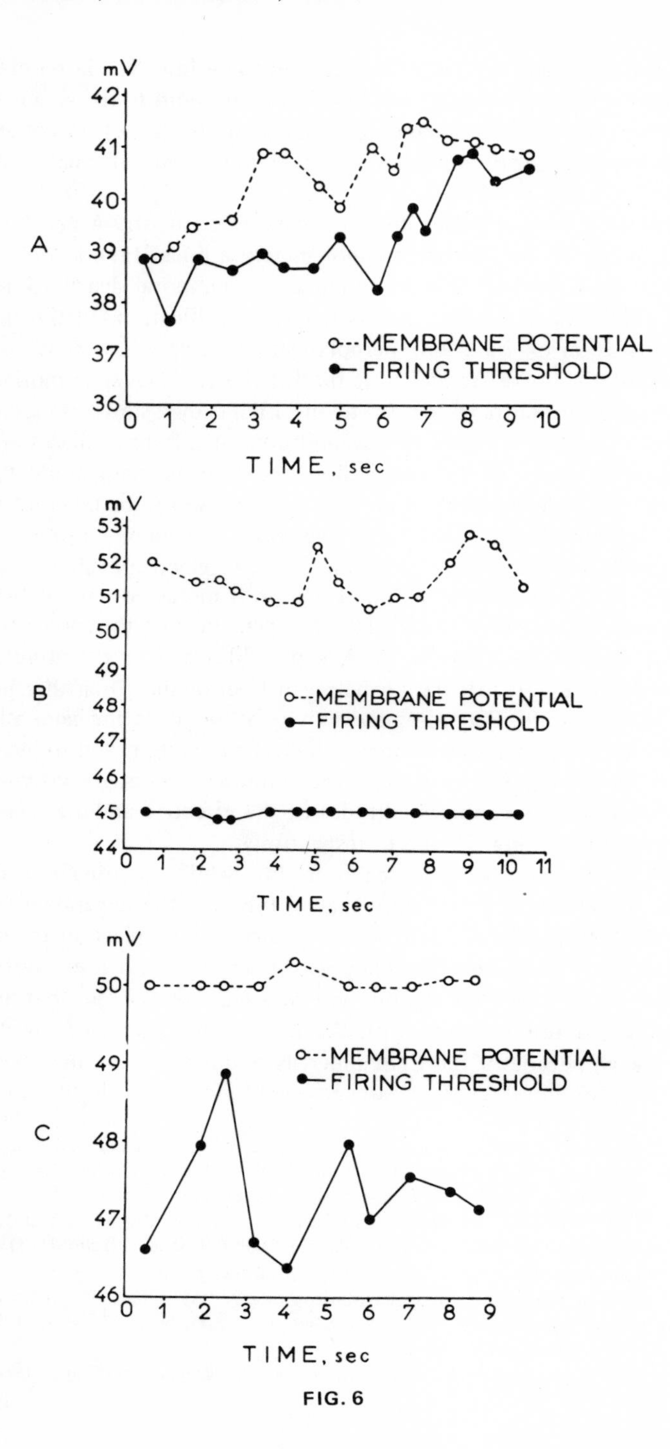

FIG. 6

when the EPSPs were of a much smaller amplitude. An increase in TMP resulting in an increase in the difference between the achieved TMP level and the firing threshold could be the possible explanation of this finding. However, along with changes of the mean value of the TMP, the average level of the threshold changed as well. The threshold "followed" the TMP level. This may be considered an "accomodation to the hyperpolarization." The electronic drift of the recording equipment was negligible when compared with these modifications of the parameters of the neuronal activity.

The threshold displayed rapid changes in addition to slow modifications. Thus, at the beginning of the experiment, when the average levels of the TMP and threshold were very close to each other, almost every EPSP ought to evoke a spike. However, a comparison of Fig. 5B and 5C showed that more than 50% of them were ineffective, although their amplitudes were up to 3–4 mV. At the same time, numerous APs were generated without any EPSP or pacemaker potential, and sometimes even during intervals where the divergence between the average levels of the TMP and of the threshold was even greater (see tenth min in Fig. 5A). These two opposite effects indicate that besides slow modifications, the TMP and the threshold undergo more rapid oscillations, and that along with their parallel changes, independent ones may take place (Fig. 6A, B, C). The amplitude of oscillations of each of the parameters reached 3–5 mV. Naturally, if the TMP and the firing threshold "diverge," even large EPSPs do not lead to a generation of the spikes. If they "converge," the probability of spike generation increases, and if they coincide, the generation of APs without any EPSP becomes possible. In the latter instance, conditions for pacemaking exist without generation of pacemaker potentials. In these conditions the recovery of the resting potential after the generation of a spike initiates the generation of a new one because the TMP and the firing threshold levels coincide again. The specific features of these types of activities in the giant neurons of *Aplysia* were described in the model of Perkel, Moore, and Segundo (1963).

Without going into particulars we must mention, however, the crucial role of the duration of the after-hyperpolarization (τ_1) and of the duration of the recovery of the threshold after the generation of a spike (τ_2) for the time-course of such a pacemaking. In the absence of the input signals, the frequency of the discharges is determined by τ_1, and in presence of them, not only by their characteristics, but as well by τ_2 and τ_1 and by their absolute values. The smaller the τ_1 and τ_2 are, and the value τ_2/τ_1 (usually $\tau_2 < \tau_1$), the sooner after the generation of a spike will EPSPs of the lowest amplitudes become effective.

In the cell described here, τ_1 and τ_2 fluctuated during the experiment. τ_1 varied from 100 to 260 msec, reaching 320 msec at the end of long spike trains.

FIG. 6 Examples of rapid modifications of the TMP and threshold. (A) Simultaneous modifications of the TMP and threshold. (B) Modification of the TMP versus a constant threshold. (C) Modification of the threshold versus a constant TMP. In all the graphs the upper curve is for the TMP.

Since an electrical stimulation was not applied in this series of experiments, it was difficult to have a precise notion about the limits of changes of τ_2 during the spontaneous activity. We use natural signals as criteria that is, the EPSPs that arrived at the soma after the generation of a spike, but which failed to elicit spike generation, though they exceeded the subsequently established threshold at "rest." During the experiment such a natural situation was encountered only a few times; however, it allowed us to establish the fact of modification of the τ_2 during the experiment. Variations of the τ_2 ranged from 50 to 90 msec, approximately.

Carpenter (1967), who investigated the giant neurons of the *Aplysia,* did not observe any changes in the firing threshold, except during periods of "bursting" when quite rapid oscillations of the threshold occurred. However, in the A neuron we have observed the phenomenon of "adaptation to hyperpolarization" in the form of a decline of threshold with the development of hyperpolarization, which to a considerable degree determined neuronal activity.

Special attention must be paid to the phenomenon of triggering spike activity by various stimuli in order to better understand the basis of activity in the intact neuron. One of the possible ways of realizing a long-lasting response has been mentioned above–pacemaking as a result of coincidence of the TMP and the firing threshold. Another mechanism is attributable to the generation of specific pacemaker potentials. Although the A neuron is not a typical pacemaker neuron, the insertion of the microelectrode at the beginning of the experiment, as well as electrical stimulation later, started spike activity via a direct involvement of the mechanism of the pacemaker. It is noteworthy that during such an activation the TMP remained, as a rule, unaltered. Some time after the insertion of the ME into the A neuron, pacemaker potentials often were not observed, and the spikes proved to be generated as a result of the onset arrival of EPSPs. Later the pacemaking could be reinstituted by prolonged subthreshold depolarization, or sometimes by single shocks.

We believe also that in the intact neuron there exists the possibility of activating pacemaker potentials by various changes in the environment, without causing a general change in the TMP. This would render a neuron's activity much more labile.

The cases when spike frequency gradually declined and vanished after impalement without changes of the TMP may represent instances during which A neurons were in a quiescent phase of the circadian rhythm before the insertion of the microelectrode.

Temporary activation of spike discharges was due to the excitation of the mechanism generating the pacemaker potentials by the inserted microelectrode. A similar temporary activation of the pacemaker may occur also against the background of spontaneous spike activity evoked synaptically. Our observations showed that the adaptation of the A neuron to the ME, being expressed as a modification of the TMP, of the firing threshold and some other parameters, manifests itself most effectively during a period of waning pacemaker activity.

The generation of pacemaker potentials elicited by electrical stimuli produced considerably shorter spike responses than the insertion of the microelectrode into the neuron. But in spite of the differing duration of the responses in these two cases, their patterns were similar. This suggests that in both cases the stimulation affected the same mechanism of generation of the pacemaker potentials, and that the duration of the responses depended on the time within which the adaptation to the acting stimuli developed. Since a mechanical injury leads to structural disorders and requires the development of "reparative adaptation," the activation in this case proved to be of a longer duration than in the case of electrical stimulation.

Apropos of this, it is intriguing to recall the research by Luco (1964) carried out on cockroaches. He has demonstrated that a stimulus can start spontaneous cellular activity in the shape of a "natural response" whose pattern remains invariable. This "natural response," starting according to the "all-or-nothing" law, apparently represents an activation of the pacemaker too. (Unfortunately, slow waves of TMP could not be recorded in those experiments as the recording was performed extracellularly).

As regards the A neurons, we may conclude that the possibility of producing "natural responses," based on the pacemaking mechanism and taking place at various levels of the TMP, is one of the basic factors responsible for the violation of the close relationship between spike frequency and the TMP level.

Evaluating the spike activity records from the soma of the A neuron, it should be noted that this neuron like many other giant molluscan neurons apparently presents a neuron-centre. The absence of a dendritic apparatus is inherent for them, and the axonal branches receive all the synaptic messages and possess quite autonomous spiking rhythms. The situation becomes more complicated by the fact that the axonal spikes, propagating in the somatofugal direction without obstacles, arrive at the soma as antidromic spikes. This is because in almost all the giant neurons having branched axons, the axonal triggering zones are separated from the soma by an indifferent zone along which the axonal spikes (A spikes) spread electrotonically (Tauc, 1966). Thus it often proves to be difficult to distinguish the A spikes from the EPSPs, when both are recorded from the soma. The passage of spikes from one axonal branch to another, depending on the time intervals and some other factors, may block the propagation of the A spike from the second branch towards the soma, though it is transmitted in the somatofugal direction (Tauc & Hughes, 1963). Consequently, in the case of recording from the soma, it is possible to evaluate with a certain confidence only the spike activity of the soma itself. The slow oscillations–if no special electrophysiological control is used for their discrimination (artificial displacement of the polarization level, etc.)–more appropriately ought to be called "somatic input signals." The question of whether the latter must be classified as local responses to the A spikes or as ESPSs is often debatable (see Fig. 8 in Tauc & Hughes, 1963).

The role of somatic spikes in the physiological state of the neuron remains, to a considerable degree, incomprehensible. As is known, the removal of the soma in the giant unipolar neurons does not affect the effectory functions of the neurons for some time. Some researchers (Saharov, 1965) favor the hypothesis that the involvement of the somatic membrane in spike activity results in an increase of RNA and protein synthesis in the soma.

Another factor violating the dependence of the firing rate on the TMP must be mentioned here. It is connected with a split of one big spike into two or more smaller ones. The giant molluscan neurons are characterized by a complex structure of their membrane which may be involved in the electrogenesis either entirely or in part only. This is reflected in the complex modifications of the spike shape.

An impalement is followed by spikes of high amplitudes (50–60 mV) which subsequently may decrease either rapidly or gradually, reaching sometimes only 2–5 millivolts by the end of the experiment ("microspikes"). Such a decrease in the spike amplitude may take place without any change of the TMP. The following three types of modifications should be distinguished: (*a*) a smooth modification of the AP amplitude; (*b*) a decrease of the AP amplitude with a corresponding expansion of a spike in time; (*c*) a decrease of the spike amplitude accompanied by a split of the spike into separate components which gradually diverge in time.

Recovery of the AP amplitude also occurred many times; in this case the changes occurred either smoothly, or by abrupt synchronization of the components. No "narrowing" of the spikes was observed. A modification of spike amplitude, accompanied by a split into separate components, is apparently caused by a desynchronization either of the axonal spikes' triggering zones or of the complex mosaic of the loci on the somatic membrane. Tauc and Hughes (1963) showed that during an antidromic stimulation, the amplitude of the spike recorded from the soma depends on the degree of synchronization of the axonal centres. However, the decrease of the AP amplitude apparently does not result from the mere desynchronization of the axonal triggering zones. Bennett and co-workers (1967a, b, c, and d) believe that the notches on the spike indicate an injury of the neuron, to which the somatic membrane is particularly sensitive. Hence the split of the spike testifies more to a desynchronization of the big loci on the electrogenic membrane of the soma, even more so because the number of the notches often exceeds the number of axonal branches. Moreover, the three types of AP amplitude modification mentioned above are far more comprehensible if considered as the results of the different patterns of the membranic mosaic.

A smooth modification of the AP amplitude, in conditions of invariant other electrophysiological parameters, suggests the possibility of switching off the activity of the elementary loci, which form the "mosaic" of the somatic membrane, along with the phenomenon of desynchronization of the big loci of the membrane. The notion of elementary loci also clarifies the process of expansion of the spike in time, which is related to its decrease and which occurs without changes in the total area of the AP. This may be interpreted as a diminution of synchronization between the elementary loci, without a complete switching of some of them, unlike the case above.

Thus, the spike recorded from the soma may be regarded as a sum of the displacements of the membrane potential caused by modification of the ionic permeability of the elementary loci which, in their turn, can be clustered. The modifications of the AP amplitude, which we have observed in our experiments, are well understood when explained by the synchronization/desynchronization phenomena between the unitary loci and between clusters of the loci.

CONCLUSION

Giant molluscan neurons integrate input signals in a complex way. The works of Tauc and other researchers have revealed the peculiarities of signal detection in these neurons. We have the impression that somatic spike activity plays a considerable role. For example, in certain conditions synaptic signals evoke independent spike activities in the axonal triggering zones; in the somatofugal direction they are transmitted to different recipients of these messages, while in the somatopetal direction, because of interaction, they produce only local responses. In other conditions the soma of the neuron proves to be involved in spike activity too, and, in its turn, evokes the generation of AP in all the axonal generating centers once more. In the latter case synaptic information coming to the axonal branches is emphasized due to the "reinforcement" of it by the soma. There are also other variants in which the "reinforcing" role of the soma in signal detection may be revealed.

Thus, the generation of somatic spikes may be regarded not only as a mechanism activating the trophic functions, but as a mechanism of detection of important messages for the neuron as well. The involvement of the soma in spike activity is determined by the spatial and temporal organization of the synaptic signals. A combination of these factors may lead to the generation of the somatic spike. Depending on the type of the neuron and on the real conditions, the somatic membrane either follows the rhythm of the axonal spikes ("executive response"), starts a "natural response," or responds only to a certain sequence of axonal spikes ("selective response").

On the other hand, it is not probable that all these responses were stable under changing conditions. One may expect that a novel stimulus, adequate or inadequate, will alter a neuronal response. This might be considered a short-term "learning," rendering the neuron better fitted to the environmental changes. From such a standpoint it does not seem to be reasonable for a neuron to react too actively for a long time to every stimulus because it would be exhausting for the neuron itself and not informative for the recipients of information yielded by that neuron. So it is more plausible that a neuron will tend to ignore the stimuli for which it has not yet elaborated a response (i.e., to the novel stimuli) as well as the stimuli of extreme strength. In this sense an adaptation to the inserted microelectrode is a convenient model of the adaptative processes in the intact neuron, as the mechanical injury represents the case when a neuron is believed to ignore the stimulus. The similar time-course of neuronal responses to mechanical injury and to electrical stimulation supports this idea (Sokolov, Arakelov, & Pakula, 1969).

The data described above obtained from experiments on the A neuron once more demonstrate the very complex organization of the electrogenic activity in giant molluscan neurons. A multifactor dependence of spike activity allows the A neuron to ignore stimuli even without changes of the TMP. From the host of factors determining the firing rate during the adaptation to the inserted microelectrode, the latent pacemaker and the "checker-board" membrane, because of their great functional mobility, are the two most efficient mechanisms in changing the pattern of response of the soma. These two mechanisms may be of great importance as well for adaptation to the natural signals in intact A neurons.

Summarizing the features of adaptation to the inserted microelectrode in the A neuron, the three most common and evident means of development of the adaptation must be mentioned. They are: (1) a gradual functional transformation of a part of the electrogenic membrane, which results in diminution of spike amplitude; (2) an inactivation of the pacemaker leading to the decline of firing rate; and (3) an increase of polarization due to which certain input signals become subthreshold, if the threshold remains unaltered.

3

HABITUATION IN THE GIANT NEURON OF A MOLLUSK DUE TO REPEATED INTRACELLULAR ELECTRICAL STIMULATION

E. N. Sokolov
A. L. Yarmizina

Moscow State University

The main problem in the study of neuronal mechanisms underlying extinction of the orienting reflex is the differentiation between pre- and postsynaptic events leading to blockade of stimulus action.

The habituation observed in giant neurons of mollusks to repeated orthodromic stimuli is a fair model of such an extinction. Bruner and Tauc (1965), using intracellular recording, demonstrated that the diminution of unitary monosynaptic EPSPs is the cause of habituation.

Holmgren and Frenk (1961) demonstrated the development of hyperpolarization, observed in the cell during the repeated orthodromic stimulation. These data were supported by Waziri, Frazier and Kandel (1965). Sokolov, Arakelov, and Levinson (1967) demonstrated the extinction of spike discharge, which resulted from the potentiation of inhibition; the latter also extinguished later. The obtained data led to the assumption that the summation of postsynaptic collateral inhibition is followed by the involvement of the presynaptic inhibition. Kandel (1967) assumed that the intensification of inhibition is due to an activation of pacemaker potentials in the inhibitory interneurons.

In all the above-mentioned studies, the participation of the postsynaptic cell (from which recordings were obtained), in the extinction processes, was not demonstrated directly. The role of the postsynaptic cell in habituation can be studied using electrical stimuli delivered intracellularly.

The present investigation is devoted to habituation in the giant neuron of the parietal ganglion of *Limnaea stagnalis* owing to repeated intracellular electrical stimulation.

METHOD

The experiments were performed on whole preparations of *Limnaea stagnalis*. Glass micropipettes filled with 2.5 *M* $FeCl_3$ were used for intracellular recording from giant neurons of the parietal ganglion. A dc amplifier MZ-3B (Nihon Kohden) was used in the experiments. The electrical activity of the neurons was recorded on film using differing gains for two beams of the oscilloscope. A "Microphot" projector was used for the visual analysis of the record. Electrical stimuli were delivered through the recording microelectrode by means of a bridge circuit. Positive electric shocks (2.5×10^{-9} amp; 60-sec duration) were applied at 60-sec intervals.

The experimental schedule consisted of: (1) recording of spontaneous activity for 20–40 min; (2) measurement of spike threshold using intracellular stimulation; (3) repeated presentations (40–50) of intracellular electrical stimuli until complete or partial habituation; (4) study of the recovery of the response after a rest-period of 5–10 min; (5) study of the recovery of the response to the initial stimulus after the application of a more intense one; (6) study of the dynamics of the response after disinhibition.

A total of 23 neurons was studied. The data presented in the paper refer mainly to one of the neurons, which was recorded for 6 hr.

RESULTS

We investigated habituation to repeated intracellular stimuli in the giant D neuron of the left parietal ganglion, which was identified among other giant neurons (A, B, C, and E) according to the schematic reconstruction of the ganglion (Yarmizina, Sokolov, & Arakelov, 1968). This neuron was a pacemaker according to the classification of Carpenter (1967).

The response of the neuron to the penetration of a microelectrode consisted of a depolarization, which developed abruptly and coincided with an increase of the amplitude of pacemaker potentials, which resulted in a high firing rate. This effect is followed by the stabilization of transmembrane potential (for details concerning the stabilization effect, see Sokolov & Pakula, this volume).

Habituation to the penetrated microelectrode was expressed by a decrease of the amplitude and frequency of pacemaker potentials resulting in decrease of firing rate. The iontophoresis of positive ions by current 2.5×10^{-9} amp* activated pacemaker potential (PMP) generation without affecting the resting potential. The amplitude of the PMP increased to such an extent that they reached the firing threshold. This resulted in an increase of firing rate. During a long-lasting iontophoresis of positive ions the activation effect of the electric current upon the pacemaker potential diminished. The amplitudes of some PMPs became lower and did not reach firing threshold. This resulted in an increase of intervals between spikes. Switching off the current evoked a depression of the pacemaker potential as indicated by a decrease of its amplitude, compared with background activity before the application of the stimulus. As a result, spike

*Stimulus duration is not specified (Ed.).

activity at first completely disappeared, and then recovered gradually up to the background level.

Some of these effects are shown in Fig. 1A, in which the on and off responses to iontophoresis of positive ions are presented. Similar phenomena were observed during the subsequent stimuli presentations. Successive presentation of

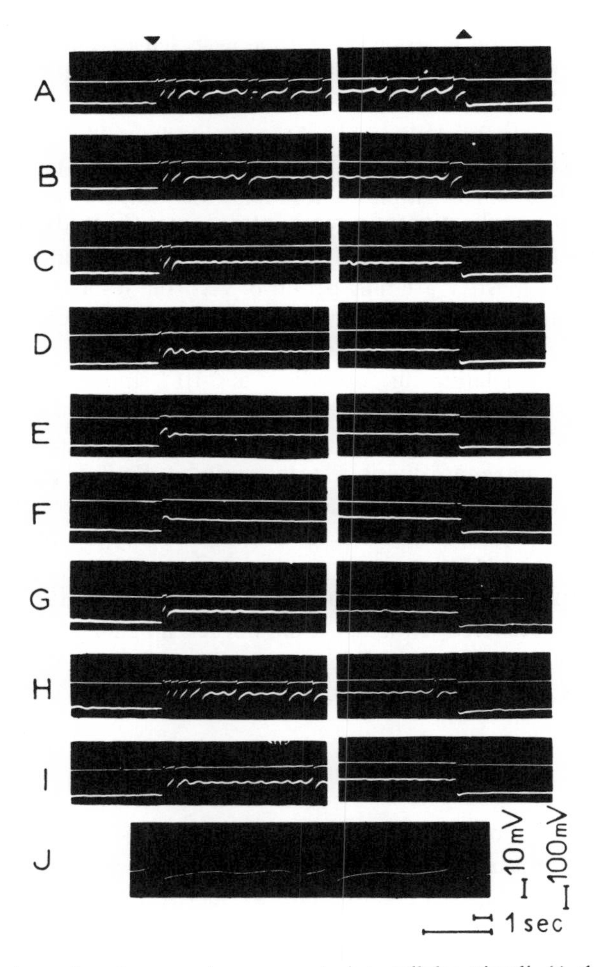

FIG. 1 The dynamics of neuronal responses to intracellular stimuli. (A through F) 2nd, 7th, 14th, 19th, 24th, and 28th presentations demonstrating the process of habituation; (G) partial recovery of the neuronal response after the 6 min period of rest; (H) activation of PMPs and increase of firing rate due to presentation of a stronger stimulus (3.5×10^{-9} amp). (I) the recovery of the response to the weaker stimulus used in the habituation series (2.5×10^{-9} amp) following presentation of an intense stimulus. Recording from one cell with low (the upper trace) and high (the lower trace) gain. Arrows indicate the switching on and off the stimulus. Shift of the potential is due to disbalance in the bridge circuit. (J) The sequence of the effective and noneffective pacemaker potentials.

the stimulus, however, activated the PMPs to a lower degree, and their amplitude gradually became reduced.

This resulted in a decrease of the number of spikes generated during the burst (Fig. 1B). The fourteenth application of the stimulus evoked spike generation only just after the beginning of iontophoresis. The response in this case consisted of merely 1–2 spikes (Fig. 1C) because the amplitude of PMPs did not reach the firing threshold. The termination of iontophoresis led to a complete depression of PMPs (Fig. 1D). The twenty-fourth electrical stimulus evoked no spikes at all. The response seen at the beginning of stimulation consisted of a group of PMPs, which did not reach firing threshold. The amplitude of PMPs dropped to zero and the iontophoresis failed to initiate the PMPs (Fig. 1E). After the elimination of spike discharges, habituation to iontophoretic stimulation consisted of a progressive diminution of PMPs, and on the twenty-eighth presentation, the response was reduced to single PMPs of a low amplitude, and represented an on-effect (Fig. 1F). After a 6-min period of rest, the response to the stimulus partially recovered; one spike, and several PMPs were evoked by electrical stimulation (Fig. 1G).

Increasing the stimulating current up to 3.5×10^{-9} amp led to recovery of the activation of PMPs, and increased their amplitude. Owing to this activation of PMPs the number of spikes during a stimulus presentation also increased. The intervals between spikes are divisible by the period of PMP. The activation effect produced by the stronger stimulus also declined when prolonged stimuli were applied. This was expressed in the reduction of the PMP's amplitude and of the firing frequency (Fig. 1H).

After the application of the stronger stimulus, the response to the original weaker stimulus recovered, although it did not reach its initial value. "Disinhibition" took place, which was also due to the activation of PMPs (Fig. 1I). Similar effects were observed in 8 other neurons.

DISCUSSION

Complete habituation in the neuron to iontophoresis of positive ions represents only one form of the dynamics of cellular responses. Some other observed phenomena were: (*a*) partial habituation, (*b*) fluctuation of the responses without a habituation, and (*c*) facilitation of the responses.

The habituation to repeated intracellular electric stimuli resulted from the weakening of the activation effect exerted by positive ions upon the mechanism of pacemaker potential generation. There exists a similarity between the weakening of iontophoretic effect due to its prolongation and habituation as a result of repeated presentations of the stimulus. One may say that habituation is a result of the summation of accomodation effects. A specific feature of the habituation to intracellular stimulation is that the current first of all fails to activate the PMPs, and hence, fails to initiate spike discharges. The process of habituation in this case seems to be dependent upon the internal structures of the neuron. The assumption concerning the mechanism of a negative feedback through inhibitory interneurons can explain only the case when iontophoresis of positive ions inhibits spike generation. Habituation, however, develops further, when no spike generation occurs and when, consequently, involvement of synaptic recurrent inhibition is less probable. Conclusive proof of this statement would be obtained from experiments on isolated cell preparations.

An important indication that the phenomena observed in these experiments are similar to habituation to orthodromic stimuli is "disinhibition" of the neuron after a stronger stimulus. The recovery of the response after the period of rest also indicates the dynamic character of the modifications occurring in the cell. The habituation described above results from inactivation of the pacemaker potential, which, according to Alving (1968), is an endogeneous event in the neuronal soma. The specific characteristic of this potential is a slowly rising depolarization (Fig 1J), which initiates spike generation when reaching the firing threshold. The PMPs are sensitive to currents, which are below the threshold of modification of resting potential. The decrease of this high sensitivity of PMPs results in effective adaptation of the neuron to repeated intracellular stimuli.

CONCLUSIONS

1. The possibility of habituation of the neuron to repeated intracellular iontophoresis is demonstrated.
2. The similarity of habituation to orthodromic and to intracellular stimuli is confirmed by the "disinhibition" of the neuron after the presentation of a stronger stimulus.
3. The participation of inactivation of the pacemaker potential in the mechanism of habituation to intracellular stimuli is shown.

SECTION II

NEURONAL DISCHARGE CHARACTERISTICS

This section contains thirteen chapters, all concerned with the extracellular recording of neuronal discharges in the rabbit. Use of the intact animal, chronically prepared, provides data which are likely to be of great relevance to the elicitation and extinction of the OR. Actually, behavioral components of the OR are not recorded simultaneously with the unit data; this is not a serious drawback, and in fact it has been achieved infrequently, if at all in the scientific world. Some chapters do, however, deal with the EEG components of the OR, and their relationship to unitary activity (Sokolova, Chapter 7 and Danilova, Chapters 14 and 15).

PART A
"Feature Detectors"

"Feature detectors" have been known for many years to exist within sensory systems. Most of the previous reports have dealt with the visual system, and have included detailed analysis of organisms as disparate as the frog and cat. The three papers in this section also deal with the visual system, in these cases, with the visual cortex of the rabbit. However, their thrust is quite different from most previous work because they are not concerned with the "structural" units of the visual world (e.g., lines, angles) but rather with three other parameters: intensity, stimulus interval, and cross-modality effects. Chkhikvadze (Chapter 4) reports units which do not exhibit a monotonic relationship between flash intensity and rate of discharge, as occurs in the optic tract, but rather a high degree of specificity such that responsiveness may be maximal to a weak stimulus. Chelidze (Chapter 5) finds neurons which behave as though they have "time receptive fields," while Polyansky and Sokolov (Chapter 6) report cells in the visual cortex that seem to be specially "tuned" to the stimulus complex of simultaneous visual and acoustic stimulation. These three chapters support the view that the sensory systems, in particular sensory cortex, performs elegant analyses of the animal's sensory world. Further, it would appear that these analyses are not subject to great plastic modification.

4

SINGLE UNIT RESPONSES TO FLASHES OF LIGHT OF DIFFERENT INTENSITIES IN THE VISUAL CORTEX OF THE RABBIT

I. I. Chkhikvadze

Moscow State University

Responses of single units in the rabbit's visual cortex depend greatly on the intensity of the applied light stimuli. Increase in the brightness of the latter results in a reduction of the latency, and changes of the number of spikes in the discharge (Gleser, 1966; Jung, 1964; Polyansky, 1965), as well as in the "pattern" of the discharge, i.e., the character of alternation of the excitatory and inhibitory phases (Polyansky, 1965).

Polyansky (1965) has divided the neurons of the visual cortex in two groups. The first group includes neurons with a simple relationship of excitatory and inhibitory processes. An intensification of the stimulus produces either an increase of the number of spikes in the discharge (neurons with a predominance of excitation), or a decrease of the number of spikes (neurons with a predominance of inhibition). The second group consists of neurons in which an increase in stimulus brightness, up to a definite level, enhances the spike discharge. A further increase in stimulus intensity results in a suppression of the spike activity instead of its intensification.

Such development of inhibition as a result of an increase in the brightness of the stimulus apparently creates conditions for a selective response of the cortical neurons to a definite intensity. Neurons with a selective response to different intensities of a flash were described in the lateral geniculate body (Beteleva, 1964) and in the tegmentum of the frog's midbrain (Mkrticheva & Samsonova, 1965).

In the present study the responses of neurons from the rabbit's visual cortex to light stimuli of different intensities were investigated. The data obtained allowed us to define neurons with a selective reaction to their "own" intensity of light flash.

METHODS

The activity of single neurons from the visual cortex of an unanaesthetized rabbit, restrained in a wooden stand, was recorded with tungsten microelectrodes with tips of 1–2 μ. The electrodes were insulated with Vinyflex and varnish, and had a resistance of 5–8 megohms.

Introduction of the microelectrode into the cortex was performed by means of a remote-control hydraulic micromanipulator. Spike activity was filmed continuously from the screen of a dual-beam "Duoscope" oscilloscope at a speed of 5 cm/sec.

Flashes of a pulse tube with intensities of 4 levels (1.25, 3.2, 5.2, 7.8 lumens) were used as stimuli. The intervals between the flashes were 1.6 sec. The pulse tube was placed 30 cm from the eye contralateral to the hemisphere under investigation. The other eye was shielded.

To test the effects of both the brightness of the flash and that of the effect of stimulus repetition on the neuronal responses, the following schedule was used: Background spike activity of the neuron was recorded for 5 sec, and for the following 10 sec, the neuronal responses to a stimulus of a selected intensity were recorded. Afterwards the spike activity was recorded for another 5 sec to evaluate the aftereffect. Then a flash of another intensity was delivered. When the whole set of intensities had been used, it was reiterated in the same order. The number of reiterations during one experiment ranged from 30 to 50. When evaluating the data, the number of spikes was counted for each 100 msec. Reference points for the number of each 100-msec period were the moments of stimulus onset.

To estimate the background activity, the record of background activity was divided into intervals of the same duration as the interstimulus ones. Furthermore, each of them was divided into periods of 100 msec.

The number of spikes per period was counted according to the same scheme as used in the stimulus presentation.

On the basis of the data processed in this way, poststimulus time (PST) histograms were plotted. They permitted comparison of responses to various stimuli with the background activity, and the study of response dynamics during the experiment.

RESULTS

Thirty neurons from the rabbit's visual cortex were investigated. Their activity was analysed statistically using the PST histogram method.

Some of the neurons (12) did not respond to stimuli (unreactive neurons); the others (18) displayed different reactions (reactive neurons). Among reactive neurons only four were selective to different levels of brightness. The majority exhibited no selectivity with respect to the intensity of the stimulus; they

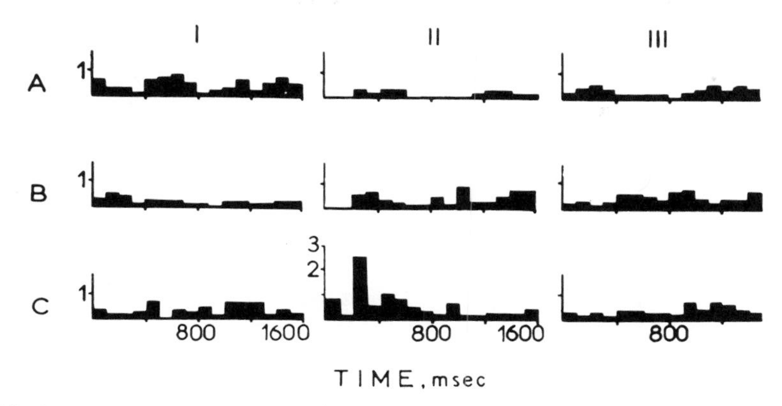

FIG. 1 Averaged poststimulus histograms for selective responses of a neuron in the rabbit's visual cortex. Three levels of intensity of a light flash: (A) 5.2 lumens; (B) 3.12 lumens; (C) 1.25 lumens; (duration of PSH = 1.6 sec). (I) histogram for the background activity; (II) histogram for the responses; (III) histogram for the background activity after stimulation. The abscissa is time in msec from the moment of stimulus application (for the background–in the case of a conventional reading of the interval). The ordinate is the averaged number of spikes within intervals of 100 msec. Histograms have been averaged for 30 stimulus presentations. This neuron (16) manifests selective sensitivity to a weak light intensity.

responded to all the intensities of the light with spike discharges of a similar "pattern," consisting of an initial spike burst of minimal latency, a subsequent 100–200 msec pause, and a secondary spike discharge.

A comparison of histograms for background activity with those obtained during stimulation showed that the neuronal response is mainly expressed not as an increase of the total number of spikes under the influence of the flash, but rather in the modification of discharge pattern. After the application of the stimulus, the number of spikes for the period analysed remains almost the same as the corresponding one for the background activity. The reactions result from the recovery to the initial background activity. The duration and magnitude of the secondary discharge exhibit a close dependence on the level of the background activity.

Neurons selective to brightness are of special interest. This property is demonstrated in reactions of neuron No. 16. The responses of this neuron to a light of minimal intensity (1.25 lumens) had a minimal latency and consisted of an initial spike discharge, an inhibitory pause (100–200 msec), and a large secondary discharge (Fig. 1C). With the increase in stimulus intensity up to 3.2 lumens, the response almost disappeared. Only a weak trace of an unstable secondary discharge remained (Fig. 1B). A stimulus of 5.0 lumens evoked a depression of the background activity (Fig. 1A).

Neuron No. 26 was similar to the one described above and responded selectively to weak stimuli. This selectivity was expressed by an increase in the

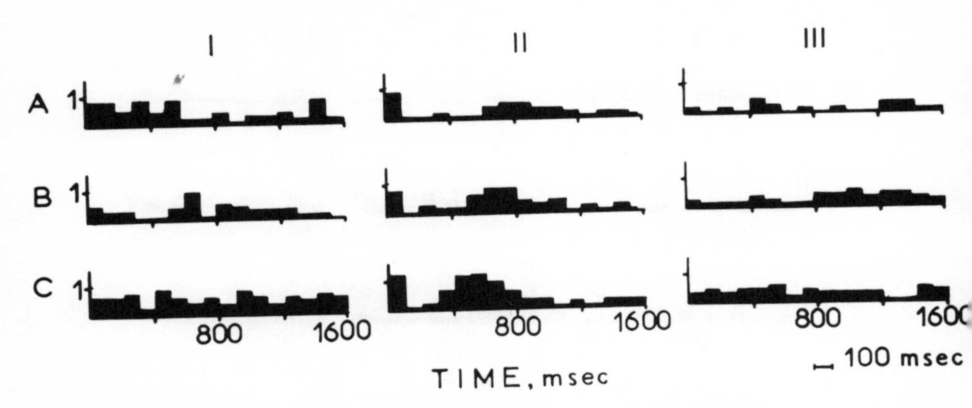

FIG. 2 In neuron 26 selective sensitivity to the minimal light intensity manifested itself in an increase of the secondary discharge. Designations are as in Fig. 1.

secondary discharge to an intensity of 1.25 lumens (Fig. 2C). This neuron responded distinctly both to the medium and the maximum intensities. In the latter case the spike discharge was, however, of a smaller amplitude and the inhibitory interval between the primary and the secondary discharges proved to be longer, the more intense was the stimulus. A decrease in stimulus intensity resulted in a reduction of this duration of inhibition.

Figure 3 shows the records of the spike activity of neurons 16 (A) and 26 (B). The stimulus of maximal intensity either does not change the background

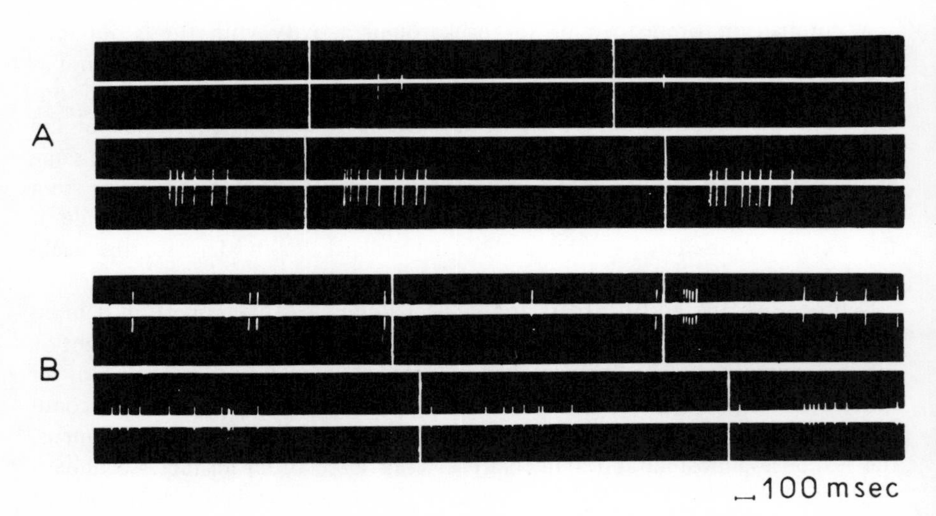

FIG. 3 Records of spike activity of neurons 16 (A) and 26 (B) for two levels of stimulus intensity (top of each pair, 5.22 lumens; bottom record of each set, 1.25 lumens). The moments of stimulus application are indicated by vertical bars. Neuron 16 did not respond to the flash of the maximal brightness. Neuron 26 responded by unstable discharge, while both neurons responded to the minimal brightness with high frequency discharges.

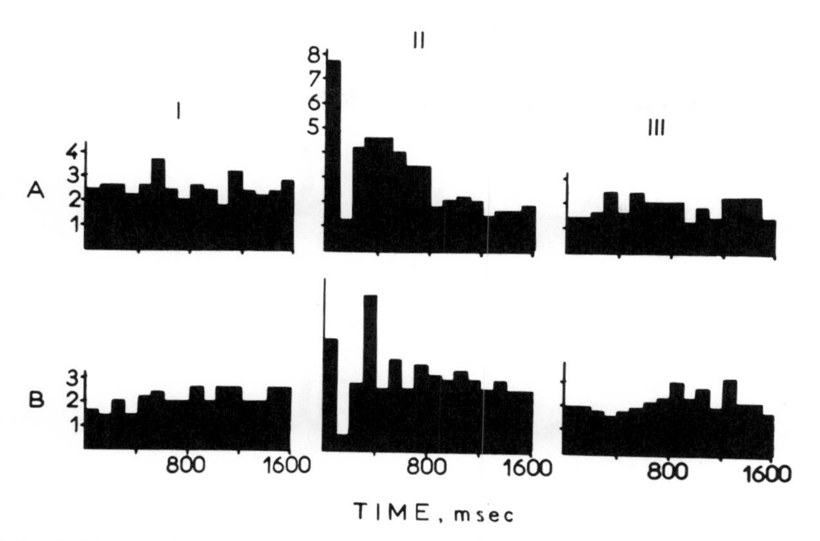

FIG. 4 Neuron 28 manifested selective sensitivity to the high levels of brightness only: (A) 5.2 lumens; (B) 3.12 lumens.

activity (A) or evokes unstable spike discharge (B). However, the low-intensity stimulus in both cases evokes a high-frequency discharge.

The selectivity of neuronal reactions with respect to intensity can be manifested in an increase of the excitatory process, as well as of the inhibitory one. For example, neurons Nos. 16 and 26 were selective to the minimal intensity, increasing their spike discharge rates, while being less responsive to the maximal intensity.

The interaction of excitatory and inhibitory mechanisms at different levels of intensity is seen in neuron 28 (Fig. 4). Stimuli of high intensities evoke responses

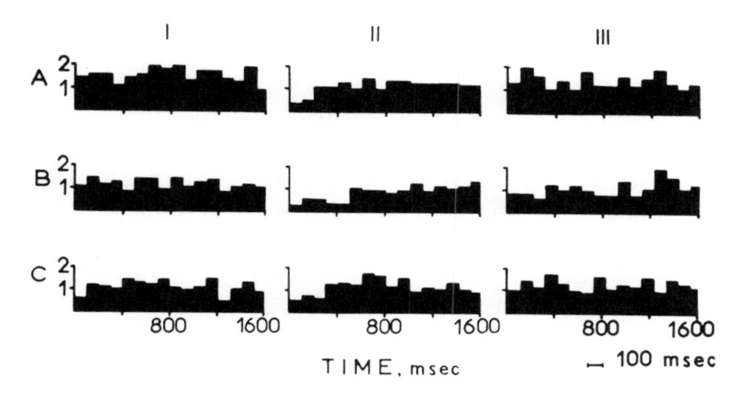

FIG. 5 Averaged poststimulus time histograms for background activity, suppressed by the stimuli of high and middle intensity: (A) 5.2 lumens; (B) 3.12 lumens; (C) 1.25 lumens. Designations are as in Fig. 1.

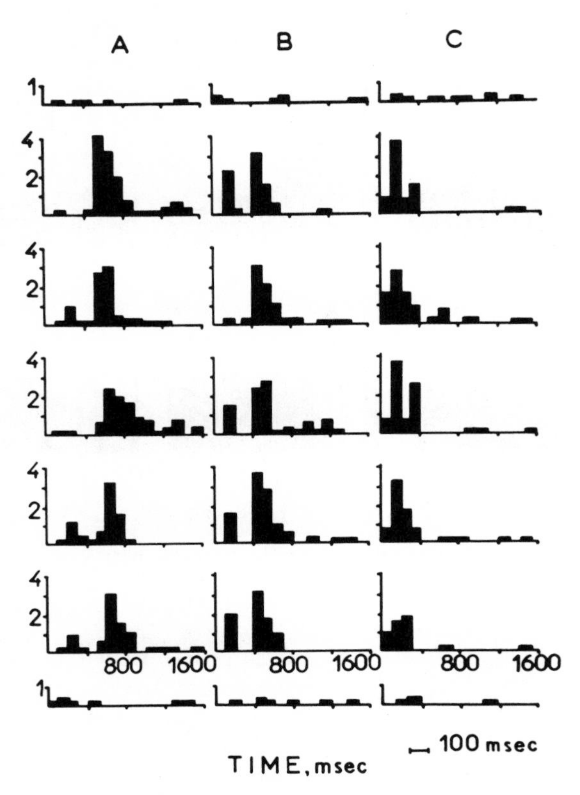

FIG. 6 A series of histograms of the neuron that displayed a change of pattern with a switch to another level of stimulus intensity. Each column corresponds to a definite level of brightness: (A) 5.2 lumens; (B) 3.2 lumens; (C) 1.25 lumens. A histogram for the background activity of this neuron is given in the beginning of each column and a histogram of the background activity after the stimulation with a conventional reading of the interval is given at the bottom of the column. Other PST histograms are obtained by successive blocks of 6–7 applications of the signal. Abscissa is time; ordinate is number of spikes during 100 msec. Explanations are given in the text.

with a well-pronounced first component; at low intensities this component is faintly expressed due to developing inhibition.

The selectivity of the reactions with respect to the maximal intensity can be manifested in an increase of the inhibitory response (Fig. 5). Thus, in the visual cortex it is possible to distinguish neurons with a maximal reaction to a certain adequate intensity of light.

Changes in the level of brightness evoke modification of pattern of the single unit discharge. This is shown in the records from cell 15, which responded to different levels of stimulus intensity by changing the pattern of the response.

The reaction of neuron 15 to stimuli of the maximal intensity (5.2 lumens) started with an inhibitory phase; this phase was followed by excitation with a close secondary discharge (Fig. 6A).

In response to a stimulus of medium intensity (Fig. 6B) the primary discharge appeared with a latency of 100–200 msec. In this case, the secondary discharge remained. The duration of an inhibitory pause separating the first and the second discharge was 100–200 msec.

In the case of the minimal intensity (1.25 lumens), the primary excitatory discharge is evoked with a minimal latency, without any sign of a secondary spike discharge (Fig. 6C). During the repetition of the stimuli, a reduction in the number of spikes was observed.

DISCUSSION

Some of the neurons in the rabbit's visual cortex demonstrated selectivity to certain intensities of light. It is assumed that such neurons deal with the mechanism of detecting specific levels of intensity as separate properties of the signal, and may be regarded as the "detectors of intensity."

Selective sensitivity to different levels of intensity is a result of unizotropic lateral inhibition among the neurons within the neuronal net which consists of units with different thresholds (Kirvlis, Pozin, & Sokolov, 1967).

The selective sensitivity to low levels of intensity is accounted for by an inhibitory mechanism which possess a high threshold. This inhibition reduces the excitation during an increase in stimulus intensity. Experiments confirmed the assumption that the active inhibition is a critical link for the selective depression of the responses to stimuli of high intensities.

5

NEURONAL REACTIONS OF THE RABBIT'S VISUAL CORTEX AS A FUNCTION OF THE INTERVAL BETWEEN FLASHES OF LIGHT

L. R. Chelidze

Moscow State University

The responses of the organism to external stimuli are associated with complex adaptive readjustments of the central nervous system. One of the manifestations of such readjustment is the effect of predicting future events on the basis of assimilating a definite sequence of signals. The conditioned reflex to the time of the action of a stimulus (Pavlov, 1952) relates to the elementary forms of prediction.

Using the registration of human motor, skin-galvanic, and EEG reactions (Fernández-Guardiola *et al.,* 1968) showed the effect of prognostication, as a result of numerous repetitions of conditioned light signals following one another at intervals of 10 sec.

In experiments on human subjects carried out by Sokolov, Chelidze and Korzh (1969) with the help of EEG and EMG recording, the extrapolation effect was obtained during the application of flashes at intervals of 1–20 sec. This effect consisted of the appearance of depression of the alpha-rhythm preceding the real action of the stimulus. In control experiments, where the flashes were applied at random intervals, the effect of prognostication was absent. The emergence of anticipatory reactions indicated the ability of the organism to memorize the interval between the stimuli and to "prognosticate" the appearance of the next signal.

In 1963, Sokolov (1963a,b) advanced a hypothesis concerning the existence of independent extrapolating neurons. In 1965 the effect of extrapolation was demonstrated by Vinogradova (1965) on some neurons of the rabbit's hippocampus which manifested discharges preceding the moment of application of rhythmical stimulus. The emergence of the extrapolation effect at the level of a single neuron of the rabbit's visual cortex was observed by Bagdonas *et al.,* (1968). Vardapetan (1967) demonstrated the ability of the neurons of the cat's auditory cortex to assimilate the periodicity of the sequence of signals. Kopytova and Rabinovich (1967), studying the conditioned reflex to time in neurons of the rabbit's motor and visual cortex, found that the majority of neurons of the cortical motor zone are able to extrapolate the appearance of a regular signal.

Realization of extrapolation effects on the neuronal level demands the obligatory participation of neuronal systems capable of measuring the duration of time intervals. Such systems can be constituted of neurons having strict periodicity of discharges, such as neurons described in the literature (Danilova, 1966; Horn, 1962, 1964). On the other hand, we can imagine neuronal systems capable of selective reaction only if the stimuli are characterized by definite time parameters. The dependence of neuronal reactions in the visual cortex upon the time parameters of stimuli is the subject of the present chapter.

METHODS

The investigation was carried out on 7 unanaesthetized rabbits. One or two days before the experiment an opening with a diameter of 5 mm was made in the skull of the rabbit; the coordinates of its center were located in the area of the best cortical evoked responses to flashes of light: 7 mm in front of the lambda and 7–9 mm laterally from the sagittal suture (Polyansky, 1963). The opening was filled with agar; a base for the micromanipulator was placed over it.

During the experiment the rabbit was tied to a stand and placed in a dark screened chamber. Spike activity was recorded by means of tungsten microelectrodes covered with veynflex and varnish. The movement of the microelectrode was regulated with the help of a distant hydraulic micromanipulator. A low-frequency amplifier was used. The recording was done from the screen of an oscilloscope with the aid of a Nihon Kohden camera on a film moving with a speed of 5 cm per sec. Flashes of a gas-discharge tube with a luminous energy of 5.2 lumens presented to the eye contralateral to the hemisphere under investigation served as stimuli.

After the end of the experiment the maximal depth of the electrode insertion was marked by means of an electrolytic lesion of the brain tissue.

The Program of the Experiments

The program of the experiments was directed at studying the modifications which take place in the neuronal responses as a result of numerous repetitions of flashes presented at a constant interval. It was important to accomplish a consecutive study of the responses of the same neurons at different intervals between the flashes. For this purpose, the flashes were presented in series. Each of the series included 20 flashes separated by a constant interval. The following intervals were used in consecutive series: 1, 2, 3, 5, 10, and 20 sec. The series were separated from one another by a 20-sec interval. The flashes were presented automatically. In some cases the intervals were fixed by the experimenter with the help of a stopwatch.

Data Analysis

During the processing of the data the number of spikes was counted for each 50 or 100 msec, depending on the frequency of the neuronal discharge. On the basis of these data, a histogram of the background activity, as well as post-stimulus time histograms (PST), were formed. In order to construct a histogram of spontaneous activity, the period of activity before the beginning of stimulation was divided into one-sec intervals, and the distribution of the spike activity for 20 intervals was averaged. Then, we formed PST histograms for single stimuli reflecting the dynamics of the neuronal spike activity after the presentation of each flash. In addition, averaged PST histograms were formed on the basis of responses to 10 or 20 flashes, which demonstrated the general pattern of the response to a light stimulation with the given interval. The beginning of the counting period coincided with the moment of flash presentation.

We likewise calculated the latencies of the first spike after presentation of the flash, as well as of the first spike in the second phase of excitation following the inhibition which separates the initial and late phases of the reaction. Finally, the total duration of the neuronal responses was measured for each stimulus presentation.

RESULTS

Responses to flashes, separated by different intervals, were recorded in 42 neurons of the visual cortex during experiments performed on 7 rabbits. We investigated mainly those neurons which manifested a pronounced response to the light stimulus. Below are the results of the processing of 22 neurons; 15 of these neurons were investigated according to the full program, whereas 7 were subjected to the action of flashes with at least three different intervals.

The background activity of the investigated neurons was of a diverse character. Both neurons with a complete absence of background activity (2 neurons) and neurons with spontaneous discharges (about 16–20 spikes/sec; 4 neurons) were encountered during the investigation. Two neurons had a "burstlike" activity in which a period of silence with a duration of 3–7 sec was followed by a high-frequency discharge that lasted 1–1.5 sec. In one case a regular discharge in the background (16–20 spikes per sec) was superseded by a "burstlike" firing during the application of flashes. The investigated neurons may be divided in two groups according to the pattern of responses to light stimulation.

1. Neurons with a multiphasic response consisting of periods of excitation and inhibition (11 neurons). A typical response of such neurons has the following pattern: Phase I, initial excitation with a latency of 20–40 msec; Phase II, inhibition lasting up to 200–300 msec; Phase III, secondary excitation. Sometimes the response includes a fourth inhibitory phase, which is longer than the second phase. The fourth phase may last up to 2 sec. The neurons with a

multiphase response are characterized by high (16–20 spikes/sec) or moderate (6–10 spikes/sec) spontaneous activity.

2. Neurons with a monophasic response that consists only of excitation or inhibition (11 neurons). The neurons with such reactions had no spontaneous activity or manifested low-frequency spontaneous discharges (up to 6 spikes/sec). The latencies of two neurons of this group were less than 100 msec. The other neurons were characterized by latencies exceeding 200 msec.

Dependence of the Neuronal Responses upon the Interval between the Stimuli

Comparison of averaged PST histograms, for stimuli presented with the different intervals, showed that among the investigated neurons existed a group in which responses depended upon the interval between the stimuli. These neurons (9 units) manifested a response of maximal duration and intensity predominantly to stimulation with one, definite, interval between the flashes. In various neurons of this group selectivity with respect to different intervals between the stimuli was observed.

Three neurons (Nos. 44, 21, and 48) exhibited a high degree of dependence upon the interval. All the three neurons were characterized by a monophasic response with a latency of about 300 msec (Fig. 1). From the figure it follows that the response of neuron No. 44 is maximal at a 2-sec interval between the flashes, and that of neuron 21, at an interval of 3 sec between stimuli. Neuron 48 showed a more complex dependence upon the interval. Its response was maximal at an interval of one sec. At intervals of 2 and 5 sec between the flashes the responses also existed, but they were weaker than at intervals of one sec. At intervals of 3 and 10 sec between the stimuli the responses were very weak and irregular, although the background activity of the neuron remained constant. Figure 2 illustrates the spike responses of neuron 44 to flashes presented at different intervals.

Selectivity with respect to interval was less pronounced, though still distinct, in the remaining 6 neurons. Among them, neurons 1 and 19 manifested maximal responses to flashes presented with intervals of 10 sec (Fig. 3). In this figure it can be seen that the response of neuron No. 19, which belongs to the group of neurons with a multiphasic reaction, reaches the maximum level at an interval of 10 sec. Neuron 1 (Fig. 3B) does not react to flashes presented at intervals of 1, 2, and 5 sec. At an interval of 3 sec, this neuron responds, but unstably and irregularly; only at intervals of to 10 sec does there appear a regular reaction consisting of one or two spikes with a latency of 300 msec. Three other neurons, which were characterized by different types of responses, exhibited a maximal reaction at the interval of 3 sec.

The responses of the other investigated neurons (13 neurons) did not depend on the intervals between the flashes. A typical representative of this group is neuron 23 with a multiphasic responses to the flash (Fig. 4). Figure 4 shows that the neuron does not change its responses as a function of different intervals

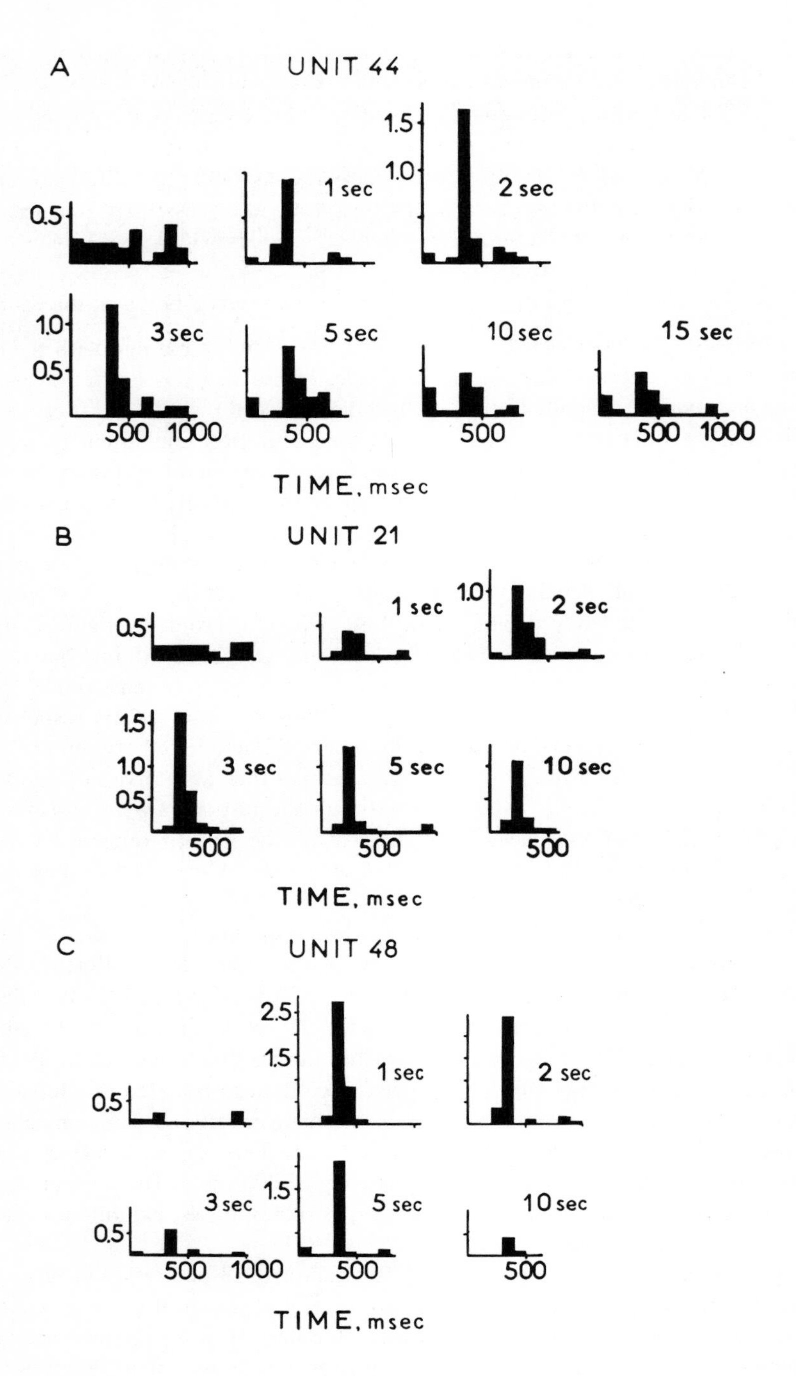
A
UNIT 44
0.5
1 sec
1.5
1.0
2 sec
1.0
0.5
3 sec
5 sec
10 sec
15 sec
500 1000
500
500
500 1000
TIME, msec
B
UNIT 21
0.5
1 sec
1.0
2 sec
1.5
1.0
0.5
3 sec
5 sec
10 sec
500
500
500
TIME, msec
C
UNIT 48
2.5
1.5
1 sec
2 sec
0.5
3 sec
1.5
5 sec
10 sec
0.5
500 1000
500
500
TIME, msec

FIG. 1

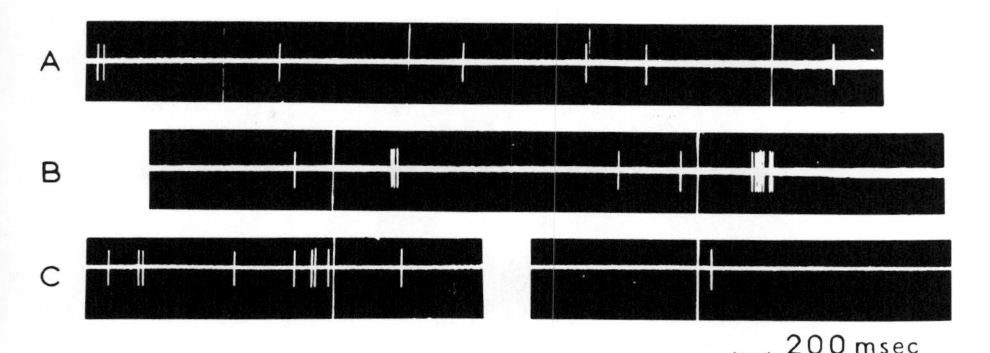

FIG. 2 Spike activity of neuron 44. Note the selective response of the neuron to the signals following with an interval of 2 seconds. (A, B, C) intervals of 1, 2, and 10 sec between the stimuli, respectively. The moments of the stimulus presentation are indicated by vertical bars.

between the flashes. The interval of one sec is an exception; at this interval the response has an incomplete form, because the duration of the reaction of this neuron is greater than the duration of the given interval.

Analyzing the formation of reactions of the neurons which exhibit selectivity with respect to intervals, it must be pointed out that these neurons did not respond to the first flashes in each series. The reactions of these neurons appear after several stimulus presentations. For example, neuron 1 starts to respond only after some period of elaboration. Thus, at the interval of 3 sec the neuron begins to react irregularly after 10 stimulus presentations, but at the interval of 10 sec a regular reaction appeared with 3 stimulus presentations.

Another type of dynamics was observed in the multiphasic selective neurons. In these cases, the reactions are present from the very beginning of application of flashes, but an elaboration is expressed in the form of a gradual intensification of reaction to the stimuli presented with an appropriate interval for this neuron.

Extrapolation Reactions

Other forms of dependence of neuronal reactions upon the time parameters of stimulation were also observed in our experiments. The most interesting is the effect of extrapolation which was clearly expressed in reactions of some neurons.

FIG. 1 (A, B, C) Averaged PST histograms for units 44, 21, and 48, respectively. Selective response to stimuli having a definite interval. Duration of the period analysed is 1 sec. The abscissa is the time in milliseconds; the ordinate is the averaged (for 20 periods) values for the number of spikes during 100 msec. Intervals between the stimuli equal 1, 2, 3, 5, 10 (and for 44, 15) sec. The first histogram in each set is for the spontaneous activity.

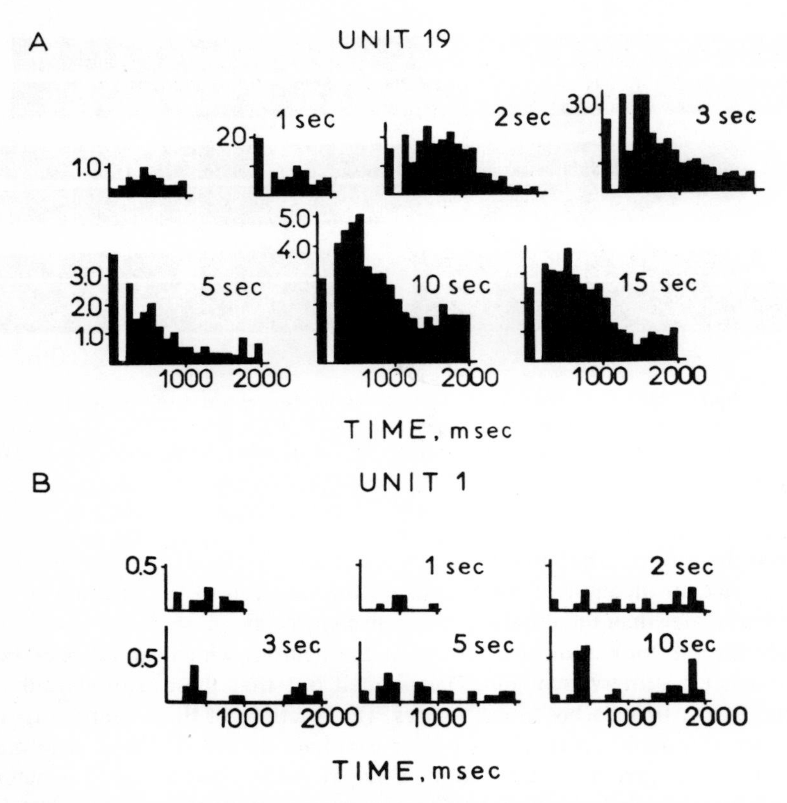

FIG. 3 (A, B) Averaged PST histograms for units 19 and 1, respectively. Rare selectivity to a 10-sec interval. The duration of the period analyzed is 2 sec. Designations are the same as in Fig. 1.

One such neuron showed an ability to extrapolate at all the intervals between the stimuli. This neuron (No. 21), which had no spontaneous activity, responded with a short burst to the flashes. During repetition of signals, the effect of extrapolation arose in the form of 1–3 spikes, appearing directly before the action of the next stimulus. At the beginning of the new series, the extrapolation effect emerged in conformity with the previous interval. Then, a transition to extrapolation in conformity with the next interval gradually developed. When the light stimulus was discontinued, a spike discharge appeared in the background from one to three times with the same interval at which the flashes previously followed one another. Figure 5, presenting PST histograms of an extrapolating neuron (No. 21), shows that with the increase of the intervals between the flashes, the interval between the response to flash and the moment of emergence of extrapolating spikes increases as well. The considerable dispersion of extrapolating spikes in time, which can be seen in this figure, is accounted for by the process of the adjustment of the neuron to a transition

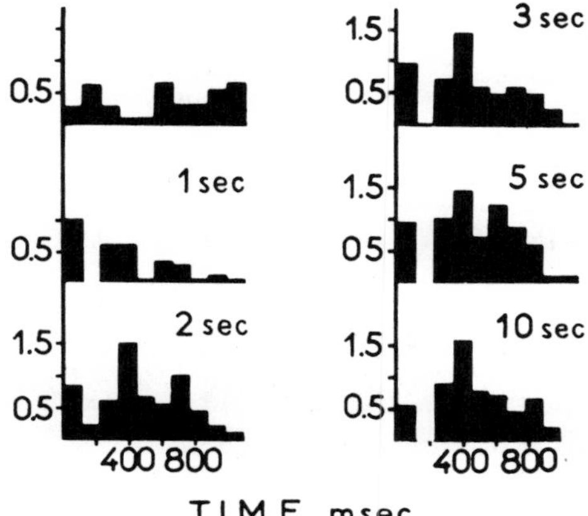

FIG. 4 Averaged PST histograms for a nonselective neuron (unit 23). The duration of the period analysed is 1 sec. It can be seen that at all intervals between the stimuli the response remains stable except for the interval of the 1 sec, at which the response is incomplete. Designations are the same as in Fig. 1.

from extrapolation of the interval of the preceeding series to that of the interval of the given series. Figure 6 demonstrates the reactions of neuron 21 to stimuli given with three different intervals. It can be seen that the extrapolating discharge emerges directly before the action of the stimulus. So we see that this particular neuron is characterized by the development of anticipatory responses

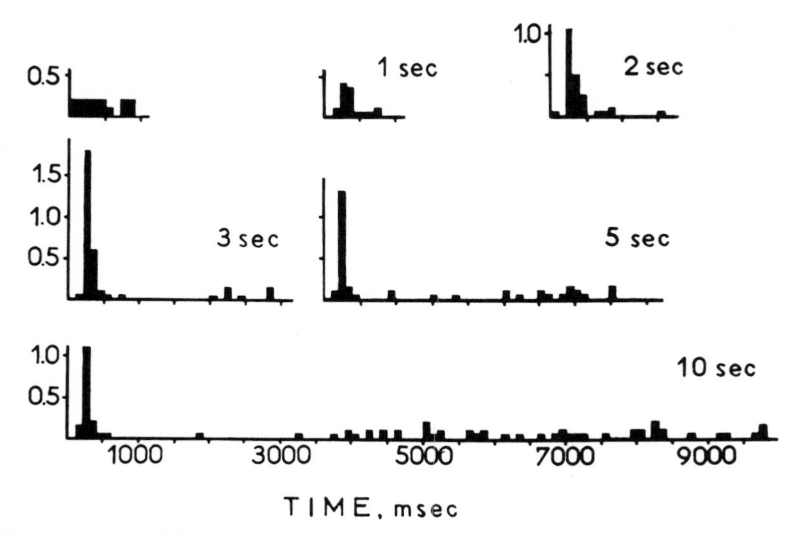

FIG. 5 Averaged PST histograms for an extrapolating neuron (unit 21). The abscissa is time in msec; the ordinate is averaged (for 20 periods) values for the number of spikes during 100 msec. Intervals between the stimuli equal 1, 2, 3, 5, and 10 sec. The first histogram is for the background activity.

FIG. 6 Spike activity of an extrapolating neuron (unit 21) at intervals 3 (A), 5 (B) and 10 (C) sec between the stimuli. It may be seen that the neuron discharges immediately before the presentation of the flash. The moments of stimulus presentations are indicated by vertical bars.

at all intervals used in our experiments. Therefore, we may consider it to be a universal extrapolating unit in the visual cortex of the rabbit.

Furthermore, we observed some neurons which showed formation of the extrapolating effect only at one definite interval between the stimuli. Among them, three neurons with a "burstlike" spontaneous firing pattern gave the extrapolating effect only at an interval of 5 sec between flashes. This reaction appeared after 4–5 flash presentations. Figure 7 shows the spike activity of one such neuron (No. 30). At an interval of 2 sec, it reacts with a group of spikes, emerging with a long latency (Fig. 7B). Further, up to the end of the interval between the stimuli, spike activity does not appear. At an interval of 5 sec between the flashes, an extrapolating response appeared immediately before application of the flashes and continued for some time after the flash presentation (Fig. 7C). Figure 8 presents the averaged PST histograms of this neuron. This figure shows that at the interval of 5 sec, the excitatory reaction is followed by a pause of about 3 sec. Then the spike activity begins to appear, its maximum coinciding with the next signal. At an interval of 10 sec, two maxima are observed: in the middle of the fifth second and in the middle of the tenth second. Analysing the reactions of this neuron, we can conclude that the neuron forms extrapolating reactions at an interval of 5 sec, does not form them at the smaller intervals, and manifests these reactions at the large intervals on the basis of the optimal one (5 sec).

Four other neurons showed the extrapolation effect at the small intervals, from 1 to 3 sec. The effect was absent at longer intervals.

DISCUSSION

The analysis of these experimental data shows that, during application of flashes with different intervals, two effects, depending on the time parameters of stimulation, may be observed. First, there were neurons reacting selectively to one definite interval between the flashes. This selectivity was expressed as an intensification of response selectivity or responses that were maximal for one definite interval between stimuli. Second, we observed the effect of extrapolation expressed in the gradual formation of anticipatory reactions to the stimuli given with constant intervals. It is evident that the ability of neurons to extrapolate might be universal (at all intervals between the stimuli), as well as selective (appearing only at a definite range of intervals). In considering these partial effects of extrapolation, there exists a danger of confusing the real extrapolating effect with an accidental coincidence of the given interval with the

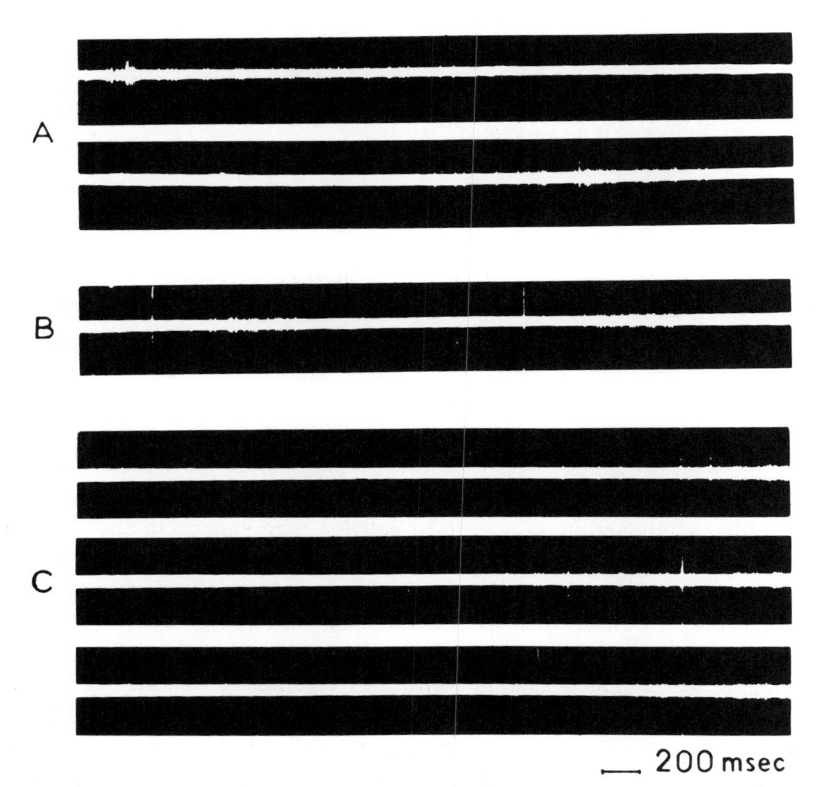

FIG. 7 Spike activity of neuron 30. Effect of extrapolation at the interval of the 5 sec. (A) Burstlike spontaneous activity; (B) monophasic reaction at the intervals of 2 sec between stimuli; (C) reactions at the interval of 5 sec. The moments of stimulus presentation are indicated by vertical bars.

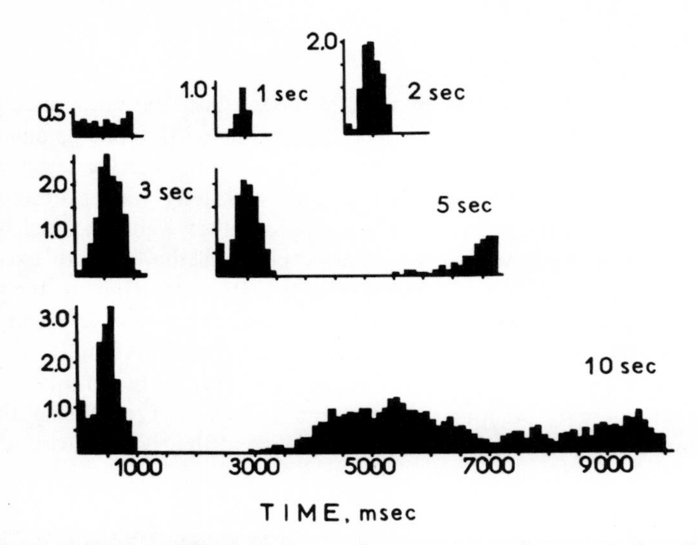

FIG. 8 Averaged PST histograms for unit 30. At an interval of 5 sec between the flashes, an anticipatory response appears. At an interval of 10 sec, two maxima are observed: in the middle of the fifth sec and in the middle of the tenth sec. Designations are the same as in Fig. 1. The first histogram is for background activity.

spontaneous periodicity in the neuronal discharge. But in the described cases no periodicity in the spontaneous activity of neurons was observed.

The data obtained may be explained with the help of the concept of a "time-receptive field." A "time-receptive field" is regarded as "a neuronal net which reacts when two excitations–immediate and delayed–coincide on one summating neuron. Depending on the time of the delay, such a neuronal net must be selectively sensitive to a definite interval between the stimuli" (Sokolov, 1966).

In order to explain the extrapolation effect on the neuronal level, it should be assumed that a circulation of the spike activity may emerge in the chain of neurons which form the "time receptive field." It is probable that the output signals of neurons of the "time receptive fields" are summated on neurons of the second order which ensure the occurrence of the extrapolation effect within a definite range of time intervals.

It may be also assumed that neurons with reactions depending upon time parameters may be regarded as "detectors of time," selectively responding to definite intervals between stimuli. The excitation of a number of "detectors of time" is summated on the extrapolating neurons. Depending upon the range characteristics of the convergent "detectors of time," the extrapolating neurons may form their reactions to a wide range of different time intervals.

6

RESPONSES OF THE RABBIT'S VISUAL NEURONS TO A SIMULTANEOUS COMPLEX STIMULUS OF SOUND AND LIGHT

V. B. Polyansky
E. N. Sokolov

Moscow State University

The development of the microelectrode technique has ensured the possibility of studying the phenomenon of integration on the level of a single neuron. Two aspects of research into this problem may be distinguished: (1), investigation of the neuronal mechanisms, which determine the elaboration of the conditioned reflex at the level of a single neuron, by reproducing the classical experiments where an indifferent stimulus is paired with an unconditioned one; (2), investigation of the mechanisms of integration of stimuli into a complex, which ensures the differentiation of the complex stimulus from the individual components of the complex. In his work, I. P. Pavlov (1952) showed in particular that after elaboration of a conditioned reflex to a complex of sound and light, the differentiation may be obtained to light and sound stimuli presented separately. The specific character of the integrative processes which take place when the stimulus complex is formed manifests itself in the fact that distruction of the cortical part of the analyser leaves conditioned responses to elementary stimuli unaffected.

Whereas a considerable number of works have already been devoted to studying the formation of an elementary conditioned reflex at the neuronal level, the neuronal mechanism of the reactions to complex stimuli has as yet not been investigated. The first stage of such research must apparently consist of studying the "unconditioned" reactions of the cortical neurons which are specific with respect to a complex stimulus. The purpose of the present work is to analyse the neuronal responses of the rabbit's visual cortex to a complex stimulus consisting of sound and light.

METHODS

The activity of single neurons in the visual cortex of a waking rabbit was recorded by tungsten microelectrodes with a tip diameter of 1–2 μ. The insertion of the microelectrode was effected by means of a hydraulic Hubel micromanipulator (Hubel, 1959). Flashes of a gas-discharge tube situated at a distance of 30 cm from the eye, contralateral to the hemisphere under investigation, as well as short clicks with a duration of 50 μsec, served as stimuli. The light and sound stimuli were presented both separately (first 10–30 applications of the light stimulus and then a similar number of applications of the sound stimulus), and jointly, as a complex of sound and light. The intensity of the flash equaled 3.12 lumens; the intensity of the click was 3.6 joules. The intensities remained constant during the presentation of the stimuli both separately and together. Two cells were subjected to the action of light and sound stimuli of several other intensities. For the sake of brevity, the intensities of the light and sound stimuli will be further designated as follows:

Intensity I: light–1.25 lumens, sound–1.4 joules;
Intensity II: light–3.12 lumens, sound–3.6 joules (this intensity was used with all cells);
Intensity III: light–5.2 lumens, sound–6 joules;
Intensity IV: light–7.8 lumens, sound–9 joules.

The interval between the stimuli in the series lasted 1–3 sec.

The spike activity was recorded from the screen of an oscilloscope on a continuously moving film. After the experiment the spikes were counted under a magnifier. Data relating to each 10 stimuli were averaged according to time periods of 50 or 100 msec from the onset of the stimulation. On the basis of these data, a table was formed, and PST histograms were formed.

RESULTS

Forty neurons of the rabbit's visual cortex were investigated in accordance with the previously mentioned program. The present study presents a qualitative and quantitative analysis of the six most typical neurons of those which were investigated. The following values were calculated for each of the six cells: the level of background activity for the period of 100 msec, as well as the number of spikes for periods of 0–100 msec, 100–200 msec, 200–400 msec, as well as for a period of 0–400 msec and 0–1200 msec from the moment of stimulus application for series of 10 successive presentations. Such time periods were selected because an analysis of the results obtained showed that the main differences in the neuronal response appeared in the initial period of 400 msec. According to Bishop and O'Leary (1936), Evarts, Corwin, and Huttenlocker (1960), Jarcho (1949), Kondratjeva and Volodin (1966), Pearlman (1963), Polyansky (1967), and Polyantsev and Serbinenko (1962), the recovery cycle of the evoked responses and neuronal responses of the cortex usually is completed in a period of 350–400 msec after the onset of the stimulus. The critical period of time, on the basis of which this ratio was calculated, was determined by us to equal 400 msec.

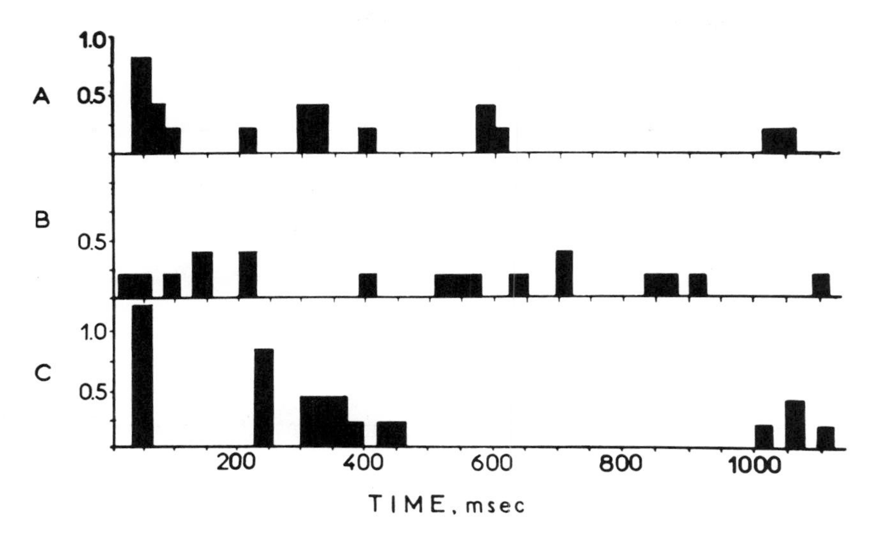

FIG. 1 Histograms for the responses of cell 120 to light (A), sound (B), and a complex of sound and light (C). The abscissa is the time in msec from the beginning of stimulation. The ordinate is the number of spikes for every 50 msec during 10 presentations of the stimulus. The intensity of the flash in all the cases was 3.12 lumens; intensity of the click was 3.6 joules.

The following values relating to the same six cells and the same time intervals (from the onset of the stimulation) were also calculated:

1. ratio between the number of spikes in the response to the light stimulus and the number of spikes in the background activity for the same interval;
2. same, for the response to the sound stimulus;
3. same, for the response to the complex; and
4. ratio between the number of spikes to the complex of sound + light and the algebraic sum of the neuronal responses to separate applications of the light and sound stimuli.

In addition, we investigated the changes of the number of spikes in the neuronal responses during successive applications of the stimuli.

An analysis of the responses of the six neurons made it possible to divide the latter in two groups: (1) neurons with a weakened response to a complex (in relation to the algebraic sum of the responses to the light and sound stimuli presented separately), and (2) neurons with an intensified response.

1. Neurons with a Weakened Response to a Complex

Cell No. 120 (Fig. 1). In the PST histogram of the neuron it can be seen that a well-pronounced response is evoked by the light stimulus. It consists of 2 discharges (one within a period of 25–100 msec, and the second within a period

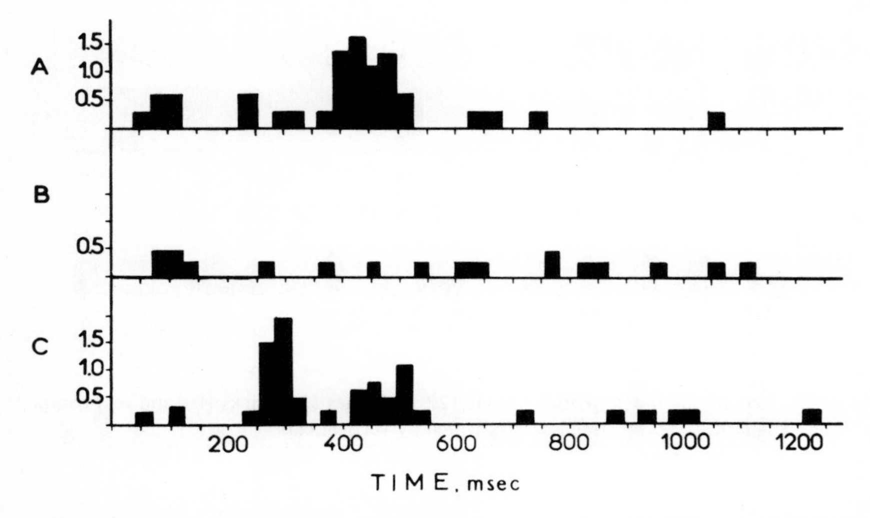

FIG. 2 Histograms for the responses of cell 121 to light (A), sound (B), and a complex of sound and light (C). Other designations are as in Fig. 1.

of 325–350 msec), between which inhibition is observed. The response to the sound stimulus is weak. A combined application of sound and light evokes a highly contrasting response with a redistribution of the spike activity (the first discharge within a period of 25–50 msec, the second within 225–250 msec) and with a very distinct period of inhibition between the discharges. Although the total number of spikes in the response to the complex is smaller than the algebraic sum of the responses to its components (Table 2), the first and second "narrow" discharges themselves exceed the algebraic sum of the spikes which arise in the same place in the responses to isolated sound and light stimuli. An essential feature of the response to a complex is its much greater stability and its more distinct character in comparison with the more diffuse responses to the light and sound stimuli presented separately.

No modifications of the response pattern were observed in the given cell as a result of repeated presentations of the stimuli.

Cell 121 (Fig. 2). The response to the light stimulus bears a distinct character (the first discharge 50–100 msec; the second, 375–475 msec), while the response to the sound stimulus is not as clear (75–100 msec). In the response to a complex stimulus, almost no primary discharge is observed, but a well formed and distinct secondary (250–275 msec) discharge, as well as a tertiary (400–475 msec) discharge (which are all separated by strongly pronounced inhibition) manifest themselves. And although the total number of spikes in the response to the complex is smaller than the algebraic sum of the response to the isolated light and sound stimuli, a redistribution of the activity in the response to the complex takes place in the given cell too. A new secondary discharge is formed; it is followed by a period of inhibition, while the tertiary discharge becomes

FIG. 3 Records of the responses of cell 122 to flashes (A), clicks (B), and to a complex of sound and light (C). The stimuli are indicated by vertical bars.

more compact. No appreciable decrease or increase of the discharge as a function of the serial number of the stimuli presentation were observed in this cell either.

Cell 122 (Fig. 3). The responses to the light and sound stimuli are of a pronounced character, the sound stimulus evoking a weaker reaction than the light stimulus. A combined application of these stimuli weakens the response, and again, as before, the maximum of the response to the complex does not coincide with the maxima of the responses to either the light stimulus or the sound stimulus. This given cell, in comparison to all other cells, manifests the greatest decline of its spike activity in the presence of a complex stimulus. In general, it must be pointed out that according to the preliminary results, the sound stimulus more often inhibits the spike activity of the neurons of the visual cortex, when acting both separately and in combination with the light stimulus. This particular cell, just as the previous ones, did not exhibit any modifications of the response as a result of repeated presentations of the stimulus.

Cell 131 (Fig. 4). The response to the light stimulus is pronounced, while the

FIG. 4 Records of the responses of cell 131 to light (A), sound (B), and a complex of light and sound (C).

TABLE 1

Cell number	Treatment[a]	Background activity spikes/100 msec	Number of stimulus applications	Number of spikes for intervals of time (msec) after stimulus application				
				0–100	100–200	200–400	0–400	0–1200
120	Background	3						
	Light (2)		10	7	0	5	12	17
	Sound (2)		10	2	3	2	7	18
	Light + Sound (2)		10	6	0	9	15	22
121	Background	1						
	Light (2)		10	4	0	33	37	75
	Sound (2)		10	8	2	4	14	36
	Light + Sound (2)		10	4	0	42	46	77
122	Background	7						
	Light (2)		10	17	3	22	42	52
	Sound (2)		10	0	12	20	32	70
	Light + Sound (2)		10	4	0	18	22	31
131	Background	6						
	Light (1)		10	1	4	19	24	67
	Light (2)		10	6	0	30	36	67
	Light (3)		10	10	10	30	50	85

	Light (4)		10	11	18	28	57	115
	Sound (1)		10	6	6	13	25	75
	Sound (2)		10	4	4	18	21	48
	Sound (3)		10	5	5	7	17	66
	Sound (4)		10	5	5	8	18	67
	Light + sound (1)		10	4	3	16	23	54
	Light + sound (2)		10	1	0	14	15	51
	Light + sound (3)		10	8	2	13	23	81
	Light + sound (4)		10	4	8	13	25	68
180	Background	7						
	Light (1)		30	13	38	34	85	111
	Light (2)		30	9	19	45	72	113
	Light (3)		30	23	10	40	73	156
	Sound (2)		30	4	5	14	23	82
	Sound (3)		30	21	0	48	69	205
	Light + sound (2)		30	21	72	72	165	225
	Light + sound (3)		30	31	37	61	129	266
200	Background	1, 3						
	Light (2)		30	1, 25	2, 0	2, 65	5, 95	14
	Sound (2)		30	0, 33	1, 97	2, 7	5, 0	7
	Light + sound (2)		30	4, 3	8, 0	0, 4	12, 7	13, 7

[a]Numbers in parentheses indicate intensity of stimulation as follows:

Light: 1 = 1.25 lumen sec; 2 = 3.12 lumen sec; 3 = 5.2 lumen sec; 4 = 4.8 lumen sec

Sound: 1 = 1.4 joules; 2 = 3.6 joules; 3 = 6 joules; 4 = 9 joules.

TABLE 2

Cell number	Time interval	Light / Background (%)	Sound / Background (%)	Sound + Light / Background (%)	(Light + sound) complex / Light + sound individually (%)
120	0–100	230	66	200	66
	100–200	0	100	0	0
(2)	200–400	83	33	150	128
	0–400	100	56	125	79
	0–1200	47	50	61	63
121	0–100	400	800	400	33
	100–200	0	200	0	0
(2)	200–400	1650	200	2100	113
	0–400	925	350	1050	82
	0–1200	625	300	640	69
122	0–100	243	0	57	23
	100–200	43	171	0	0
(2)	200–400	157	142	128	43
	0–400	150	114	79	29
	0–1200	62	83	36	25
131	0–100	16	100	66	57
	100–200	66	100	50	10
(1)	200–400	160	108	133	53
	0–400	100	104	96	47
	0–1200	93	104	75	24
(2)	0–100	100	66	16	30
	100–200	0	66	0	0

	200–400	250	108	117	13
	0–400	150	87	62	26
	0–1200	93	66	70	35
(3)	0–100	160	83	133	50
	100–200	160	83	33	32
	200–400	250	58	108	35
	0–400	208	70	96	33
	0–1200	118	91	112	37
(4)	0–100	183	83	66	39
	100–200	300	83	133	45
	200–400	233	66	108	53
	0–400	237	75	104	33
	0–1200	159	93	94	37
180	0–100	128	57	300	161
	100–200	271	71	1028	300
(2)	200–400	321	100	228	54
	0–400	293	82	582	172
	0–1200	134	96	268	117
(3)	0–100	329	300	443	70
	100–200	143	0	528	370
	200–400	285	342	437	69
	0–400	296	247	460	90
	0–1200	185	244	267	74
200	0–100	96	25	330	240
(2)	100–200	154	150	615	201
	200–400	102	104	15	7
	0–400	114	96	244	116
	0–1200	84	42	82	65

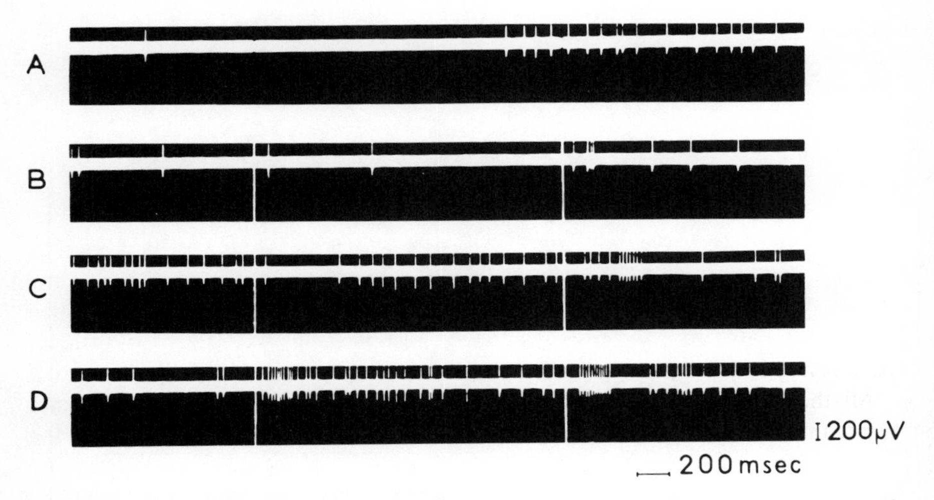

FIG. 5 Records of the responses of cell 180 to flashes (B), clicks (C), and a complex of sound + light (D). (A) Background neuronal activity.

response to the sound stimulus is not. The response to the complex of sound and light resembles in its pattern the response to light. This cell was investigated during the action of sound and light stimuli (both separately and in a complex) of different intensities (Tables 1 and 2). An increase in the intensity of the light stimulus results in a greater number of spikes during almost all the time periods (except an interval of 100–200 msec for the second intensity). A sound of the 1st intensity does not evoke any reaction whatever. Sounds of higher intensities inhibit the reaction only insignificantly, and the stimulated activity does not differ much from the background activity (Table 1).

The response to a complex of sound and light at all intensities of the stimulus is smaller than the algebraic sum of its components, but at each intensity the pattern of the neuronal response is highly characteristic and differs from others (Table 1). This particularly relates to the periods of 0–100 msec and 100–200 msec. It can be seen that in these columns there is no gradual growth of the number of spikes as a function of the increase of the stimulus intensity. Apparently, a "struggle" between excitation and inhibition is taking place, and owing to this, the pattern of the response sharply changes (it is probable that this facilitates the discrimination of one complex from another). In this given cell, like in all other cells, the response does not change when the stimulus is presented repeatedly.

2. Neurons with an Intensified Response to a Complex

Cell 180 (Fig. 5). The response to a flash (Fig. 5B) consists of a primary discharge (80–100 msec), a period of inhibition (130–170 msec), and a new neuronal discharge (180–320 msec). The overwhelming majority of spikes is observed during the period between 200 and 400 msec (Table 1). Practically no response to the sound stimulus manifests itself (Fig. 5C), except perhaps a slight inhibition of the background activity during the first 100–200 msec. A certain increase of the number of spikes is observed in the response to the light stimulus in this cell as a result of repeated stimulus presentations, but the pattern of the response remains quite stable.

The first applications of the complex of sound and light evoke a strong neuronal reaction; the maximum of the response clearly shifts in the direction of 100–200 msec (Fig. 5D, Table 1), while within the period of 200–400 msec the activity declines and inhibition develops. The response to the complex has a distinct pattern (differing from that of the response to light) and is quite stable: As a result of repeated applications, it increases only to a small degree (to 5%).

All the facts described above relate to the second intensity of the sound and light stimuli. If the intensity is increased (Tables 1 and 2), the ratio between the response to the complex and the algebraic sum of the responses to the isolated light and sound stimuli decreases in all the time periods, except the period of 100–200 msec (37%). The response to the complex within the period of 400 msec is very close to the algebraic sum of the responses to the light and to the sound stimuli (90%).

Cell 200 (Fig. 6). The response to the light stimulus consists of a primary, secondary, and tertiary discharge, but the number of spikes in it is low. The response to the sound stimulus is represented by a weak primary discharge. The number of spikes in the neuronal response diminishes after a repetition of the sound and light (Fig. 6B–F); this is particularly apparent in the case of the sound stimulus.

The very first applications of the complex of sound + light cause a very strong response which differs from the response to the light stimulus, as well as to the sound stimulus. A new configuration of the response may be clearly seen: a small primary discharge followed by distinct inhibition and a very powerful secondary discharge which is followed by a new period of inhibition. This response decreases with the repetition of the complex, and becomes stabilized by its thirty-third to thirty-fifth application; but even during the thirtieth application of the complex the total number of spikes proves to be greater than the algebraic sum of the responses to the light and sound stimuli. Tables 1 and 2 show that the greatest increase of the response is observed within the period between 0 and 200 msec. Subsequently, the response to the complex becomes comparable to the algebraic sum of the responses to the sound and to the light; within the period between 200 and 400 msec the activity becomes markedly inhibited, and during the whole period of 1200 msec the total number of spikes is smaller than the algebraic sum of the responses to the light and sound stimuli presented separately. Thus, we see an obvious redistribution of the neuronal activity—a contrasting character of the response pattern.

In the given cell no signs of an elaboration of a conditioned reflex are observed after 30 combined applications of the sound and light stimuli; the application of

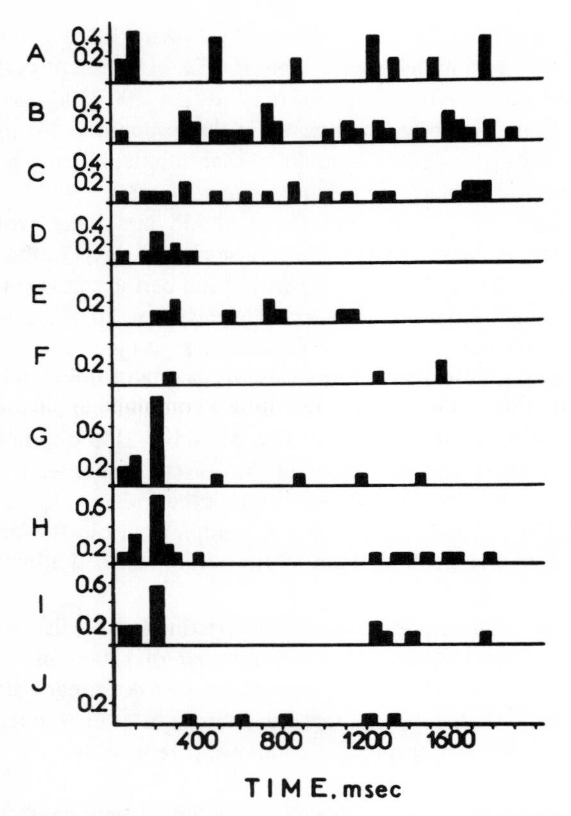

FIG. 6 Histograms of the responses of cell 200 to light (B–C), sound (D–E–F), and a complex of sound and light (G–I). (A) background neuronal activity; (B) averaged responses of the cell to the first 10 applications of the light stimulus; (C) the same for the 11th to 20th applications of the flash; (D) responses of the cell to the first 10 clicks; (E) the 11th to 20th applications of the sound; (F) the twenty-first to thirtieth applications of the sound. The extinction of the response to the sound is clearly seen. (G) the first to tenth applications of the complex of sound and light; (H) the eleventh to twentieth applications of the complex; (I) the twenty-first to thirtieth applications of the complex; (J) the tenth test of the sound, immediately after the applications of the complex were stopped; the sound does not evoke any response similar to that evoked by the light stimulus or by the complex.

clicks after 30 complex stimuli does not result even in a single reproduction of a response which is similar to the response to the light stimulus or to the complex (Fig. 6J). On the contrary, the response to the sound proves to be even weaker than during tests before the application of the complex.

All this probably indicates that the response to the complex is highly specific in this cell.

DISCUSSION

An analysis of the results obtained leads to the conclusion that in some neurons of the rabbit's visual cortex the responses to separate applications of sound and light stimuli, as well as to their combined applications, greatly differ. This difference relates mainly to the initial period of the response (up to 400 msec), whereas their later components are fully comparable with the responses to the sound and light stimuli presented separately, or with their algebraic sum.

As stated previously, the basic difference between the response to the complex and to its components consists of the distinct character of the response, in a shift of the maximum of the discharge to some other place than in the response to separate components of the complex, and in the emergence of new inhibitory phases (cells 120, 180, 200). At the same time a combined application of sound and light evokes in most neurons (cells 120, 121, 122, 131) a smaller response than the algebraic sum of their responses to these stimuli when applied separately.

The change of the response indicates that the action of a complex of sound + light is to some degree specific, although in some cases the sound manifests also an unspecific action (cell 200).

Of all the neurons analyzed in the present paper two are of particular interest–neurons 180 and 200. These neurons are characterized by a sharp increase of activity, and by a very strong and contrasting response to the complex of sound and light (within a period of 400 msec).

Moreover, in cell 180 the sound stimulus, which by itself does not evoke any response, displays a very powerful action on the neuronal response when applied in combination with a light stimulus. This is in accordance with the data of Buchwald, Hull and Trachtenberg (1967) obtained in experiments with neurons of the cat's frontal cortex. Rutledge and Duncan (1966) and Dubner (1966) also pointed out the facilitating action of two combined stimuli. However, whereas the facilitation may be accounted for by a summation of the responses to the complex of sound and light, the responses of cells 180 and 200 can be explained neither by the algebraic sum of the responses to these stimuli during their separate action, nor by an additional influence exerted by the orienting reflex (see Table 1 and the description of cell 200).

Inhibition plays a highly essential role in the formation of a new discharge pattern (Fig. 5G–H). This inhibition develops within the period of 50–300 msec and is apparently of a recurrent character (Andersen & Eccles, 1962; Fuster, Creutzfeldt, & Straschill, 1965; Krnjevic, Randic, & Straughan, 1964; Polyansky, 1965; Supin, 1967). It may be assumed that the excitation, which is evoked by the sound, limits and "stresses" this inhibition. Then the increase of the discharge may be explained (cells 180 and 200) either by a strong facilitation effect after the inhibition or by postanodic exaltation (Andersen & Eccles, 1962; Andersen & Sears, 1964; Supin, 1967).

A comparison of the responses to stimuli of different intensities in two cells was made in our experiments. In one of these cells (No. 131) an increase of the intensity, which evoked no change in the pattern of the neuronal response as a whole (the response to the complex was smaller than the algebraic sum of the responses to the light and sound stimuli acting separately), exerted certain influence on the redistribution of the number of spikes within periods of 0–200 msec from the moment of the stimulus application. In the other cell (No. 180) an increase of the intensity reduced the ratio between the response to the complex and the algebraic sum of the responses to its components. However, at the given intensity the period of 100–200 msec stands out in bold relief (ratio, 370%). Apparently, in the case of the given cell an increase of intensity leads to a greater difference between the response to the complex and those to its separate components within a short time interval (100–200 msec). Besides, according to the results obtained by some investigators (Li & Chou, 1962; Polyansky, 1965; Supin, 1967), an increase in the intensity of the stimulus to a certain point results in a more rapid development of inhibition (in comparison with excitation), and the number of spikes in the response diminishes. Therefore, although we related the given cell to the group of neurons having an intensified response to a complex (such a response manifested itself at the second intensity of the stimuli, according to our observations, of stimuli of moderate, and optimal intensity), it can be seen that the division of the neurons into groups with intensified and weakened responses to a complex is of a rather relative character.

An analysis of the responses of all six neurons to a complex, shows that in some of them (cells 120, 180, and 200) the following specific features are clearly distinguished:

1. Strongly pronounced distinctive effect and greater stability of the response to a complex in comparison with the responses to its components (light and sound).
2. Redistribution of the neuronal activity (as regards the time and the number of spikes) in the response to a complex in comparison with the responses to its components.

Such a great difference between the responses to a complex and those to its components cannot be explained by an orienting reaction. In neurons 120 and 180 the response to the complex was stable and did not become extinguished, while in neuron 200 the response to subsequent applications of the complex decreased (Fig. 6) to a certain stable level; but even in this case the response to the complex was greater than the algebraic sum of the responses to its components.

The responses of these cells to the complex can hardly be accounted for by an elaboration of a temporary connection, since these responses arise with the very first applications of the complex; after 30 tests of the complex the application of the sound stimulus alone (Fig. 6J) does not produce a response similar to that which is evoked by the complex or by the light stimulus.

All this allows us to assume the existence, in the rabbit's visual cortex, of neurons which specifically react to a complex stimulus.

Further research in this direction will help to elucidate the role played by such "detectors" of complexes in the formation of a conditioned reflex to a complex stimulus.

PART B
Neuronal Behavior in Selected Brain Loci

The chapters in this section provide data on different regions of the rabbit's brain, but many use the same technique of applying repetitive stimulation and testing for dishabituation and also spontaneous recovery. Sokolova (Chapter 7) finds a correlation between EEG desynchronization and neuronal discharges in the motor cortex, but only for units which are activated both by sensory and reticular formation stimulation. Also, the discharges of units in the reticular formation are correlated with the state of the EEG. Relationships between the electrocorticogram and neuronal discharges in the non-specific thalamic system are reported in two papers by Danilova (Chapters 14 and 15). The extreme vulnerability of thalamic spike bursts to sensory stimulation is evident in both chapters. Within the sensory systems, Skrebitksy (Chapter 8) finds plasticity in visual cortex cells. However, this does not only take the form of a decrement with repeated stimulation, but also of potentiation. As the author notes, the potentiated neurons could participate in extinction of the OR if they were cells which inhibited those closer to the effector side of the organism. Whatever models are brought forth, however, these findings suggest that even if the visual cortex does contain many finely tuned "feature detectors," there also exist cells which seem to have other functions. At a lower level of the visual system. Dubrovinskaya (Chapter 9) finds response stabilization to repeated visual stimulation, not widespread habituation. She emphasizes the importance of depth of recording within the tectum, believing habituating cells are to be found only ventrally within the superior colliculus. Working with auditory stimulation, Bagdonas (Chapter 10) found stabilization to characterize inferior colliculus neurons; little habituation was found. Beteleva (Chapter 13) examines the responses of lateral geniculate cells to nonvisual stimulation, finding that acoustic and reticular formation stimulation have similar effects upon neurons in this visual nucleus. She suggests that the acoustic effects upon the visual system are mediated by the reticular formation. Vinogradova (Chapter 11) adds a major contribution to the literature in reporting extensive investigations of the properties of hippocampal neurons. The importance of finding neurons whose response to sensory stimulation is systematically modified by stimulus repetition, and which exhibit other characteristics of the orienting reflex, cannot be overemphasized. This chapter is likely to stand for a long time as a basic work for those interested in the orienting reflex, whether or not the theoretical model survives the vagaries of science. It should also be pointed out that the data presented represent activity from only fields CA 2 and CA 3. Neurons of the caudate nucleus behave quite differently from those in the hippocampus (Vinogradova and Sokolov, Chapter 12). Although neurons in the caudate exhibit selectivity to sensory stimulation of various modalities, they do not exhibit the dynamics found in hippocampal neurons; caudate cells exhibit discharge modification, but only after a very large number of stimulus repetitions.

7

MICROELECTRODE INVESTIGATION OF THE ACTIVATION REACTION IN THE MOTOR AREA OF THE RABBIT'S CORTEX

A. A. Sokolova

USSR Academy of Medical Sciences

Investigations carried out in recent years showed that most neurons of the motor area of the rabbit's cortex are polysensory, i.e., respond to stimulations of different modalities (Buser, Imbert, 1961). These data were first obtained on encéphale isolé preparations of the cat. Subsequently, they were confirmed for both cats and rabbits on preparations immobilized by means of muscle relaxants (Kotlyar & Shulgovsky, 1966; Vasilevsky, 1965) as well as on unrestrained waking rabbits (Shulgina, 1967; Voronin, 1966; Voronin & Skrebitsky, 1967). The possibility of conducting microelectrode investigations without applying special immobilizing and narcotic agents (i.e. in conditions approximating chronic experiments) makes the rabbit most suitable for studying the neuronal responses to different stimuli. Besides, simultaneous EEG recording makes it possible to control, to some extent, the animal's state (the degree of its wakefulness).

Our aim was to investigate experimentally in waking rabbits the responses of the neurons of the cortical motor area to stimuli of different modalities and to correlate them with the arousal reaction in the EEG.

METHODS

The rabbit was fixed in a stand in an extended position. The microelectrode was inserted with the aid of a miniature micromanipulator fastened on the bone of the skull by means of dental cement. The activity of the neurons in the motor area was recorded extracellularly from the point where stimulation by an alternating current evoked a movement of the contralateral foreleg. Simultaneously an EEG of the same area was recorded through the electrodes implanted in the bones of the skull, as well as an EMG of the corresponding leg by means of superimposed electrodes (for details see Sokolova & Lipenetskaya, 1966). An electrocutaneous stimulation of the rabbit's forepaw, contralateral to the recording electrode, was used as adequate stimulation (i.e., coming to the recorded area along the lemniscal ascending pathways). The threshold intensity of the stimulation was determined by the appearance of a motor reaction. In addition, we applied light and acoustic stimuli which were considered as inadequate ones for the motor area of the cortex. The presence (or absence) of an activation reaction was ascertained in the simultaneously recorded EEG. In the given investigation an EMG was recorded in order to control the presence or absence of the animal's movements, since, as shown by the investigation, a motor reaction substantially changes the character of the response to the stimulus. This article presents data obtained on 219 neurons recorded during 75 experiments in 15 rabbits.

RESULTS

According to the responses evoked by electrocutaneous stimulations, the investigated neurons may be divided into the following three groups: neurons with an acceleration of activity, neurons with an inhibition of activity, and nonreactive neurons. It should be noted that in most cases neurons of the cortical motor area responded to stimuli with variable, unstable reactions. In view of this, it was necessary to determine the criterion of the presence of a reaction, and the criterion of a reacting neuron. We agreed that the term "neuronal reaction" should imply at least a 50% change of the background frequency of discharge and the term "reacting neuron" a case when the given neuron responds to not less than two stimuli out of the three applied. With this approach, only 124 neurons (out of 219) could be related to any of the previously mentioned groups (to the remaining neurons the electrocutaneous stimulation was applied less than three times). Besides, the group of nonreactive neurons included also a number of neurons whose responses were of an indistinct character.

Quantitatively the neurons were distributed among the groups as follows:

neurons responding with activation:	48	(38%)
neurons responding with inhibition:	22	(18%)
nonreactive neurons and neurons with indistinct responses:	54	(44%)
Total:	124	(100%)

Investigation of the responses of these neurons to stimuli of different modalities was carried out separately for each of the groups mentioned above.

A comparison of the presence of a neuronal response with the presence of an activation in the EEG showed that *nonreactive* neurons, as a rule, do not respond to stimuli of other modalities either, irrespective of the modifications observed at this time in the EEG.

Both mono- and polysensory units could be recorded among the neurons which become inhibited. No distinct dependence between the activation in the EEG and the neuronal response was established in these neurons.

The responses to stimuli of different modalities were investigated in 37 activatory neurons (out of 48); in 11 cases less than 3 stimulations of one modality were presented, or the stimulation was accompanied by a motor reaction which prevented a correct evaluation of the character of the response to the stimulus.

Six neurons (out of 37) were monosensory, i.e., responded only to electrocutaneous stimuli, without reacting to light and acoustic stimuli, regardless of the

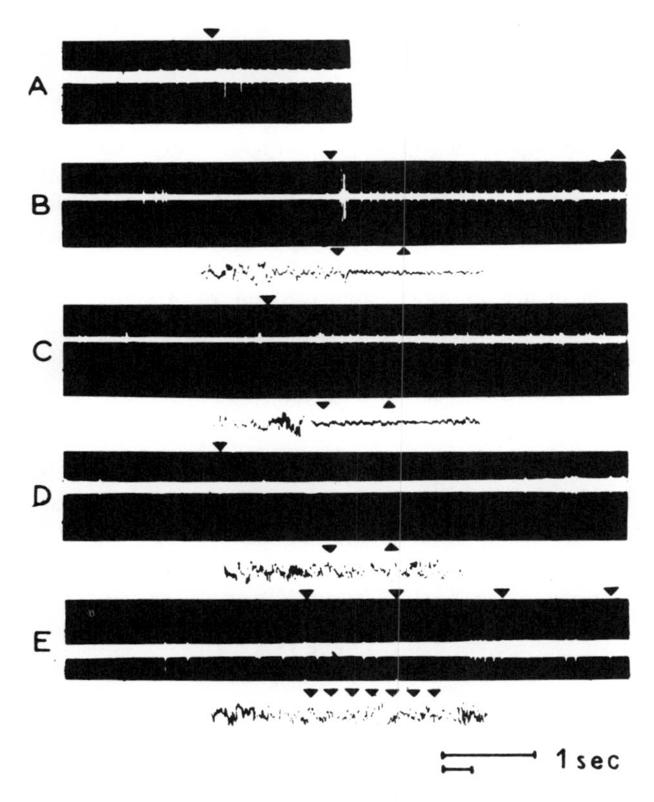

FIG. 1 Correlation between reactions of a polysensory activatory neuron from the rabbit's motor area and EEG activation. (A) The neuronal response to an electrocutaneous stimulus. (B) The neuronal and EEG responses to an acoustic stimulation. (C) The neuronal and EEG responses to a high-frequency light stimulus. In B and C a distinct activation reaction is seen. In D and E the activation reaction is absent. In B, C, D, and E the upper trace is an extracellular record of the neuronal activity; the lower one is an EEG from the motor area of the rabbit's cortex; bipolar recording. The upper time calibration is for the neuronogram; the lower one is for the EEG. (▽) onset of stimulus; (△) offset of stimulus.

presence of an activation reaction in the EEG. The response to electrocutaneous stimulation in these neurons, as a rule, consisted of a high-frequency burst of impulses.

Thirty-one neurons (out of 37) proved to be polysensory, i.e., responded not only to an electrocutaneous stimulation, but also to stimuli of some other modalities (one or two). As a rule, the responses to all stimuli were of a tonic character. A high correlation between the response of the cell and the activation in the EEG was established in this group of neurons; a distinct acceleration of the neuronal activity was observed only in combination with a strongly pronounced activation in the EEG.

Figure 1 presents an example of such a polysensory neuron. It can be seen that an acceleration of spike activity in response to electrocutaneous, acoustic, and light stimuli (Fig. 1A, B, C) is accompanied by activation in a simultaneously recorded EEG (Fig. 1B, C). If stimuli of the same modality do not evoke activation, no acceleration of the neuronal activity is observed (Fig. 1D, E).

Table 1 shows the correlation between the presence of a neuronal response and the presence of an activation reaction in the EEG for all the acoustic and light stimuli applied in the course of recording the given neuron. As it can be seen, in 18 cases (out of 19) the presence of both reactions coincides. A similar calculation made for the whole group of polysensory activatory neurons showed a 90–100% coincidence in most cases (28 out of 31). In the case of three neurons this percentage ranged from 60 to 80, which practically corresponded to full absence of such a coincidence.

A number of additional observations likewise testify to the existence of a correlation between the acceleration of the spike activity and EEG arousal:

1. In some cases simultaneous recording of the neuronal activity and of the EEG showed that the activation in the EEG is retarded in comparison with the beginning of the stimulus. Thus, in Fig. 2A the activation in the EEG begins 1.5 sec after the onset of the acoustic stimulus. Simultaneous recording of the

TABLE I

The Relationship between Responses of a Single Polysensory Neuron to Sensory Stimulation (Light, Sound) and the EEG Activation Reaction in the Motor Cortex

No. Stimulus Presentation	1	2	3	4	5	6	7	8	9	10	11	12	13	14	15	16	17	18	19
Neuronal Response	+	+	–	–	–	+	+	–	–	–	+	–	+	–	–	–	–	+	+
EEG Activation	+	+	–	–	+	+	+	–	–	–	+	–	+	–	–	–	–	+	+
Correspondence	+	+	+	+	–	+	+	+	+	+	+	+	+	+	+	+	+	+	+

+ indicates presence of neuronal response or EEG activation; also, correspondence between these two reactions

– indicates absence of neuronal response or EEG activation

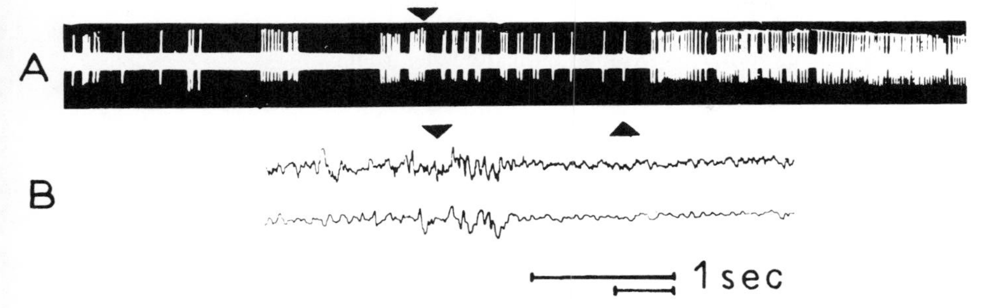

FIG. 2 Retardation of the onset of the neuronal activatory response, as well as of the onset of the EEG activation relative to the moment of presentation of the acoustic stimulus. (A) the neuronal response to an acoustic stimulus; (B) simultaneous record of the EEG from the rabbit's cortex. Designations are the same as in Fig. 1.

neuronal activity shows that its acceleration starts 1.5 sec after the onset of the acoustic stimulus (Fig. 2B).

2. In some cases the EEG activation was observed only as an off-effect of acoustic (or light) stimuli. Simultaneous recording of a polysensory neuron showed that its activation starts when the acoustic stimulus is switched off.

3. As a result of repeated applications of the same tone, activation in the EEG becomes gradually extinguished. During simultaneous recording of the neuronal activity there is a gradual decline in the intensity of the response (until its disappearance) which takes place in parallel with the extinction of the activation in the EEG. When the tone is changed, the activation reaction in the EEG reappears; at the same time the distinct reaction of acceleration of the neuronal activity in response to the acoustic stimulus recovers.

The fact that polysensory neurons usually are more reactive to sound than to light stimuli may be explained by the existence of a correlation between the acceleration of the activity of the polysensory neuron and the presence of activation in the EEG. A comparison of the behavior of neurons which may be regarded as bisensory according to their responses (responding only to electrocutaneous and acoustic stimuli), with simultaneous EEG records, shows that a light stimulus did not evoke any activation.

Proceeding from the generally accepted concept concerning dependence of the EEG arousal on activity of the diffuse activating system, or unspecific system of the brain stem (Jasper, 1949; Moruzzi & Magoun, 1949), it can be assumed that the activatory reactions of the neurons described above are determined by the excitation of the midbrain reticular formation.

In order to verify this assumption, it was necessary to demonstrate that these neurons also respond, with an acceleration of the spontaneous activity, to a direct stimulation of the reticular formation of the brain stem. A special series of experiments was performed to prove that.

The stimulation of the midbrain reticular formation (RF) was performed in this series of experiments through implanted electrodes.

Two forms of stimulation were applied: stimuli of a high frequency (100 per sec) that evoked an activation in the EEG, and stimulations by impulses of low frequency (6 per sec) that did not evoke any activation in the EEG. The methods were described in detail in another work (see Kalinin & Sokolova, 1968). Among the 53 neurons investigated in the course of 18 experiments, 20 units were found to belong to the group of polysensory neurons responding with activation. A correlation of the presence of an activation in the EEG with the presence of a reaction of acceleration in neuronal activity in response to stimulation of the RF was carried out for the aforementioned 20 neurons. Figure 3A shows a distinct activation in the EEG during a stimulation of the RF by a high-frequency current. It can be seen that an appreciable increase in the frequency of the neuronal discharge (from 6 per sec to 12 per sec) corresponds to the moment of activation in the EEG. Figure 3B presents the moment of stimulation of the RF by a low-frequency current. As may be seen, this stimulation does not evoke an activation in the EEG and is not accompanied by a change in the frequency of the neuronal activity (6 per sec in the background and 6 per sec after the stimulation). The same phenomenon manifested itself in all the 20 neurons investigated in this series. They responded with acceleration of activity to a stimulation of the RF by high-frequency impulses. A stimulation of the RF by impulses of low frequency–if it was not accompanied by an EEG activation–did not result in an acceleration of the neuronal activity (Fig. 3B). It should be added that the neurons which did not respond to the applied afferent stimuli neither reacted to a stimulation of the RF.

The data obtained confirm the assumption that the investigated neurons of the motor area of the rabbit's cortex respond with an acceleration of activity to a direct stimulation of the midbrain reticular formation. Thus the assumption of

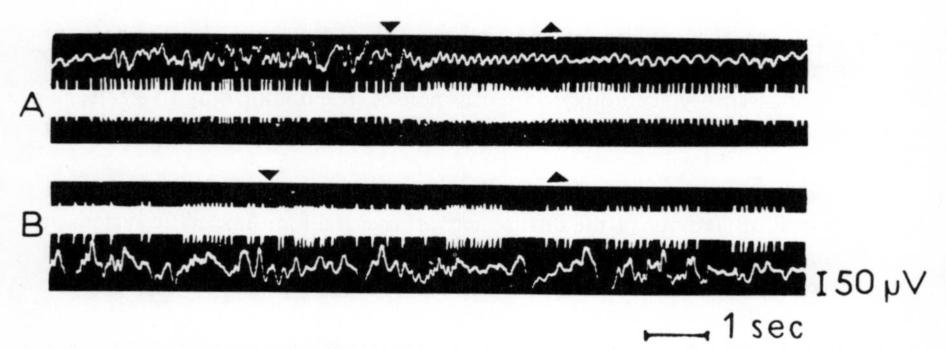

FIG. 3 Modifications of the activity of a polysensory activatory neuron from the motor cortex during a stimulation of the midbrain reticular formation. (A) High-frequency (100 per sec) stimulation evoking activation in the EEG (from the visual cortex). (B) Low-frequency (6 per sec) stimulation evoking no activation in the EEG.

the dependence of these responses on the excitation of the diffuse activating system of the brain stem becomes more substantiated.

From this it follows that neurons of the brain stem reticular formation must possess the same properties as the neurons of the motor area described above.

In order to verify this, a series of experiments, recording the neuronal activity of the midbrain reticular formation, was carried out. The recording was done in the same conditions in which the neurons of the motor area were investigated, i.e., in experiments on waking and unrestrained rabbits. In this series of experiments extracellular recording of neurons of the reticular formation was combined with simultaneous EEG and EMG recording (see the description of the methods), and changes evoked by the presentation of light, acoustic, and electrocutaneous stimuli were investigated. Altogether, activity of 54 neurons of the reticular formation of the midbrain was recorded.

As regards the character of the responses to electrocutaneous stimuli, the investigated neurons were divided as follows: neurons responding with activation (42), neurons responding with inhibition (7), and nonreactive neurons (5). From these figures it follows that most of the recorded neurons of the reticular formation responded with activation. The greater number of these neurons (27 out of the 33 subjected to investigation, i.e., 80%) proved to be polysensory, i.e., they responded both to electrocutaneous and acoustic stimuli (27 neurons), or to stimulations of all the three modalities (7 neurons). In 6 neurons the response to the acoustic stimulus was of an inhibitory (2 neurons) or indistinct character (4 neurons). A study of the correlations between the presence of an activatory reaction of the cell and the presence of an activation in the EEG in response to a given stimulus showed that, for the group of neurons which respond with activation, the number of coincidences of these two reactions amounted in most cases to 90%. Only in the base of 2 neurons was this percentage considerably lower, i.e., no distinct dependence between the investigated reactions was observed.

Figure 4 presents an example of a neuron of the reticular formation which responds with an increase of spontaneous activity to an acoustic stimulus evoking EEG activation (Fig. 4A), and which manifests no response to the acoustic stimulus that does not evoke activation in the EEG (Fig. 4B). The general character of the tonic response of a reticular neuron and the correlations with the activation in the EEG are very close to those which were described above for neurons of the cortical motor area.

Consequently, as was expected in planning the given series of experiments, it proved possible to record, in the midbrain reticular formation, a considerable number of polysensory neurons similar—in the pattern of their response and their dependence on the activation in the EEG—to the neurons described by us in the motor area. For convenience, these polysensory neurons, which respond with tonic activation, will be called "unspecific" neurons.

In the light of the facts described above it was interesting to establish whether such neurons are encountered in a specific area, namely, in the area of primary

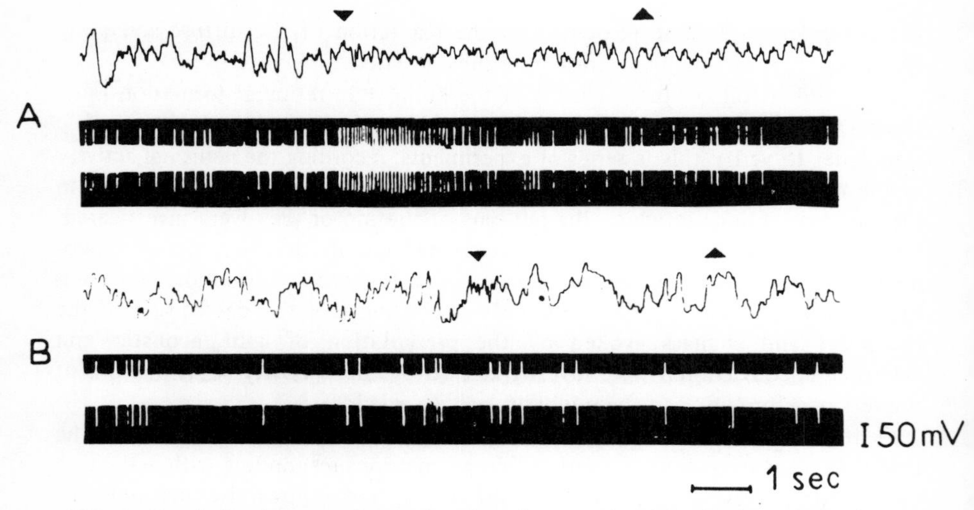

FIG. 4 Correlation between the acceleration of neuronal activity in the reticular formation in response to an acoustic stimulus and the activation in the EEG. (A) Activation in the EEG from the visual area related to the acceleration of the neuronal activity in the reticular formation in response to acoustic stimulation. (B) Lack of both EEG activation and acceleration of the neuronal activity in response to an acoustic stimulus.

sensory projection. With this aim in view, we investigated the neurons of the rabbit's visual cortex under the same conditions as the neurons of the motor area. Altogether 55 neurons were recorded in the course of 20 experiments on 5 rabbits (these experiments were carried out with L. I. Gapich and V. G. Skrebitsky). 18 of these neurons responded neither to acoustic nor electrocutaneous stimuli, 34 neurons were inhibited by acoustic (and electrocutaneous) stimuli, and 3 neurons became activated in response to an acoustic stimuli. In the case of the last 3 neurons, a well-defined connection between the EEG activation and the neuronal response was established. Figure 5 presents an example of a response of one of these neurons to an acoustic stimulus evoking EEG activation (Fig. 5A) and the case when the sound did not evoke any distinct activation reaction (Fig. 5B). It can be seen that in the first case an obvious acceleration of spike activity takes place, while in the second case there is no neuronal response. Figure 5C shows the response of this neuron to a light stimulus which is accompanied by an activation in the EEG. An acceleration of the neuronal activity, though not very pronounced, can be seen. Figure 5D shows that the acceleration of the neuronal activity corresponds approximately to the fifth flash. On the simultaneously recorded EEG it may be seen that this moment corresponds to the appearance of activation in the EEG.

Thus, an investigation of the visual cortex from the point of view of the presence of unspecific neurons in it showed that only a small number of such neurons are encountered in this area (3 out of 50, or 6%).

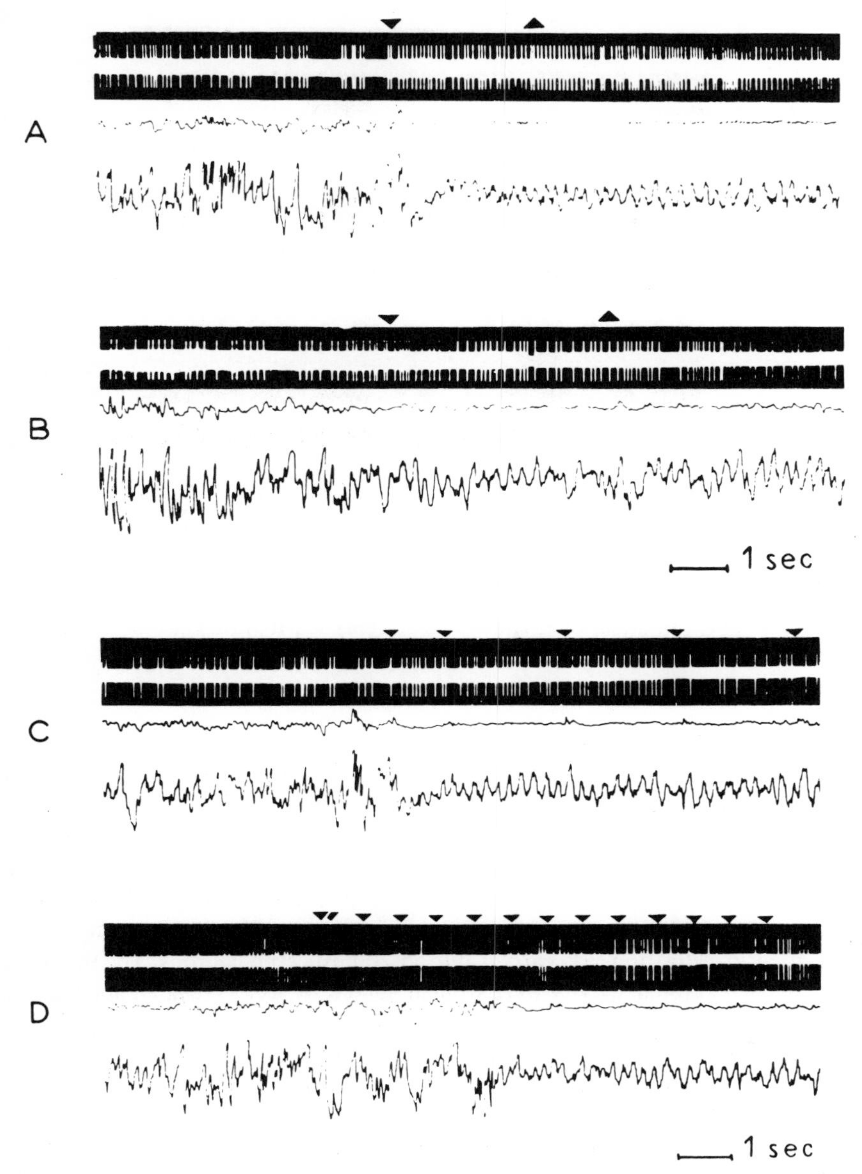

FIG. 5 Correlation between the acceleration of the activity of a polysensory neuron from the rabbit's visual cortex and the EEG activation. (A) Acceleration of the neuronal activity in response to an acoustic stimulus accompanied by an EEG activation. (B) Lack of both the acceleration of neuronal activity and the EEG activation to an acoustic stimulus. (C) Diffuse acceleration of the neuronal activity during an EEG activation to a light stimulus. (D) The acceleration of the neuronal activity starts during the 5th flash, simultaneously with the EEG activation. (A, B, C, D) the upper EEG records from the motor cortex, the lower ones from the visual cortex.

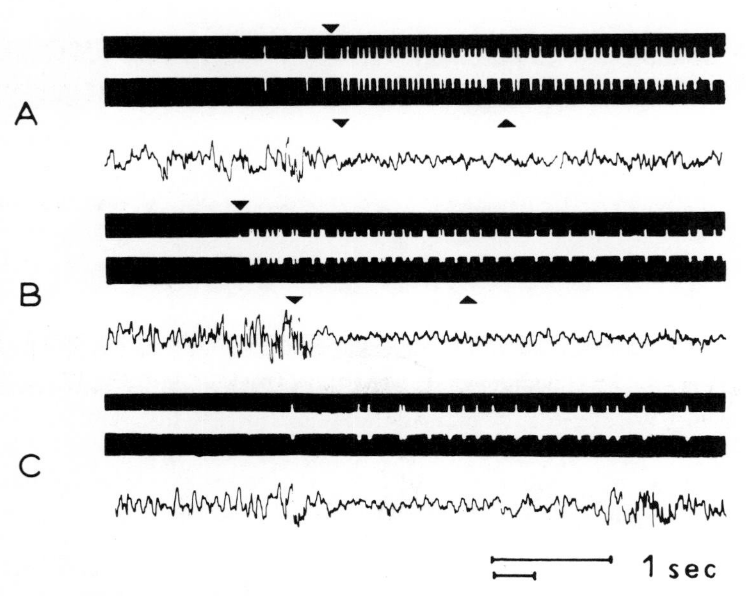

FIG. 6 Different degrees of acceleration of the neuronal activity depending on the intensity of EEG activation. Designations. (top record) activity of the neuron in the motor cortex; (bottom record) EEG from the visual cortex.

Since it was found that an acceleration of the activity of unspecific neurons emerges in response to all stimuli which evoke an arousal reaction in the EEG, the following question arose: Is there any difference between arousal reactions both on the cellular and EEG level to the different stimuli.

In order to answer this question, a comparison was made between the response of the unspecific cells of the motor area mentioned previously and EEG modifications in various cases of an arousal reaction to different stimuli.

Such an investigation was carried out on 20 unspecific neurons. The modifications of the neuronal activity were studied in conditions of repeated applications of light and acoustic stimuli and correlated with the activation in the EEG.

Certain differences in the reaction of acceleration of the spike activity, coinciding with the onset of the EEG arousal reaction, could be observed at different levels of activation in the EEG. Figure 6 shows three cases of such an acceleration of activity in a single cell. The frequency of the discharges in all the three cases proved to be different: 16 per sec, 14 per sec, and 6 per sec (if the frequency is taken for the first sec after the onset of the reaction) (Fig. 6A, B, C). The arousal reaction, which was observed on the simultaneously recorded EEG, was expressed in a desynchronization of activity in the sensorimotor cortex and in the appearance of a theta rhythm in the visual area. The frequency of the theta rhythm in the EEG varied correspondingly to the three variants of acceleration of neuronal activity. In the first case its frequency was 7 per sec, in the second 6 per sec, and in the third 4 per sec (Fig. 6A, B, C). Thus, the investigation revealed parallel differences in the frequency of the spike discharges and in the frequency of the theta-rhythm on the EEG. A more detailed analysis of the neuronal response discloses some other phenomena in the dynamics of the frequency of the neuronal discharges; thus, one observes a regular decline in the frequency of the discharges from the onset of the response to its end (Figs. 1 and 6). A gradual decrease in the frequency of cellular discharges is also observed when the response becomes extinguished as a result of repeated presentations of the stimulus (with simultaneous extinction of the activation reaction in the EEG; Sokolova & Lipenetskaya, 1966). Similar regular changes of the frequency were revealed by an investigation of the rabbit's theta rhythm. These phenomena allow us to assume that the differences in the frequency of the neuronal discharges are associated with different intensities of the arousal reaction.

Apparently, the intensity of the arousal reaction or its "level" varies in different cases, being reflected in the frequency of the unspecific cell discharges (just as in the frequency of the theta rhythm in the EEG of the rabbit).

DISCUSSION

At present no general agreement exists in the literature concerning the existence of a correlation between the neuronal activity and the activation reaction in the EEG. A number of authors (Akimoto & Creutzfeldt, 1957; Akimoto, Saito, & Nakamura, 1961; Creutzfeldt, & Jung, 1961; Vinogradova, 1965) state that such dependence between the activation reaction in the EEG and the acceleration of activity in some cells of the visual area does exist. There are still more definite statements concerning the existence of a connection between the activation reaction in the EEG and the tonic response of an activated neuron in the cat's associative cortex (Saito, Maekawao, Takenaka, & Kasamatau, 1957). Some authors point out that the activity of some cells of the rabbit's motor area becomes accelerated during an arousal reaction in the EEG, whether spontaneous or evoked by the action of a dc stimulation (Creutzfeldt, Fromm, & Kapp, 1962; Jung, Kornhuber, & Da Fonseca, 1963). But there exists quite an opposite point of view, according to which the neuronal activity is in no way associated with the activation reaction in the EEG (Jasper, 1964; Ricci, Doane, & Jasper, 1957; Woldring & Dirky, 1950). Finally, according to some authors, the activity of neurons (of the pyramidal tract) declines at the moment of arousal in the EEG (Evarts, 1963; Okuda, 1963 Whitlock, Arduini, & Moruzzi, 1953). In the research mentioned above the reactive cells were not investigated from the viewpoint of their mono- or polysensory character. A correlation with the activation reaction was performed either for all cells recorded or only for the so-called "pyramidal" neurons. Apparently, this selection of neurons is responsible for the fact that no dependence was detected. According to our data, a definite dependence between the neuronal activity and the arousal reaction in the EEG manifests itself only in the case of the group of polysensory neurons which respond with activation.

In our experiments the acceleration effect was also obtained for the same group of neurons during stimulation of the midbrain reticular formation (RF). This effect was described for the cells of the visual area by Akimoto and Creutzfeldt (1957), Fuster (1961), Narikashvili, Arutyunov, and Moniava (1965), and Skrebitsky and Bomstein (1967); for some cells of the motor area it was described by Krupp and Monnier (1964). These investigations, however, did not correlate the effect of stimulation of the RF with the type of the responses to sensory stimuli.

The established correlation between the activation in the EEG and the acceleration of activity of a definite neuronal group led to the assumption that this group of neurons is associated with the diffuse activating system of the brain stem. This assumption was confirmed by a similarity of these neurons with the neurons of the reticular formation (both with respect to the response pattern and the correlation with the activation in the EEG). As regards neurons of the reticular formation, there exist in the literature very definite statements according to which their activity becomes accelerated during the activation reaction (Bell, Buendia, Sierra, & Segundo, 1963; Duensing & Schaefer, 1963; Machne, Colma, & Magoun, 1955; Mollica, Moruzzi, & Naquet, 1953). However, these investigations were carried out predominantly on cat's encephalé isolé preparations. In view of this, it seemed necessary to verify these data on the neurons of the rabbit's reticular formation without apply any relaxants or narcotic agents.

During the investigation of changes taking place in the neuronal activity of this group of neurons during EEG activation, differences were observed in the frequency of the arousal discharges, apparently connected with different levels of the activation. A gradual decline in the frequency of the discharges of the activatory cells, with habituation to the stimulus, was described for neurons of the reticular formation (Horn & Hill, 1964), for neurons of the visual cortex (Skrebitsky & Gapich, 1963), and for neurons of the hippocampus (Vinogradova, 1965). No systematic comparison with the course of extinction of the EEG activation was made in the previously mentioned investigations, but the general course of changes in the frequency of the spike discharges generally coincide with the data obtained in our experiments.

The aforesaid correlations of the frequency of the discharges of the unspecific cells with the frequency of the rabbit's theta rhythm corroborate the opinion that the frequency of the theta rhythm in the EEG of the rabbit depends on the level of excitation (Anokhin, 1963; Voronin & Kotlyar, 1962). Apparently the frequency of the neuronal discharges of the group under investigation, just as the frequency of the rabbit's theta rhythm, may serve as an additional indicator of the level of the activation.

Thus, a microelectrode investigation of the activation reaction revealed the existence of a definite group of neurons which are more closely connected with the unspecific activating system of the brain stem than other cortical neurons. These cells, characterized by a number of properties, in particular by the polysensory and tonic character of their responses, are designated "unspecific" neurons by us.

The neurons of the classical sensory pathway (medial lemniscus) and the surrounding neurons of the reticular formation are classified in a series of investigations carried out by Darion-Smith and Yokota (1966) as specific, or lemniscal and unspecific, or reticular. These terms are used also in a number of other investigations (Carreras & Anderson, 1963; Vasilevsky, 1965). The propositions advanced by Jung *et al.* (1963) differ somewhat from those mentioned above; these authors distinguish specific and unspecific neuronal responses and consider these two kinds of responses to be recorded in one and the same neuron.

Thus far we failed to detect in our records any responses of the "specific" type (according to Jung) in the neurons which were related by us to the group of the unspecific type.

According to the available data, the unspecific neurons are represented in various structures of the brain to different degrees. They comprise the majority of neurons of the reticular formation (65%) but are relatively seldom encountered in the visual cortex (6%). The presence of a considerable number of such neurons in the motor area of the rabbit's cortex (25%) indicates that the latter resembles the unspecific or associative areas as regards its functional properties. These data accord with the opinion of Buser and his co-workers, which is based on investigations of evoked potentials and according to which the motor area of the rabbit may be regarded as an associative one (Cazard & Buser, 1963).

The fact that a correlation with the activation reaction in the EEG is not revealed in all neurons, but only in a definite group of cells (namely, in polysensory neurons with tonic activatory reactions) leads to the assumption that these neurons form a special system that is represented, to a greater or lesser degree, in different parts of the brain.

8

PLASTIC PROPERTIES OF THE VISUAL NEURONS IN THE ALERT RABBIT

V. G. Skrebitsky

USSR Academy of Medical Sciences, Moscow

When studying the electrical activity of single neurons, investigators come across the peculiar phenomenon of systematic changes in cellular responses to rhythmically applied stimuli with stable parameters. Although the responses of most, if not all, neurons of the higher divisions of the nervous system are not stable, and to some degree change from one stimulus presentation to another, it is still possible to single out certain cells with responses that display pronounced and definite transformations as a result of a repeated action of the stimulus. The behavior of these cells reproduces, in a highly simplified form, some plastic responses of the nervous system.

Two basic tendencies in transformation of the neuronal responses may be distinguished: a weakening of the responses during repeated applications of the stimulus, and the intensification of these responses. The first kind of transformation was described by Hubel, Henson, Rupert, and Galambos (1959) in the auditory cortex, by Vinogradova and Lindsley (1963) in the visual cortex, by Vinogradova (1965) in the hippocampus, by Huttenlocher (1961) and Horn and Hill (1964) in the reticular formation of the brain stem. The second kind of transformation was detected by Voronin (1966) in the sensorimotor cortex, and by Bell, Sierra, Buendia, and Segundo (1964) in the mesencephalic reticular formation.

Attempts to explain the plastic properties of central neurons by constructing hypothetic neuronal chains consisting of elements with *a priori* assumed properties (Sokolov, 1965) are presently of growing interest for modelling various forms of the activity of the central nervous system.

METHODS

The experiments were performed on alert rabbits slightly restrained in a special stand. The microelectrodes were inserted into the brain with the help of a remote hydraulic micromanipulator, fixed on the skull of the rabbit. A detailed description of the micromanipulator was given in another article (Skrebitsky & Voronin, 1966). The neuronal activity was recorded extracellularly with glass micropipettes filled with a 3 *M* solution of KCl. Flashes of a gas-discharge tube with a duration of 75 msec were used as light stimuli.

RESULTS AND DISCUSSION

The present paper summarized observations of the responses of visual cortex neurons in alert rabbits to rhythmical flashes of light. We shall consider here only those neurons whose response patterns change; the only factor which determines a change in their responses is a repeated application of the stimulus. On the basis of this criterion the following four main types of neurons may be established:

1. Neurons with potentiated responses. This type is characterized by the appearance of responses in previously nonreactive neurons, and/or by an increase of the number of spikes during repeated applications of the stimulus.

Figure 1A presents a record of the activity of a neuron with such properties. It can be seen that the number of evoked impulses steadily increases from the first flash to the subsequent ones and reaches its maximum in response to the 4th–5th stimuli. In this case, the response develops as postactivation potentiation, described for spinal neurons (Eccles, 1957; Kostyuk, 1959).

Figure 1B shows a record from a neuron whose response to a light stimulus does not appear from the beginning of the light stimulation, but, as a rule, only at the third or fourth flash, irrespective of the stimulation frequency (compare the left and right parts of the record). In this case we apparently deal with a summation of subthreshold influences, or with an increase of the effectiveness of the synaptic transmission, which leads to the emergence of a response when the

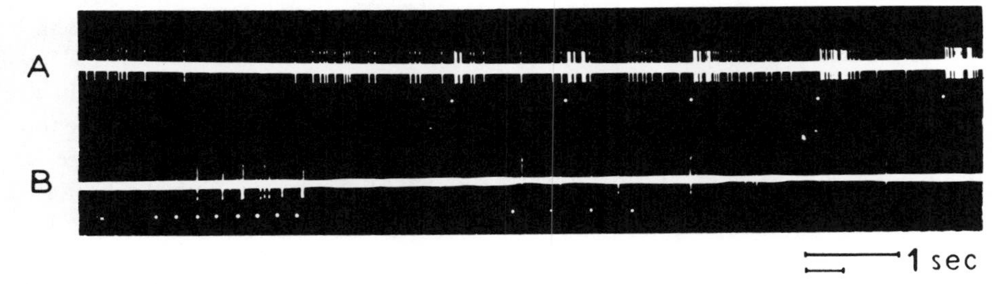

FIG. 1 Neurons with potentiated responses. Responses to flashes of light become intensified as a result of repeated applications of the stimulus. (A and B) records of activity of two different cells. Dots indicate the light stimulus presentation. Details are given in the text. Time calibrations are 1 sec for A (the upper one) and for B (the lower one).

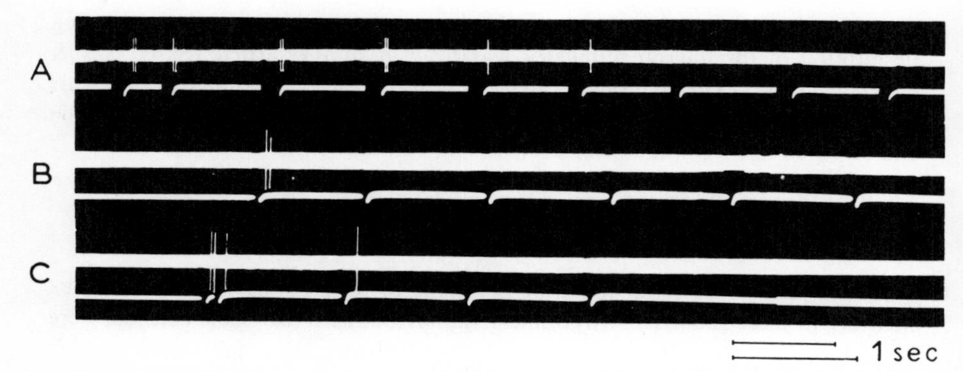

FIG. 2 Neurons responding to the first stimuli: (A) first neuron; (B and C) second neuron. In (A), (B), and (C), the upper beam denotes activity of the cell; the lower beam shows light stimulus artifacts A thickening of the lower beam on (C) is the sound presentation. Details are given in the text. The calibrations are 1 sec for A (the upper one) and for B and C (the lower one).

threshold level is reached. This case may be also regarded as one of the forms of potentiation.

2. *Neurons responding chiefly to the first stimulus.* This type of response is opposite to the first type. Figure 2 presents two records from neurons representative of such a time-course of responses.

Figure 2A illustrates a case when the cell responded with two spikes to the first four flashes, with one spike to the 5th and 6th flashes, and completely ceased to respond to all subsequent flashes. Figure 2B represents a record of the reaction of another neuron with almost the same type of response. It is noteworthy that the spontaneous activity of these cells is, as a rule, very low, which probably testifies to a high firing threshold. Neurons of this type are similar in a way to "attention neurons" (Hubel *et al.,* 1959; Vinogradova & Lindsley, 1963). However, their distinctive feature is that they do not react to any new stimulus, but only to the first flashes, i.e., prove to be modally specific. In Fig. 2C, for example, it can be seen that the neuron does not respond to an acoustic stimulus (a tone).

3. *Neurons exhibiting an increase in the duration of primary inhibition.* In a number of cases the response of visual cortex neurons to a flash began immediately with an inhibitory phase which was not preceded by a neuronal discharge—primary inhibition (Skrebitsky & Voronin, 1966). Figure 3A shows a typical example of cessation of the neuronal spontaneous activity evoked by a flash. As a result of presentation of a series of light flashes, the primary inhibition may increase and become more protracted and complete. In the right set of Fig. 3B it may be seen that a short inhibitory phase appears in response to the first flash, and is followed by a series of spikes. In response to the second flash, the duration of the inhibition increases, but the inhibitory phase is still interrupted

by two spikes. In response to the third flash, protracted and complete inhibition is observed; it is followed by a group of spikes (a rebound).

It should be noted that the inhibitory phases of the neuron under consideration are correlated with the activation of another cell (Fig. 3B). This cell responds to flashes of light with a short latency (in Fig. 3B the first spike in the response of this neuron is indicated by arrows), and is possibly responsible for the development of primary inhibition directly or through other interneurons.

An intracellular investigation of the inhibition of visual neurons revealed inhibitory postsynaptic potentials with a short latency (Skrebitsky & Voronin, 1966). The absence of initial activation, along with a short latency, indicate that it is not recurrent inhibition of the Renshaw type, but direct collateral (Eccles, 1965b) or parallel inhibition (Sokolov, 1965).

4. Neurons exhibiting a decline or primary inhibition. This type of responses is opposite to the above-described type. The example presented in Fig. 3C illustrates a gradual reduction of the duration of the inhibitory phase from 600 msec, in response to the first flash, to 100 msec in response to the sixth and seventh flashes. As far as we know, such a "habituation of the inhibitory response" was never described before in the visual cortex. It is one of the most pronounced dynamic changes of neuronal reactions in the visual cortex.

Considering the character of changes in the responses of the afore-mentioned four groups of neurons, one can observe a definite correlation in their activity. Proceeding from the concepts of certain forms of interaction between individual neurons (reciprocal inhibition, Jung, 1964), and assuming the possibility of a postactivation potentiation in the cortical neurons, it is possible to construct very simple schemes which approximate the real properties of the recorded neurons.

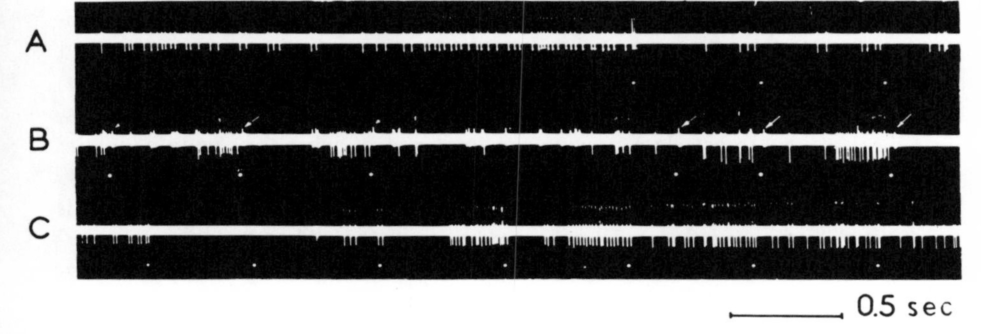

FIG. 3 Change of the duration of primary inhibition: (A, B, and C) records of activity from three different neurons. (A) Example of primary inhibition. (B) Potentiation of the inhibitory pause as a result of repetition of the stimulus. (C) Decline of primary inhibition. Dots indicate the light stimulus presentation. Arrows in B indicate the first spikes in the responses of another neuron recorded with the same electrode. Details are given in the text.

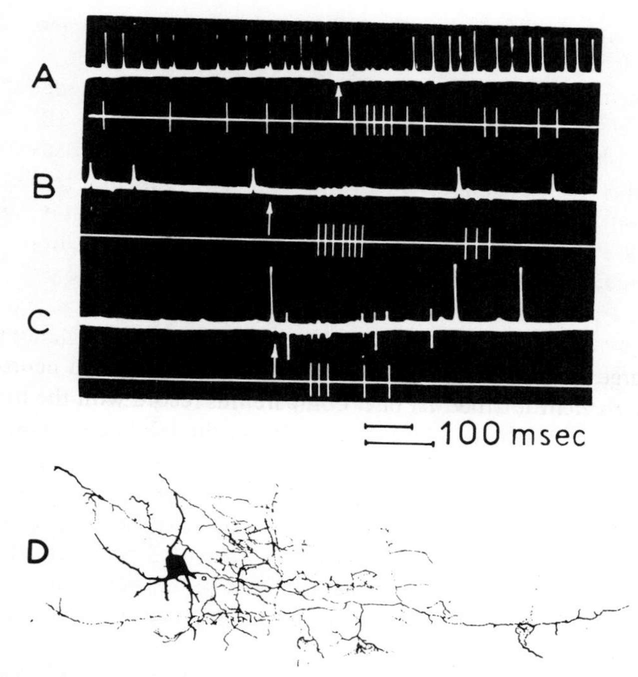

FIG. 4 Hypothetic inhibitory neurons. Lower records in (A, B, and C): schematic representations of the activity of hypothetic inhibitory cells presented on the upper oscillograms. Records B and C are retouched. Arrows are presentation of the light stimulus. (D) Stellate neuron from the visual cortex forming nests in which other cells are located; on the drawing they are seen in the shape of shadows (after Shkolnik-Yarros, 1965, with a slight modification). Details are given in the text. Time calibrations are as follows: upper, A record; lower. B and C records.

One of the necessary conditions for constructing such schemes is the supposition that special inhibitory neurons, similar to those which have been described for the spinal cord, hippocampus, and cerebellum, exist in the cortex (Eccles, 1964).

We have no direct data proving the existence of inhibitory neurons in the visual cortex of the rabbit. But some facts give ground to such a possibility. Thus, the works of Krnjevič *et al.* (1964) showed that cortical inhibition is effected by neurons located in the cortex itself. Probably, these are the interneurons which discharge during periods of inhibition in other cells. The histological work of Shkolnik-Yarros (1965) demonstrated the existence, in the visual cortex, of stellate cells with short densely ramifying axons which form a plexus of the "neat" and "basket" type. The bodies of the pyramidal neurons which are located in these plexes possess numerous synaptic contacts with the axonal terminals. An analogy with the stellate neurons of the cerebellum and hippocampus functioning as inhibitory neurons, along with data concerning the postsynaptic inhibition in the neurons of the visual cortex (Skrebitsky & Voronin, 1966), allows us to assume that certain stellate cells of the visual cortex perform the function of inhibitory neurons (Shkolnik-Yarros, 1965; Skrebitsky & Shkolnik-Yarros, 1967).

Figure 4D shows a stellate cell whose axon forms nests around the bodies of the pyramidal neurons. One can see a striking analogy between this figure and the well-known schemes and drawings of inhibitory neurons in the cerebellum and hippocampus (Andersen, Andersson, & Lomo, 1963).

The same figure presents records of the activity of two neurons, made with the help of one microelectrode (Fig. 4A, B). On the records it can be seen that the inhibitory phase of the response of one neuron to a light stimulus is correlated with a discharge of another. It may be assumed that this second neuron acts as an inhibitory element for the first one. Compare this record with the oscillogram presented in Fig. 3B, where, as mentioned above, the inhibition of one neuron also corresponds to the activation of another. However, it is self-evident that a more precise answer to the question concerning the mechanisms of formation of inhibitory reactions requires further experiments with a comparison of the latencies and durations of the discharges, as well as of other parameters of the responses manifested by the inhibitory and "inhibited" neurons.

The schema in Fig. 5 summarizes some of the afore-mentioned properties of neurons of different groups and the forms of their interaction. This figure is based on slightly modified records from neuron discharges presented in this chapter (Fig. 1A, B; Fig. 2).

Cell 1 is a neuron with a potentiated response, having its own afferent input which does not change in the conditions of the given experiment (aff[1]). The

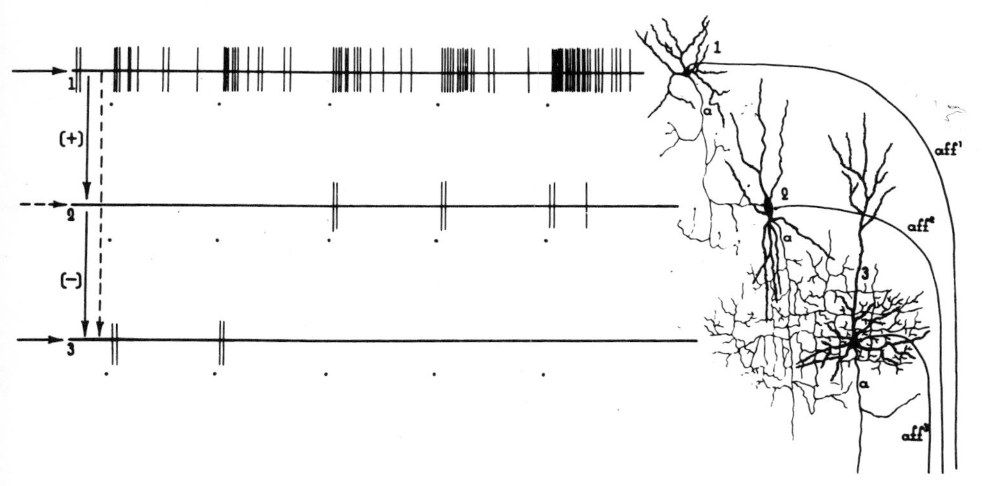

FIG. 5 A hypothetic scheme of the interaction of neurons. Explanations are given in the text.

axon of this neuron contacts with the body of cell 2 which may have a direct afferent input (aff^2) or may not, and which possesses a high threshold of generation of a spike potential. Cells 1 and 2 are presented in the figure in the form of stellate neurons. Cell 3 is a pyramidal neuron with its body located in the plexus formed by the axon of cell 2. It is supposed that cell 3 has a direct activating afferent input (aff^3) and that cell 2 acts as its inhibitory neuron. Under the action of a rhythmic light stimulus the discharges of the "potentiated neuron" 1 are intensified. Neuron 2, possessing a high threshold of spike generation, begins to respond to afferent signals coming to it through the direct afferent input (aff^3).

The purpose of the scheme is to demonstrate the possible forms of interaction of the neuronal responses and to explain one of the most interesting properties of certain neurons–that of responding only to the first stimulus. However, it is possible to simplify this scheme by assuming that it is cell 2 that proves to be the potentiated neuron. In this case, the participation of neuron 1 is not necessary. Finally, neuron No. 1 may act as an inhibitory element for cell 3 (this connection is not shown in the drawings of the neuron); in this case neuron No. 2 may be excluded from the scheme.

Strictly speaking, this scheme presents a certain detailed rendering of Sokolov's (1965) interpretation of the phenomenon of "habituation" on the level of a single neuron as a result of potentiation of the inhibitory neurons ("elaborated inhibition").

There exists, however, another essentially different interpretation of this phenomenon, which proceeds from the fact that habituation is inherent in all divisions of the nervous system and manifests itself even in spinal reflexes with the participation of one interneuron. According to this interpretation, "habituation" on the neuronal level is a universal phenomenon, whose actualization does not require any neuronal nets, and which develops as a result of desensitization of the membrane to a repeated dose of the transmitter, depletion of the transmitter, or its inactivation by enzymes (Sharpless, 1964).

At present it is difficult to say which of these two interpretations is preferable, especially since no direct facts confirming or disproving the assumptions mentioned above have been obtained thus far on cortical neurons. It seems to us, however, that the first approach contains greater possibilities for plastic reconstructions and for coordinated work of the elements of the neuronal nets.

9

DYNAMICS OF CHANGES IN THE RESPONSES OF SUPERIOR COLLICULI NEURONS OF THE UNANESTHETIZED RABBIT DURING REPEATED LIGHT STIMULATION

N. V. Dubrovinskaya

Moscow State University

Microelectrode investigations carried out in recent years have revealed some appreciable differences between neurons as regards the characteristics of their reactions. The method of repeated stimulus presentation proves to be highly valuable for ascertaining the complete characteristics of each neuron, and provides a dynamic classification of neurons (Vinogradova, 1965; Vardapetian, 1967; Sokolov, 1967). This test permits the classification of cells according to the stability of their responses. Neurons with different properties are united into functional groups which ensure the effective functioning of various reactions of the central nervous system. A hypothetic scheme of the neuronal structure of the orienting reflex at the cortical level (Sokolov, 1963) presupposes the existence and interaction of elements with stable and various forms of unstable reactions, which accomplish operations connected with the reception of information, comparison of the signal with the standard, anticipation of the future significance of the stimulus, and elaboration of an "order." The existence of such neurons in the cortex was experimentally proved (Vinogradova & Lindsley, 1963; Vardapetian, 1967; Sokolov *et al.*, 1967).

Since the orienting reflex develops as a multilevel reaction, which involves various brain structures in a complex interconnection, it was interesting to investigate the dynamic properties of neurons in structures other than the cortex. The object of our research was the superior colliculus of the rabbit, which is known on the one hand as a specific subcortical relay, and on the other as a reflex center involved in the realization of orienting reactions (startle reflex, according to Sepp). It was assumed that a study of the behavior of superior colliculi neurons during repeated stimulus presentation would allow us to draw certain conclusions concerning the participation and role of this structure in the general organization of the orienting reflex.

METHODS

Of the 100 investigated neurons of the superior colliculus we selected 8 representatives of the three most typical groups, which were established by us (Dubrovinskaya, 1967) on the basis of the pattern of their responses to a flash of light. The investigation was carried out on unanaesthetized, restrained rabbits prepared for chronic experiments. Extracellular recording of potentials from the neurons of the colliculus was performed by tungsten microelectrodes (tip diameter 2–4 μ, resistance 5–7 megohms); slow-wave and spike activity were recorded simultaneously. A short flash produced by an FD-1 stimulator with a brightness of 5.2 lumens presented to the animal's contralateral eye at intervals of 1–1.6 sec served as a stimulus. The number of stimulus presentations varied from 40 to 190. Sound signals produced by the same stimulator were used as disinhibiting agents. The position of the electrodes was marked by passing an electric current of 40 μA for 60 sec through the microelectrode. Both the spontaneous and evoked activities were taken into account during the processing of the data

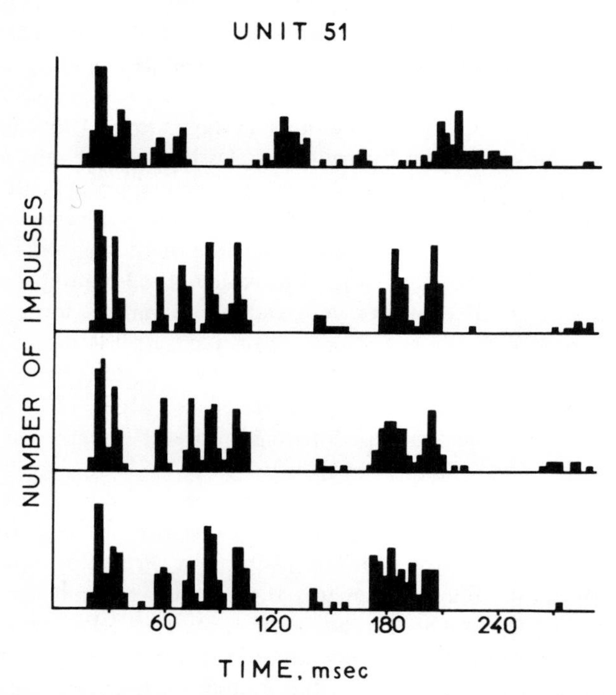

FIG. 1 Change in the response of unit 51 during repeated stimulation. Abscissa, time in msec; ordinate, average number of spikes in the response. From top to bottom: 1st–10th, 20th–30th, 50–60th, and 80th–90th applications of the flash. Stabilization of the pattern of response is demonstrated; it is particularly pronounced in the 3rd and 9th blocks (rows 2 and 4).

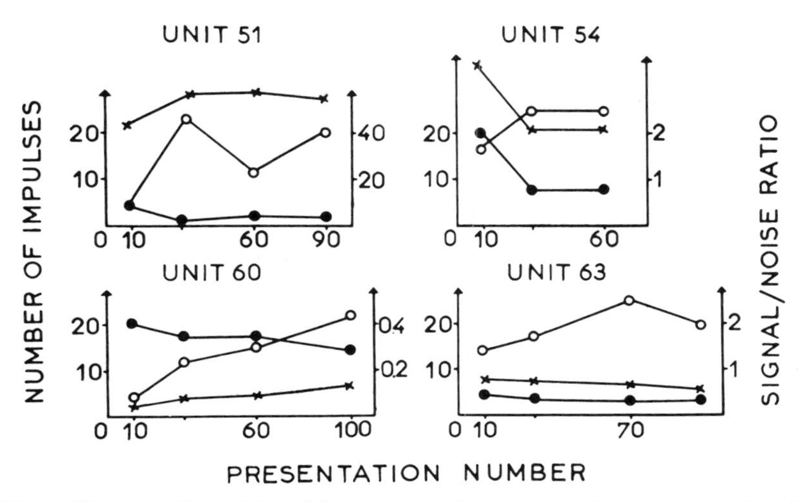

FIG. 2 Change in the activity of four neurons during repeated applications of the flash. Unit 51 is the 1st group; units 54 and 63 are the 2nd group; unit 60 is the 3rd group. Abscissa, stimulus presentation number; left ordinate, average number of spikes per unit of time; right ordinate, signal-to-noise ratio. Solid circles are spontaneous activity; crosses are evoked activity; open circles are signal to noise ratio.

obtained. In order to ascertain the characteristics of the spontaneous activity, we counted the number of spikes for 400 msec before the onset of each stimulus in blocks of 10 flashes, and then deduced the mean value for each block. The responses were evaluated according to poststimulation histograms, where the average number of spikes during 10 applications was calculated either for a period of 3 msec or 20 msec after the onset of stimulation. We also calculated the average number of spikes for each block during 400–1000 msec after presentation of the flash.

RESULTS

The effects of repeated stimulus applications are described below for each group of neurons separately.

1. Cell 51 is a typical representative of a group of neurons with an excitatory type of response. The latter consists of 7–10 short (6–12 msec) bursts of impulses that are evoked consistently. The total duration of the discharge is 285 msec. The fact that this neuron belongs to the group which responds to a flash with a monophasic excitatory reaction is confirmed by tests with paired and rhythmic stimuli, which did not reveal any signs of inhibition in the responses.

Repeated applications of flashes did not result in appreciable changes of the response in the direction of an increase or decrease of the average number of

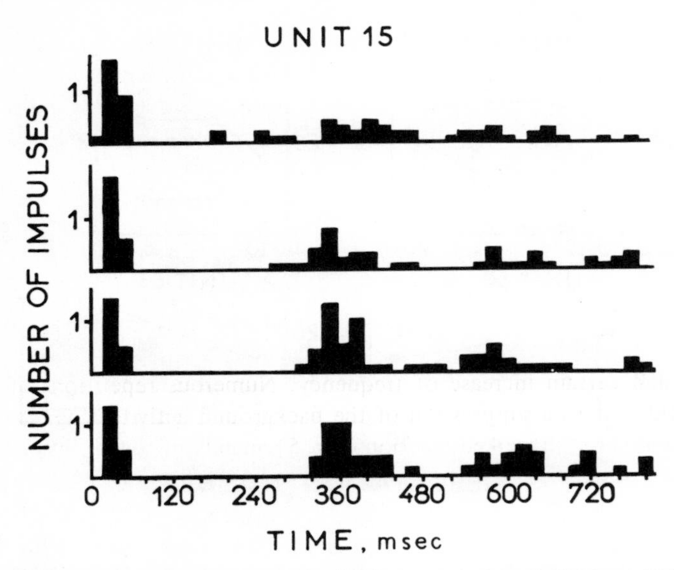

FIG. 3 PST histogram for responses of cell 15 to successive application of blocks of 10 flashes. From top to bottom: 1st–10th, 10th–20th, 20th–30th, and 30th–35th applications. A modification of the primary and secondary discharges and an increase of the inhibitory interval are demonstrated in the figure.

spikes, but led to its stabilization in the form of a more distinct alternation of bursts of impulses and periods of silence (Fig. 1).

The following factors also contributed to a greater expressiveness of the response in the post-stimulation histogram: (1) a suppression of spontaneous activity, as a result of repeated applications of the flash (15 spikes during 500 msec at the beginning of the test and 0.6 spikes for the same period during the application of the 9th block of stimuli, Fig. 2, cell 51); (2) a certain reduction—in the same conditions—of the dispersion of initial spike latencies from 8 msec in the first block to 3 msec in the ninth.

2. Neurons with a triphasic response constitute the most numerous group. In view of this, we selected four representatives of this group for purposes of investigation. The response of these neurons consisted of initial and secondary spike discharges separated from each other by an inhibitory interval which could be identified by the absence of spikes, as well as by the results of application of paired and rhythmical flashes.

The behavior of these neurons in our particular experimental situation proved to be uniform: there was a gradual weakening of both the background and evoked activity (Fig. 2: 1, 2, cells 54, 63) during the excitatory phase. The latter may develop at the expense of the initial discharge (cell 54), as well as of the secondary one (cell 63). A peculiar form of the decrease of the response was a reduced probability of the initial discharge. Thus, in cell 20 this index changed from 1 in the first block to 0.4 after 180–190 applications of the flash.

However, a full disappearance of the response was not observed in a single case. The above-mentioned weakening of the evoked activity was accompanied, just as in cell 51, by certain signs of stabilization of the response in the shape of its emphasized configuration due to the prolongation of the inhibitory pause (cell 15, Fig. 3) and to a reduced dispersion of the latencies, with the mean value remaining constant. Thus, in cell 20 (Fig. 4) the latency varied from 19 to 23 msec during the first 10 applications of the flash but equaled only ±1 msec in the nineteenth block.

3. Cell 60 is representative of a small group with a bi-phasic, inhibitory–excitatory response. The presentation of a single flash results in a protracted (517 msec), complete inhibition of spontaneous activity with its subsequent recovery and certain increase of frequency. Numerous repetitions of the flash (up to 100) led to a suppression of the background activity (Fig. 2, cell 60-1) and weakening of the evoked reaction (Fig. 5), mainly of the inhibitory phase. It can be seen in the figure that with the increase of the number of the flashes, an increasing number of spikes begins to appear against the background of the inhibitory pause, leading, as it were, to the recovery of the spontaneous activity. The application of an extra stimulus–a sound–evokes the inhibitory response in its initial form. With further repetition of the flashes, the same reduction of

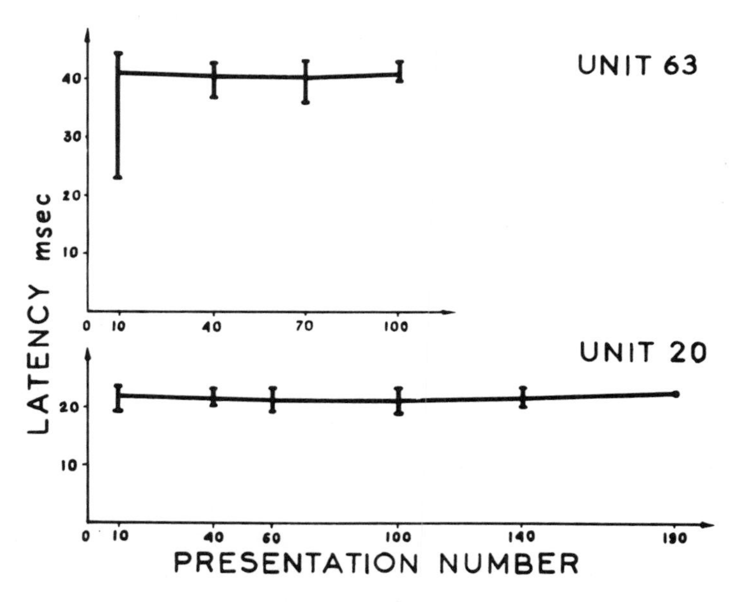

FIG. 4 Dependence of the average latency of responses of two neurons on the stimulus' presentation number. The figure demonstrates a decrease in the dispersion of the latency values at the end of the test, with the mean value remaining constant. Abscissa, stimulus' presentation number; ordinate, latency of the response in milliseconds.

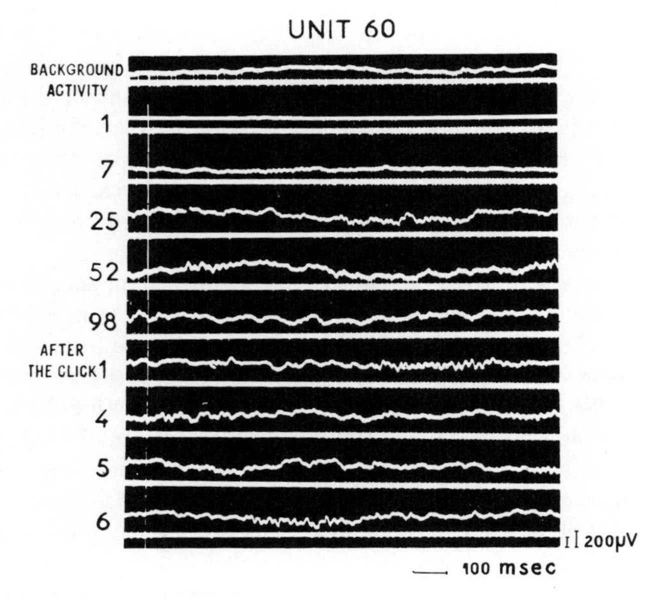

FIG. 5 Extinction of inhibition in unit 60. Upper line, slow activity; lower line, spike activity. The moment of stimulus presentation is indicated by a vertical bar. Numbers at the left of the records indicate fresh presentation number. The figure shows a gradual filling of the inhibitory pause with spikes and a temporary recovery of the response after the disinhibiting action of the click.

inhibition arises after a much smaller number of stimulus presentations. No signs of stabilization of the response were observed in this neuron.

DISCUSSION

When analysing the behaviour of representatives of the three neuronal groups in conditions of repeated stimulation we established three types of effects (1) a constancy of the response with signs of stabilization; (2) a weakening of the response with signs of stabilization; (3) a weakening of the response without any signs of stabilization.

The first of these effects, apparently, emphasizes the response, while the third masks and equalizes it with the background activity. But in the second type of effect the weakening of the response is accompanied by an opposite phenomenon–its stabilization. How, then, can this phenomenon be explained?

It was shown that in the course of the experiment the evoked and background activity exhibited a decline. As can be seen in Fig. 2, at definite moments these two processes may develop not only in parallel (unit 54), but also in different directions (units 51, 60). However, the effectiveness of the response is determined by its level over the noise (i.e., spontaneous activity). In view of this, in order to interpret the dynamics observed, it is expedient to consider the signal-to-noise (S/N) ratio. The S/N ratio was calculated as the ratio between the average number of spikes in the response and in the background activity for the same time period. (Coefficients calcualted for separate phases of the response did not differ in their dynamics from the general coefficient; therefore the value of the latter will be used by us further on.) We also took into account the fact that the S/N ratio is more than 1 for responses of the excitatory type and less than 1 for responses of the inhibitory type, and that, consequently, a change of this coefficient is of opposite significance for the excitatory and inhibitory reactions. The responses of all three groups of neurons were subjected to such processing. It proved that for all cells, the S/N ratio increased during the first 30–70 applications of the flash (Fig. 2). In the case of neurons of the first and second groups, this meant that during the first minute of flash presentation, a peculiar "accentuation" of the response developed, consisting of an increase of the contrast between it and the background activity together with a stabilization in the shape of a reduced dispersion of the latencies and the appearance of a more distinct phasic character. Thus, the significance of the dynamic changes in the responses of the first two groups of neurons proved to be similar, in spite of their different manifestations.

On the contrary, an increase of the S/N ratio in cell 60 (a representative of the third group) means a gradual approximation of the response to the background activity, i.e., its true decline.

Taking into account the aim of the present investigation, it is necessary to confront the afore-described changes of the neuronal responses with the laws of development of the orienting reflex. The first, but not the only factor indicating that the neuron belongs to the system of the orienting reflex is the extinction of its response. According to our data, only the dynamics of neuron 60, with a gradual decrease of inhibition as a result of repeated stimulations, which is eliminated by an extra stimulus, complies with this condition. A similar "extinction of inhibition," but one which took place after 6–15 applications of the signal, was described in the hippocampus (Vinogradova, 1965), cortex (Skrebitsky, 1966) and tectotegmental area (Horn & Hill, 1966). In our case this process is somewhat delayed. Apparently, cell 60, not being by itself a detector of novelty, is included in the same system with the latter.

By its functional characteristics, this neuron may be related to the efferent neurons of the orienting reflex, since its function is to suppress activity during the first applications of the stimulus and to gradually weaken the inhibitory regulation with the repetition of the stimulus. A change in the situation (click dishabituation) again puts these blocking mechanisms into operation.

In most neurons of the colliculus, the fact that no extinction of the responses takes place contradicts the results obtained by Horn and Hill (1966), who observed such extinction in 88% of the neurons investigated by them in the rabbit's superior colliculus. The following question naturally arose: Do our data not result from differences in methods? Our investigation was carried out on unanaesthetized animals in conditions of chronic experiments, which are adequate for studying the phenomenon of extinction. The only drawback was that the rabbits were investigated in a prone position, being fixed in the stand, which evoked protracted activity from the proprioceptors. However, this fact could not substantially influence the results obtained, since in the experiments of Arden (1963), and Horn and Hill (1966), extinction was observed in conditions of acute experiments with the use of a stereotaxic apparatus, anaesthesia, and artificial respiration.

Let us now consider the factor of intensity of the signal applied. In the experiments of Horn and Hill (1966), as well as Bell et al. (1964), in the course of which the phenomenon of extinction of neuronal responses was observed, weak light signals were used—modifications of the background illumination, geometrical figures, etc. As is known from the works of Sharpless and Jasper (1956), and Arden (1963), the intensity of stimulation and the rate of extinction are inversely proportional. It might be assumed then that the flash applied by us (5.2 lumens × sec) was too bright for experiments of this kind. However, the effect of extinction could be achieved even in response to electrical stimulation (Scheibel & Scheibel, 1965). Therefore it seems to us that the factor of intensity is by no means responsible for the results obtained.

Now it remains only to consider the role of the frequency of stimulus presentation which is emphasized in the works of Altman (1961), Webster, Dunlop, Simons, and Aitkin, (1965), Fernandez-Guardiola, Toro-Donoso, Aquino-Cias, and Guma-Dias (1965), Horn and Hill (1966). It is possible that the interval between the stimuli used by us (about 1 sec) was insufficient for obtaining the effect of extinction, even more so because the neurons of the colliculus often show a facilitation and intensification of the responses during the action of rhythmic stimuli. But the constant character of the latency indicates that no activation of the neuronal responses was observed in these conditions. Besides, it was demonstrated that the interval chosen by us exceeded the duration of the transitory process of the colliculi neurons measured in experiments with paired flashes (Dubrovinskaya, 1967).

Taking this into consideration, we may conclude that the absence of the extinction effect in the responses of the superior colliculi neurons in the course of repeated light stimulation is a property inherent in these neurons, and does not result from experimental methods.

The following two considerations corroborate our results and explain their divergence from the data of Horn and Hill:

1. The localization of the electrodes. It seems to us that the divergence between our data and those of Horn and Hill may be explained to some degree by the different location of the investigated neurons. An examination of a picture, presented by the authors, of the brain section showing a microelectrode placement, showed that most of the recorded neurons (multisensory!) are

located below the colliculus proper and belong to the area of the tegmentum and central gray matter–nonspecific structures of the midbrain.

In our case the relations prove to be inverse. Most neurons investigated by us are localized in three upper layers of the structure. They form a superficial neuronal complex (Viktorov, 1966) consisting of monosensory neurons of a specific type which participate in the specific visual function of the superior colliculus.

2. Accordingly, the neuronal responses in our experiments usually exhibited a distinct specific pattern, while in the experiments of Horn and Hill they consisted of tonic increase or decrease of the spontaneous discharge. The drawing of a parallel between the pattern of response and the functional characteristics of the neuron is quite justified, since the works of some authors (Vinogradova & Lindsley, 1963; Vinogradova, 1965; Cross & Green, 1959; Sokolova, 1966; Scheibel & Scheibel, 1965) reveal a correlation (which is sometimes emphasized by the authors and sometimes not) between the multisensory property, the tonic type of response in the shape of an increase or decrease in the spontaneous frequency, and the possibility of response habituation. In the light of these data it is no wonder that the responses of the monosensory neurons of the specific type are not extinguished (our data), while multisensory elements with a tonic character of response exhibited habituation in the course of 15–20 repetitions of the signal (data of Horn and Hill).

Apparently, the accentuation of the response during the first two minutes of the action of repeated stimuli has a definite physiological significance. It may be assumed that the developing stabilization contributes to the fine analysis of the given signal on higher levels of the visual analyser, as well as to the formation of efferent commands addressed to the muscles of the eyes, neck, and trunk.

10

DYNAMICS OF THE NEURONAL RESPONSES IN THE RABBIT'S INFERIOR COLLICULUS DURING REPEATED APPLICATIONS OF A SOUND STIMULUS

A. Bagdonas

Moscow State University
and *USSR Academy of Sciences*

The development of the microelectrophysiological technique has made possible a closer approach to the ascertainment of the neuronal mechanisms which determine the elaboration of a temporary connection (Bureš & Burešova, 1967; Jasper, Ricci, & Doane, 1958; Kamikawa, McIlwain, & Adey, 1964; Morrell, 1963; Hori, & Yoshii, 1965; Hori *et al.*, 1967). In all these investigations the electrical neuronal responses were elaborated in the classical form, by pairing two stimuli applied to different receptors or structures of the nervous system. However, plastic changes which are due to the formation of nervous connections can be investigated also with the help of repeated applications of a single stimulus (Sokolov, 1958, 1964). The dynamics of the neuronal responses, which develop with the repetition of the stimulus, serves in this case as an indicator of synaptic functional readjustments. A classification of neurons according to their dynamic properties (Vinogradova, 1965; Vinogradova & Lindsley, 1963) allows the establishment of intermediate processes between the extreme points of these dynamic modifications (Sokolov, 1967).

Investigations of neurons of the auditory cortex carried out predominantly on cats were confined to the properties of neurons as detectors of various parameters of the signal (Gershuni, 1967; Katsuki, 1964; Kiang, 1965; Nelson, Erulkar, & Bryan, 1966; Rose, Brugge, Anderson, & Hind, 1967; Evans & Whitfield, 1964; Suga, 1964, 1965). The character of the dynamic changes in different parts of the auditory system has usually been studied as habituation of the evoked potentials (Altman, 1960; Dunlop, Webster, & Day, 1964; Simons, Dunlop, Webster, & Aitkin, 1966; Webster *et al.*, 1965; Kratin, 1965, 1967; Hernandez-Peon, 1956). The dynamic characteristics of neuronal responses in the auditory cortex of a cat were investigated by Hubel (Hubel and coauthors, 1959) who discovered "attention" neurons, and by Vardapetian (1967) who suggested a dynamic classification of neurons according to the stability of their responses during repeated stimulus presentations.

In the present work an attempt is made to investigate the dynamic properties of neurons in the inferior colliculus of the unanaesthetized rabbit.

METHODS

The experiments were carried out on 12 unanaesthetized rabbits each weighing 2.5–4 kg.

The experimental animal was fixed in a special stand. A trephine opening with the center located 2.5 mm laterally from the sagittal suture and 2 mm anterior to lambda was made under local anaesthesia in the bone of the skull over the inferior colliculus. The dura was removed and the trephine opening was filled with a 3–4% agar in a physiological solution. Then a Plexiglas base was mounted over the trephine opening, and a hydraulic manipulator was fixed on this base during the experiments.

Spike activity of the neurons was recorded with tungsten microelectrodes covered with vinyflex and acetone varnish (resistance of the microelectrode–2–5 megohm, tip diameter–1.2 μ). An UBP 1-01 amplifier with a cathode follower was used; recording was continuous during the investigation of the neuron using a Yauza-10 tape recorder. Subsequently, the record was reproduced on a film moving at a speed of 5 cm per sec.

A click, a flash of light, and a simultaneous complex of a click plus flash were used as stimuli. In order to generate a click, one flashing tube, enclosed in a box, was connected with a transmitting loudspeaker. The discharge of a second tube served as a light stimulus. The energy supplied to both flashing tubes was changed in the ratios of 1:2, 1:4, 1:6, 1:14; these four levels were designated as the first, second, third and fourth intensities, both for the clicks and flashes, as well as for the complex of sound plus light.

The investigation was usually conducted in the following sequence:

1. Recording of the background activity for 20–60 sec.
2. Presentation of 10 clicks at intervals of 5–10 sec.
3. Presentation of clicks of different intensities.
4. Application of paired clicks with intervals of 10 to 300 msec.
5. Presentation of a series of rhythmical clicks with a frequency of 0.6 per sec. (from 30 to 300 applications) and with 1–8 omissions each lasting 10–60 seconds. Clicks presented with a frequency of 10 per second during 10–30 seconds served as a disinhibiting factor.
6. Presentation of a series of flashes (30–50 applications) followed by a complex of a click and a flash (30–50 applications) and then again by flashes alone (20 sec) also presented with a frequency of 0.6 per sec. At the end of the experiment the background activity was again recorded.

In the course of processing, poststimulation time histograms (PST) were formed. For this purpose, the interval between the stimuli was divided into

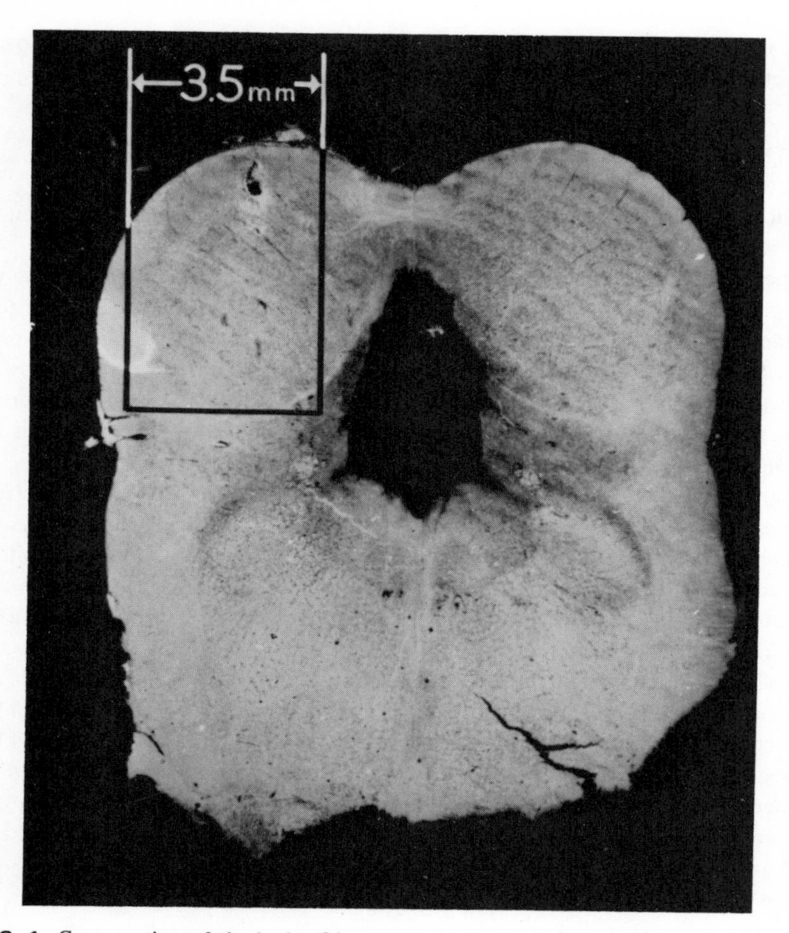

FIG. 1 Cross section of the brain (80 μ thick) at the level of the inferior colliculus. The approximate area from which neurons were recorded is outlined by a rectangle. A coagulated point is seen near the surface of the left colliculus.

periods of 50 msec each, in which the number of spikes was counted. When processing the background or the omissions of signals, we first divided the record into sections equalling the interval between the stimuli and then into periods of 50 msec. Corresponding periods of 10 successive applications were summed and the average number of spikes in each section was determined. The total number of spikes in the interval between the stimuli was also counted and the latency, as well as the duration of the inhibitory pause, were measured. In addition, graphs were formed to illustrate the dependence of the signal-to-noise ratio (correlation between the number of spikes in the response and that in the remaining part of the interval between the stimuli) as a function of successive stimulus applications.

We determined the depth of insertion of the microelectrode tip according to the scale of a micromanipulator, considering that the inferior colliculus is located at the depth of 5.5–6 mm from the brain surface, as shown in the atlas of Monnier and Gangloff (1961). The area of recording could be established also on the basis of the neuronal responses to sound stimuli (whistle, rustle, claps, clicks). The localization of the microelectrode was verified by histological means. In order to mark the tip of the microelectrode, a direct current of 40 μA was passed through it for 1–2 min. Immediately after coagulation, the brain was fixed in a 10% solution of formalin. Two weeks later brain sections 80 μ thick were made on a freezing microtome.

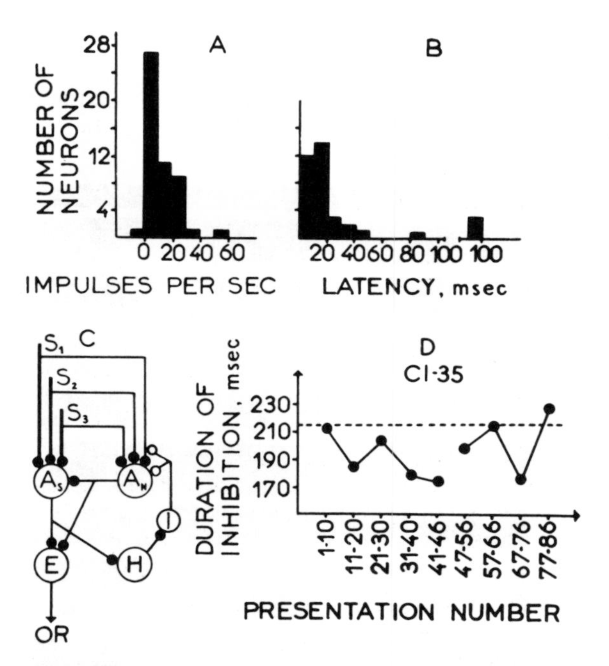

FIG. 2 (A, B) Distributions of the neurons according to the frequency of their background activity (A) and according to the latency of responses (B).

Ordinate–number of neurons; absicssa, frequencies of the background activity (A) and latencies (B). The column to the left of the zero (A) indicates a neuron without any background activity. (C) A hypothetic scheme explaining partial habituation in the afferent specific neuron (A_s) and complete habituation in the efferent (E) and nonspecific afferent (A_n) neurons during the action of an adequate stimulus.

S_1, S_2, S_3, stimuli differing in one of the parameters; OR, output of a component of the orienting reflex; H, neuron of the hippocampus; I, interneuron. The solid circles designate excitatory synapses; the open circles denote inhibitory synapses. (D) Dependence of the duration of the inhibitory phase on the number of successive applications (in blocks) of rhythmical clicks.

The dashed line gives duration of the inhibitory phase in the case of random stimulus presentations (average of 10 single applications).

RESULTS

Activity of 51 neurons of the inferior colliculus was recorded. Twenty-seven of them were stimulated by more than 200 rhythmical clicks during the process of recording, 45 by more than 100 clicks, and the remaining by less than 100 clicks. Only 10 neurons were investigated according to the relatively full experimental program described previously. All the recorded neurons were located at a depth of about 4.5 mm from the surface of the colliculus. In Fig. 1 the approximate area of recording is outlined by a rectangle and in the main corresponds to the central nucleus of the inferior colliculus.

General Characteristics of the Neuronal Responses

Most of the recorded neurons (54%) possessed a spontaneous activity of less than 10 spikes per sec. The number of neurons with a higher background activity was much smaller (Fig. 2A). In 70% of the cases the background activity was random while in all other cases (30%) it was characterized by group discharges.

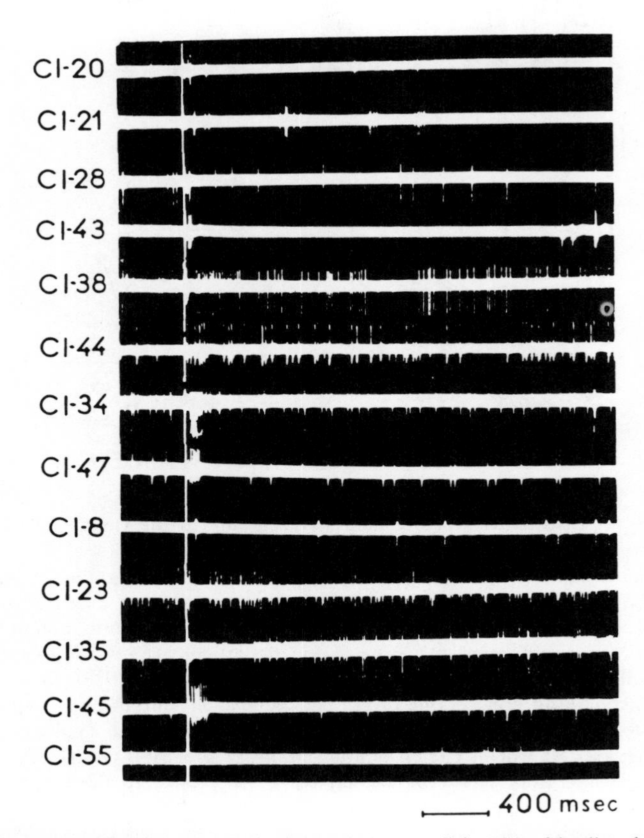

FIG. 3 Examples of responses of 13 various cells to a click. The white line drawn through all the records indicates the moment of stimulus application; to the left of it, the number of the cell is given in each record.

The responses to a click were highly diverse (Fig. 3) which is due mainly to the correlation between the phases of inhibition and excitation. In most cases (34%) neurons of the excitatory–inhibitory type were recorded; in these neurons the initial phase of excitation is superseded by inhibition which may sometimes be followed by a secondary phase of excitation with a subsequent (second) phase of inhibition. The percentage of such neurons is apparently higher than that mentioned above, since the division of the record into periods of 50 msec masks the structural character of responses of a short duration. The response of cell IC 34 shown in Fig. 5 seems monophasic but the use of 10 msec analysis periods reveals a polyphasic character (Fig. 4C).

The response may begin directly with inhibition of the background activity which may be followed by a phase of excitation (14% of all neurons). It should be noted that the probability of recording a response that begins with inhibition proves to be the greater, the higher the frequency of the neuronal background activity. However, there are some exceptions to this rule. The duration of the inhibitory phase in neurons of the types mentioned previously may vary from 100 msec (and less) to 1000 msec. Fourteen neurons (28%) responded to a click with monophasic excitation; here too a great diversity of responses was observed–from 1 spike to scores of spikes in a response. The fourth group (24%) includes neurons which do not respond to a click, although most of them responded to other sounds during the experiment (rustle, whistle, claps). Some of the neurons, which at first did not respond to click, began to react to it when it was repeatedly presented.

The latencies of the neuronal responses varied from 8 to 200 msec, but in the majority of neurons they did not exceed 20 msec (Fig. 2B). If the neuron responded with inhibition, the time necessary for the discontinuance of the background activity was regarded by us as the latency period.

Figure 4 (A, B, and C) shows a modification of the response depending on the intensity of the click, as well as the recovery cycle of neuron IC 34 which responded with excitation and with subsequent slight inhibition. The number of spikes during a period of 600 msec following stimulus application increased with the increase of the intensity (Fig. 4A, B). Such a dependence was observed in all six neurons recorded in accordance with this program.

The phasic type of response, even with a slight inhibitory pause, complicates the recovery cycle of the neuron (Fig. 4C). The response to the second click depends on the phase upon which it falls. If it is the phase of excitation, then the response proves to be almost fully reproduced (compare the histograms at intervals of 10 msec and 170 msec); if it is the phase of inhibition, the response is reduced at longer intervals (240–280 msec).

Dynamics of the Responses in Conditions of Repeated Stimulus Presentations

1. General remarks. As stated above, 45 neurons received more than 100 rhythmical clicks each, and some neurons up to 300 such applications. Although the neuronal responses themselves were quite variable, and their magnitudes

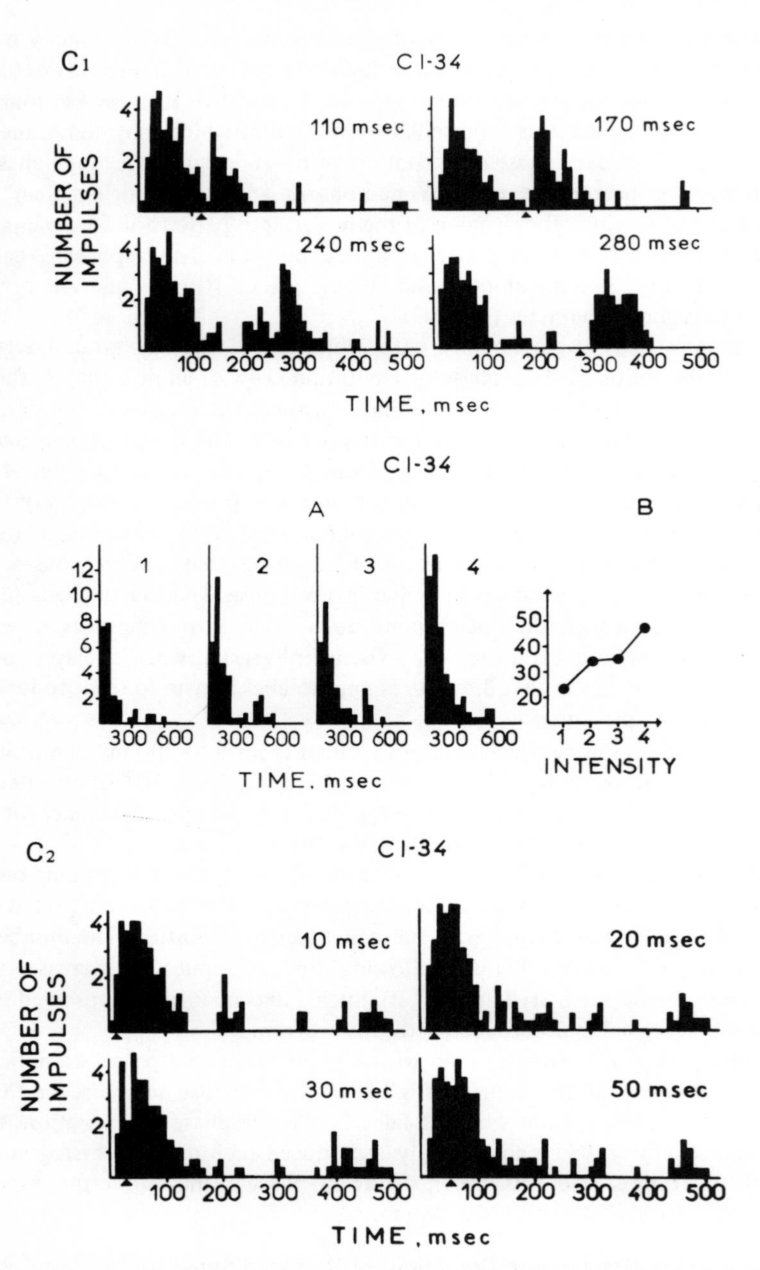

FIG. 4 Cell CI-34. A. PST histograms of responses to clicks of 4 increasing intensities. The Ordinate is number of spikes (average of 3 applications); abscissa is time. The intensity is designated by a figure above each histogram. (B) Graph showing a change in the number of spikes (average of 3 applications) depending on the intensity.

changed from stimulus to stimulus, they were always in evidence and did not disappear as a result of numerous repetitions of the stimulus. However, statistical processing showed that with the repetition of the stimulus, some directed modifications of the responses still take place. A partial extinction was observed most frequently. The first applications of the stimuli always evoked a stronger reaction than subsequent stimuli. This was observed in all cells, except neurons which intensify their reaction, and a few neurons with highly variable responses. In some neurons, this intensification of the response to the first applications of the stimuli, in comparison with the subsequent ones, was more pronounced, and less pronounced in some.

The rate of extinction of the responses varied in different neurons. Thus, it proves possible to single out neurons which present extreme cases of dynamic modifications and between which the bulk of neurons with intermediate characteristics is distributed. One of these extreme poles is formed by neurons with a minimal dynamic variability (stable neurons), the other by neurons with a high dynamic variability.

2. Stable neurons. We failed to observe absolutely stable neurons, because the first responses were invariably stronger than the subsequent ones. But since the extinction of the responses was only slightly expressed, they can be related to the group of stable neurons. Cell IC 34 may serve as an example of neurons belonging to this group (Fig. 5). The histograms show that the responses to the first 10 applications of the stimulus are just a little stronger than to the 231st–240th rhythmical clicks. Manifesting only slight variations, the responses remain constant during the whole period of stimulation. Neither omissions (after 100 and 180 applications), nor rhythmical clicks with a frequency of 10 cps (histogram 231–240) exert any influence on the responses. This stability can be also seen in Fig. 4 (A, B, C), which presents the recovery cycles and the dependence of the response of the same cell on the intensity of the click. Neurons with relatively stable responses comprise more than 70% of all the responding units, including neurons with variable individual responses, and all the neurons with inhibitory reactions.

3. Neurons with a partial extinction. Not in a single case could we observe a full extinction of the response. Therefore, this group includes those neurons in which the effect of partial extinction is of a more pronounced character. Cell IC 45 is a typical representative of this group of neurons (Fig. 6B, C). The records clearly demonstrate a gradual decline of the response. Statistical processing

---

The 4 points of the graph correspond to the 4 PST histograms in A. (C) Cycles of response recovery to paired clicks.

Ordinate, average number of spikes during 3 applications for each period of 10 msec plotted on the abscissa. The arrows under the abscissa indicate the moments of the application of the 2nd stimulus of the pair (the first click coincides with the beginning of the histogram). The figures above the histograms indicate the intervals between the paired clicks.

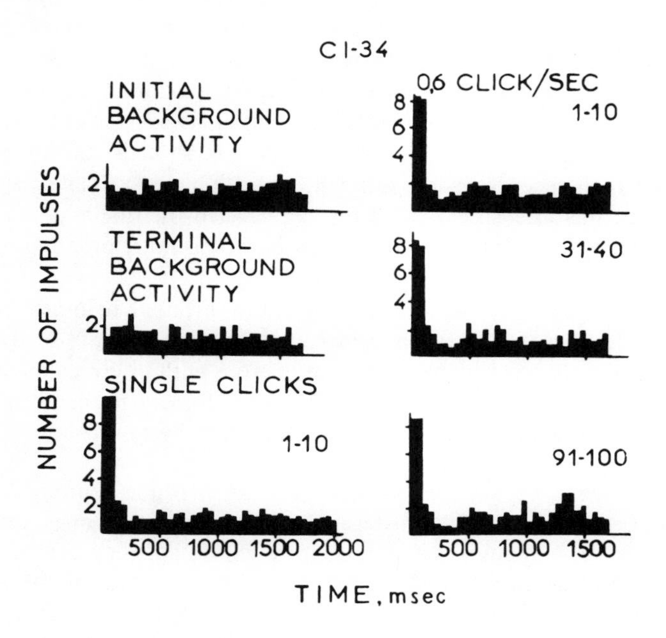

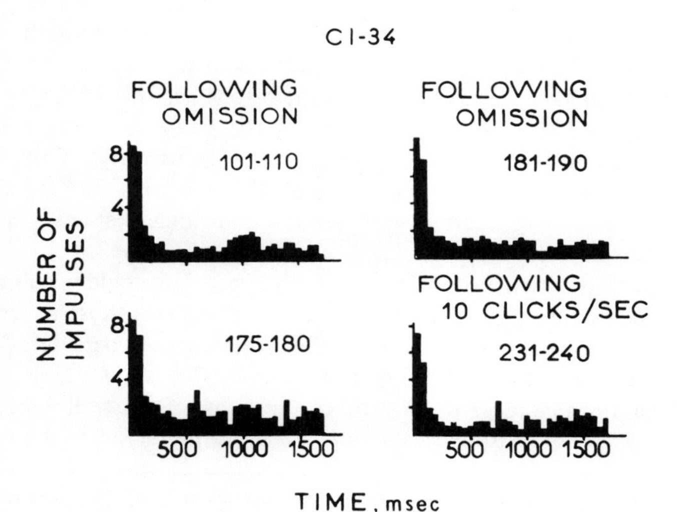

FIG. 5 Histograms of the background activity and PST histograms of the responses of a stable type from neuron CI-34 to clicks. The numbers of the successive applications are shown above each histogram. Ordinate, average number of spikes in each 50-msec period plotted on the abscissa. The moment of the stimulus presentation coincides with the beginning of the histogram.

showed that the total number of spikes during the interval between the stimuli decreases, which is mainly due to a decline in the background activity (Fig. 6C). After an omission, the response recovers to a certain degree, but as a result of further stimulation, the number of spikes shows a still greater decrease. A second omission also brings about recovery of the response. The average number of spikes in the response before the omissions is smaller than after them. Both the response itself and the background activity become weakened. However, the signal to noise ratio in the course of the repeated clicks presentation gradually increases (Fig. 6B). During random application of the clicks, S/N ratio proves to be lowest (in the figure this level is indicated by a dotted line). With the transition to rhythmical clicks this index proves to be somewhat higher, and it increases considerably with the repetition of the stimuli. After an omission, the S/N ratio markedly declines and rises to a higher level during further repetitions. From this it follows that the response increases with the increase of the noise (i.e., background activity). When the stimulus is repeated, the noise becomes suppressed and the response itself decreases. The S/N ratio, however, rises at the expense of a more rapid process which is connected with the suppression of the background activity.

Unit IC 39 (Fig. 7 and Fig. 8) presents a transitional type between the two extreme types of neurons–stable ones and those with a partial extinction of the response. The latter is characterized by a partial extinction of the responses which develops during a rather short period of time. The first applications of clicks evoke a strong reaction which becomes weakened by the 10th application (Fig. 7). During the action of the first 50 rhythmical clicks, the response remains rather variable. After an omission of the signal the response becomes considerably stronger (Fig. 7, No. 51). This disinhibition of the partial extinction is still more pronounced after a second omission of the spinal (No. 101). But the previously mentioned effects manifest themselves with particular clarity during a rhythmical repetition of the complex of click plus flash (Fig. 8). The given neuron exhibits no reaction to the light stimulus. The first applications of the complex evoke a strong reaction which becomes extinguished by the fourth application; after that it remains on the level achieved during all the subsequent applications of the complex. A flash after combination of a click and flash begins to evoke a weak reaction which becomes rapidly extinguished. A repeated presentation of the complex of flash plus click again evokes a strong reaction, which proves to be partially extinguished by the third application of the complex; it then remains on a constant level.

4. Neurons with a full extinction of the responses. As mentioned previously, we were not able to record a complete extinction of the response to a click in a single case. It proved difficult to evaluate the action of the light stimulus, since in our experiments each flash was accompanied by a weak sound. Apparently however, in some cases the light stimulus itself evoked reaction. Figure 6A presents PST histograms of cell IC 38. Whereas a click evokes a reaction of a strongly pronounced phasic type, the response to rhythmical flashes presented

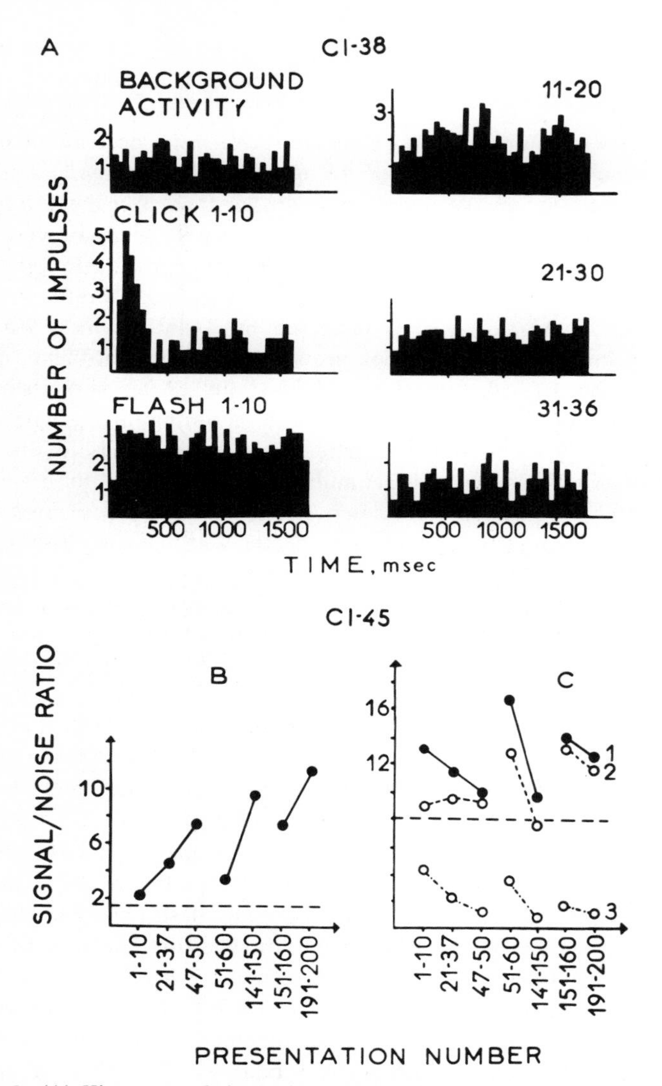

FIG. 6 (A) Histogram of the background activity and PST histograms of the responses from cell CI-38 to clicks and flashes.

Designations, the same as in Fig. 5. (B) Modification of the signal to noise ratio (correlation between the number of spikes during the first 100 msec after the stimulation and the number of spikes during the remaining part of the interval between the stimuli) in successive blocks of rhythmical clicks (cell CI-45).

Dashed line–level of the signal-to-noise ratio during the random application of clicks (1–10) single applications). The interruptions in the graph indicate the intervals in stimulation. (C) Changes in the total number of spikes (1), in the number of spikes per response (during the first 100 msec after the click) (2), and in the number of spikes in the remaining part of the interval between the stimuli (3), in successive blocks of stimulus applications. Dashed line–level of background activity during a period equal to the interval between the stimuli.

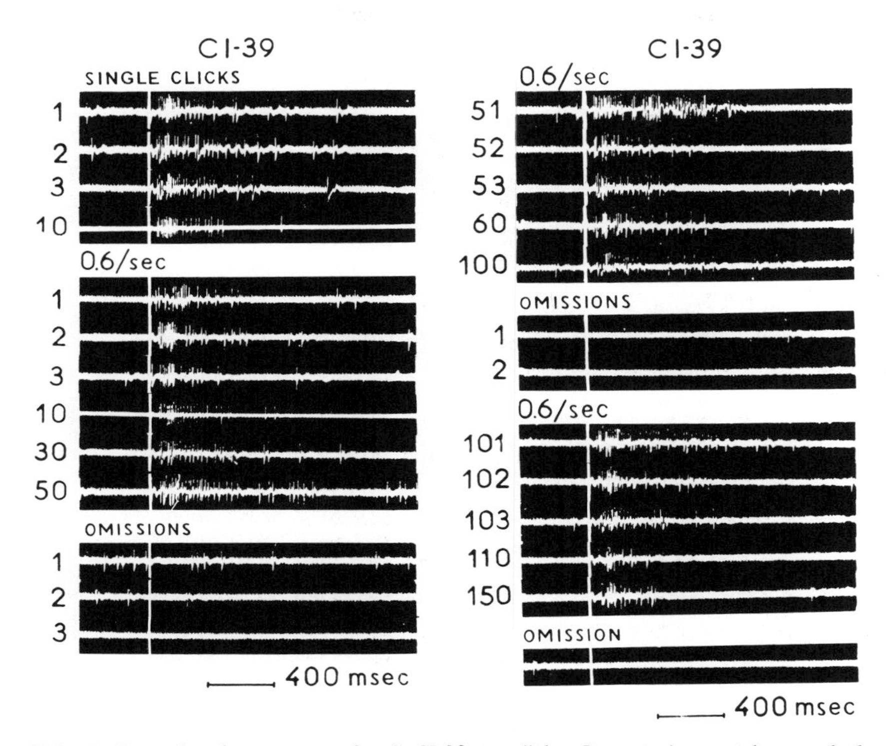

FIG. 7 Records of responses of cell CI-39 to clicks. Presentation numbers and the numbers of application are indicated on the left-hand side. In the case of omissions (after 50, 100, and 150 applications) the mark of the expected stimulus is conventional (the place where it would have been presented).

with a frequency of 0.6 per second bears a tonic character; this response becomes gradually and completely extinguished, reaching the background level during subsequent repetitions (compare the histogram of the background activity with the histograms of 1–10, 11–20, 21–30 and 31–36 successive applications of rhythmic flashes).

5. Neurons with an intensified response. The next type of modification, which is characterized by an intensification of the response as a result of repeated presentations of the sound stimulus, was observed only in one case. Figure 9 presents records of reactions of cell IC 31. The first five applications of single clicks, following one another in a random order, evoked no response whatever. The first applications of rhythmic clicks (0.6 per second) evoked a response consisting of one spike; this response persisted during subsequent applications and increased to two spikes. After 10 applications a 30-second interval of silence was introduced, and during this interval not a single spike emerged. No response was observed to the 11th and 12th applications (the first applications of rhythmical clicks after the omission). The first response (consist-

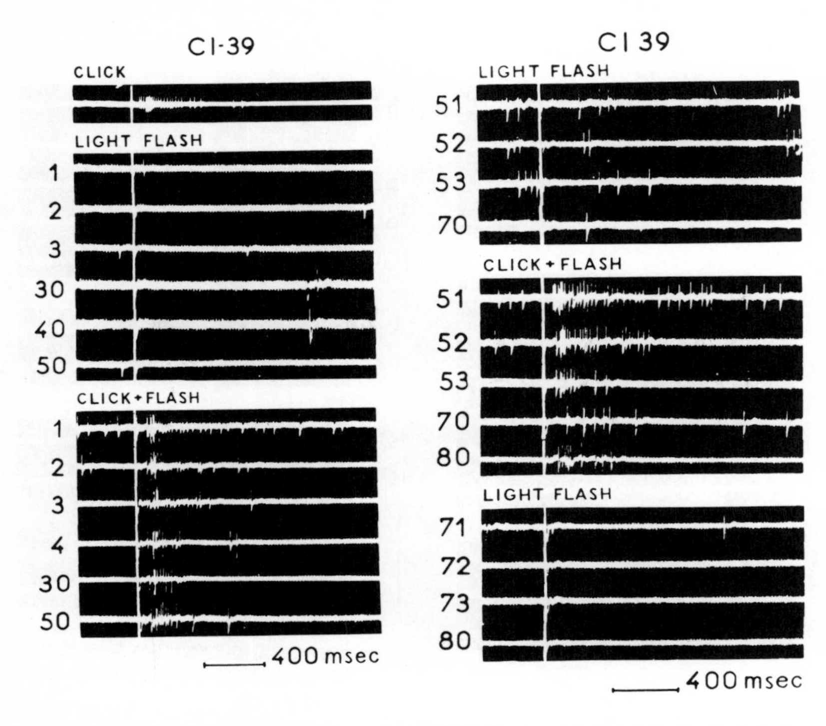

FIG. 8 Records of responses of the same cell CI-39 to a click, to rhythmical flashes and to a complex of click plus flash. Presentation numbers are shown on the left-hand side.

ing of 1 spike) emerged to the 13th application and was observed in all subsequent applications, showing an increase to 3–4 spikes. Attention should be paid to the long latency of this response (about 230 msec) which varies insignificantly and does not undergo any directed change as a result of stimulation.

6. Absence of an elaboration of a conditioned reflex to time and of a conditioned reaction to a complex of sound plus light. As mentioned above, series of omissions of the stimulus were introduced during rhythmic stimulations. Forty-nine neurons were obtained having 1 to 8 such series (from 5 to 20 successive omissions in each series). However, not in a single case could we observe any significant manifestations of a conditioned reflex to time either in the form of an increase of spike activity during stimulus omission, or as a response which occurs at the scheduled time of stimulus presentation. Cell IC 39 may again serve as an example (Fig. 7). Seven spikes emerge in response to the first omission after 50 applications of rhythmical clicks, 1 spike to the second omission, and no spikes at all to the third. However, the increase of activity during the first omission is apparently due to the afteraction of the response to the fiftieth application of the click, which proved to be of a longer duration than usual. During omissions which were made after 100 and 150 applications, no increase of activity took place at all.

During the demonstration of a partial extinction of the responses of cell IC 39 (Fig. 8) to a complex of click plus flash, it was observed that whereas prior to the combination with the sound stimulus the flash did not evoke any response, after such combination a weak reaction began to emerge. But this case, as well as other similar cases, may not be regarded as a true conditioned reflex, since the weak subthreshold sound which accompanies each flash of the tube could itself increase the excitability of the auditory system.

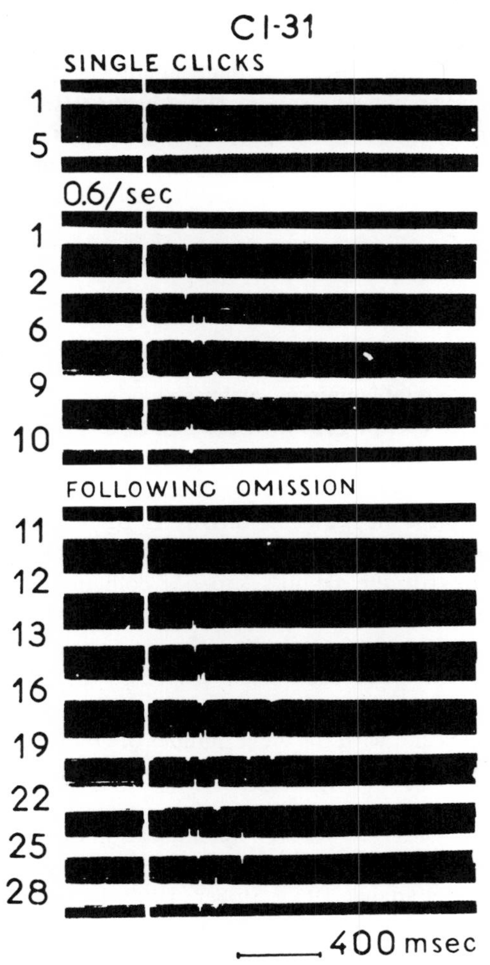

FIG. 9 Records of responses of a cell with an increasing reaction to single and rhythmical clicks. After 10 applications of the rhythmical click a 30-sec interval of silence was interposed.

7. Connection between the dynamic changes and the type of response. In the examples described before, we considered the dynamics of excitation during numerous repetitions of the clicks. No such dynamic changes were observed in neurons which responded with inhibition: Inhibition was relatively more stable than the excitation. Figure 2D presents a graph showing modifications in the duration of inhibition (cell IC 35 of the inhibitory type) depending on the sequence of the clicks. These modifications did not assume any definite direction during stimulation. Therefore, all neurons, except cell IC 55, which respond to a click with an initial phase of inhibition, may be related to the group of stable neurons. As to cell IC 35 (the record of its response is presented in Fig. 3), it differed from other cells by a very long initial inhibition (from 1000 to 2000 msec). During rhythmical stimulation, the inhibition accumulated. This was reflected by the fact that the recovery of the background activity (which in the course of stimulation was practically suppressed after the first click applications) required less time during the first series of omissions (about 3 sec) than during the subsequent omissions (5 sec and over). This specific feature gives grounds to single out this cell as a neuron with a partial intensification of the inhibitory reaction.

DISCUSSION

The neurons were recorded in the central nucleus of the inferior colliculus where the fibers of the lateral lemniscus, which extends to the medial geniculata body, partially terminate (Sepp, 1959). In the course of evolution, the inferior colliculus develops later than the superior colliculus, but their functions are apparently similar. The difference between them is that they belong to the different analyser systems. In view of this, it is particularly interesting to compare the responses of their neurons. Dubrovinskaya (1967) carried out an investigation of neurons of the rabbit's superior colliculus, and showed that according to the types of their responses to a flash, the neurons of the superior colliculus may be divided approximately into the same groups as the neurons of the inferior colliculus. In both cases their responses are of a quite diverse character, which is connected with the complexity of the morphological organization of these structures and with the great number of connections between the tectum and other divisions of the central nervous system (Tarlov & Moore, 1966).

In cases of specialized forms of behavior (echolocation of bats, hydrolocation of dolphins) there is observed a relative increase of the size of the inferior colliculus (Morgane & McFarland, 1965; Slijper, 1962; Friend, Suga, & Suthers, 1966). It may be assumed that in the course of evolution both the inferior and superior colliculi become more complex and turn into integrating centers, below level of which the initial processing of information about the incoming signals has already taken place.

Considering these dynamic changes, we have to point out that our work to some degree contradicts the works of Horn and Hill (1964, 1966) who demonstrated complete habituation of responses in most neurons of the midbrain tectum and tegmentum. This contradiction is apparently explained either by a difference in the intensity of the stimuli presentations, or by the fact that the afore-mentioned authors recorded the neurons not only of the specific structures, but also of the adjoining unspecific areas.

Various combinations of excitatory and inhibitory influences of different origin converge on a neuron. In other words, the work of a neuron is determined not only by its inherent properties, but also by the net into which it is included, as well as by the place which is occupied by the given neuron in this net (Sokolov, 1965, 1967). Due to this, not only a great variety of the types of neuronal responses should be expected, but also a great variety of dynamic modifications of these responses. The following several types of neurons with different modifications of their responses as a result of numerous applications of clicks may be established:

1. Neurons manifesting no responses both at the beginning and in the course of stimulation (11 neurons, or 22%).
2. Stable neurons (32 neurons, or 64%): (*a*) neurons with a stable inhibitory response (6 neurons, or 20%, example in Fig. 2D); (*b*) excitatory and excitatory inhibitory stable neurons with a slight partial habituation (16 neurons, or 32%; example in Fig. 4); (*c*) excitatory and excitatory-inhibitory neurons with highly variable, but persistent responses (10 neurons, or 20%).
3. Neurons with a pronounced partial habituation (5 neurons, or 10%; example in Fig. 6B, C).
4. Neurons with an increasing response (2 neurons, or 4%): (*a*) 1 neuron with accumulated inhibition (description in the text); (*b*) 1 neuron with intensified excitation (example in Fig. 9).

We failed to observe, in any of the investigated neurons, a complete habituation to clicks. The most frequently observed dynamic change consisted of a partial extinction accompanied by a stabilization of the response. Two extreme cases of a partial extinction may be singled out in the classification mentioned above: neurons with a minimal extinction of the response (relatively stable neurons) and neurons with a pronounced partial extinction of the response. The essence of this partial extinction is that the first applications of the stimulus evoke a stronger and longer reaction than its subsequent applications. The degree of extinction and its rate vary from neuron to neuron, but after the extinction reaches a definite level, the response then remains stable. An interval in the stimulation or an extra stimulus lead to a recovery (disinhibition) of the response. It is important that this effect of partial extinction is accompanied by an increase of the signal to noise ratio. Similar phenomena were described in neurons of the superior colliculus (Dubrovinskaya, 1967), as well as of the lateral geniculate body (Beteleva, 1967). Apparently, this is an expression of the interaction of specific and unspecific influences on the level of a single neuron. The unspecific influence becomes more or less rapidly extinguished, while the specific one stably persists. The unspecific character of the habituating component is proved by the fact that the partial extinction is similar in its temporal parameters to a complete extinction of the responses of the reticular neurons (Scheibel & Scheibel, 1965) and to a complete extinction of the responses of the specific neurons to inadequate stimuli (Skrebitsky & Bomstein, 1967).

The following question arises: Why is the phenomenon of complete habituation not observed on the level of the midbrain, especially since the afferent manifestation of different components of the orienting reaction becomes extinguished after the first 10–20 repetitions of the stimulus? A study of neurons in different central sensory structures showed that a complete habituation to specific stimuli is either absent or is encountered in very rare cases (Vardapetian, 1967; Vinogradova & Lindsley, 1963; Beteleva, 1967; Dubrovinskaya, 1967). This leads to the assumption that the main defect of the functional models of habituation (Horn, 1967; Fernandez-Guardiola *et al.*, 1967; Hernandez-Peon, 1960) is that they assume the possibility of a complete disappearance of the responses in the afferent link to be the result of repeated presentations of specific stimuli. A selective extinction in the afferent systems, particularly in those which are connected with different components of the orienting reflex, is possible only if the afferent neurons continue to transmit information. The basic effect evoked by numerous repetitions of the stimulus in afferent neurons is a partial habituation, whose mechanism may be described as follows: Excitation coming along the classical sensory pathway to the specific afferent neuron (A_s in Fig. 2C) reaches the nonspecific neuron (A_n) along the collaterals; the latter extends its processes both to the afferent (A_s) and efferent (E) neurons. The latter, being connected with a realization of definite components of the orienting reflex, discharges only when summary excitation comes from the specific (A_s) and unspecific (A_n) neurons. When the stimulus is repeatedly presented, a system which blocks the unspecific excitation comes into operation. This blocking may be performed by the cortex–hippocampus system. Vinogradova (1965) showed that, as a result of the first applications of a stimulus, most neurons of the hippocampus become inhibited. When the stimulus is repeated, a recovery of the activity takes place. Some neurons exhibit an increase of activity which is timed to the moment of the action of the stimulus. On the other hand, Redding (1967) and others, applying direct electrical stimulation, showed that the hippocampus exerts inhibitory influence on the activating reticular formation of the midbrain. Thus, it may be assumed that with the repetition of a stimulus, a neuron of the hippocampus (H) begins to block the unspecific neuron through the system of the interneurons (I). The reaction of the efferent neuron, and corresponding component of the orienting reflex, become gradually extinguished, since the summation of the specific and unspecific excitation coming to the efferent neuron declines. Finally, the reticular neuron proves to be fully blocked.

The excitation continues to arrive through the afferent neuron, although now it cannot alone evoke a reaction of the efferent neuron. Thus, a full extinction of the reactions of the efferent and unspecific afferent neurons is accompanied by a partial extinction of the reactions of the afferent neuron. The given scheme also explains the complete habituation of the specific neuron to the inadequate stimulus. When some parameter of the stimulus is changed (for example, the frequency of the tone), the orienting reaction reappears; it becomes extinguished again when the modified stimulus is repeated. The emergence of a reaction as a result of a change only in one parameter of the stimulus is due to the fact that simultaneously with such a change, the channel of maximal excitation becomes displaced both in the classical pathway and in the unspecific pathways (in the scheme this is designated as S_1, S_2, S_3). It should be further assumed that the blockade of the unspecific excitation is selective with respect to the stimulus by which it was evoked.

11

THE HIPPOCAMPUS AND THE ORIENTING REFLEX

O. S. Vinogradova

Moscow State University
and
USSR Academy of Sciences
Pushchino on the Oka

The limbic system, and particularly, its largest division the hippocampus, are still "mute zones" of the brain in spite of the numerous attempts made in recent years to ascertain their functions. The concept of olfactory reception as the leading function of this area ("rhiencephalon") has been fully discarded during the last decades. The participation of the hippocampus in the regulation of vegetative functions ("visceral brain"–McLean, 1954) has not been fully confirmed either. Likewise, the opinion that the limbic system is a regulator of sleep ("hypnogenic limbic system"–Hernandez-Peon, 1963) cannot, apparently, be considered comprehensive although numerous available data show that the hippocampus can exert an inhibitory influence on the activating system of the brain. The progressive development of the hippocampal system, which manifests an absolute and relative growth in the course of evolution (Brodal, 1947, and others) and reaches its highest level of development in primates (particularly in man) should be indispensably accompanied by a corresponding development of its functions. Thus, the functions of the hippocampal system can hardly be reduced to the regulation of the vegetative processes, olfaction, or sleep.

An analysis of clinical and experimental data shows that among all the diverse functions which are attributed to the hippocampus, its participation in inhibition of the orienting reflex and in the processes of memory are consistently implicated (see, for example, the reviews by Douglas, 1967; Meissner, 1966; and Vigouroux, 1959).

The concept of participation of the hippocampus in the suppression of the activation reaction (or orienting reflex), which arose as a result of electrophysiological and behavioral investigations, served as a starting point for the present work. We attempted to find correlates of the orienting reflex, with its complex dynamics, in the responses of hippocampal neurons.

METHODS

Extracellular recording of neuronal activity was performed in the course of experiments on chronically prepared, unanaesthetized rabbits, slightly restrained in a special box. Tungsten microelectrodes, with tips having a diameter of 1–3 μ and a resistance of 4–10 megohms in saline, were used; the microelectrodes were inserted with the aid of a Hubel remote hydraulic micromanipulator. Neurons of the dorsal hippocampus in fields CA_2 and CA_3 (Lorente-de No, 1934) were recorded.

When the microelectrode was passing through the dorsal hippocampus, we found two zones having a high probability of unit discharges, one 300–700 μ from the surface of the hippocampus, the other 2000–2500 μ (near its lower border). These levels correspond to fields CA_2 and CA_3 in the investigated area indicating that the majority of the records were obtained from hippocampal pyramidal cells. Calculations of the depth to which the electrode was inserted were verified after the end of a given series of experiments by means of electrolytic lesions or marking by iron ions through a steel microelectrode in sections of 10–20 μ. The following stimuli were applied: clicks, pure tones, light flashes, changes of general illumination, and air puffs on the animal's muzzle. The stimuli were usually presented in series at constant intervals of 2.75 or 5 sec.

Recording was performed on a Yausa-10 tape recorder, with subsequent filming of discharges. On the records thus obtained, the periods between the stimuli were divided in periods equalling 125 msec; the number of spikes was determined for each of these periods. In order to establish the level of spontaneous activity, an equivalent number of periods of background activity was counted without stimulus presentation. On the basis of the numerical matrix obtained for a number of presentations of the same signal we formed (1) individual post-stimulation (PST) histograms or graphic tables showing the density of the spikes; (2) averaged histograms for a number of stimulus presentations; (3) graphs indicating the dynamics of changes in the responses based upon the total number of spikes for each period between the successive stimuli. For the purpose of evaluating the action of long-lasting stimuli, we formed cumulative curves for the same periods of 125 msec during the action of the stimulus, as well as for periods of the background activity of equal duration before and after the presentation of the stimulus. We also used automatic counting of the number of spikes for 10-sec periods with the help of a "Volna" pulse counter.

RESULTS

A. Two Types of Responses of Hippocampal Neurons

Neurons with the following characteristics constitute the majority of reactive units in fields CA_2 and CA_4 of the dorsal hippocampus:

1. They possess high rates of irregular spontaneous activity (average frequency, 15–30–60 per sec).
2. They respond to all stimuli applied, usually without preference to a specific modality or to an optimal range of stimuli within one modality.
3. The type of response is uniform, irrespective of the stimulus. (The principle of frequency coding in the output of a multimodal neuron does not appear).
4. Their responses possess a relatively long latency (70 to 300 msec); the precise values are difficult to establish due to the presence of high rates of spontaneous activity. Also, responses are tonic in nature; in the case of a short signal (1–3 msec), they may persist for 10–12 sec, while in the case of long-lasting stimuli, the response may last tens of seconds.
5. With repetition of the signals, the responses gradually decline; this is expressed in a shorter duration of the discharges, the initial parameters of the response remaining the same. A complete disappearance of the response takes place after 10–20 presentations of the signal.
6. Any change in stimulus parameters or in the conditions of its presentation leads to the recovery of the initial response (see Section 2).

The previously mentioned features relate to 80% of the reactive neurons of the dorsal hippocampus. This percentage is high enough to draw the conclusion that these features characterize the whole system of hippocampal neurons.

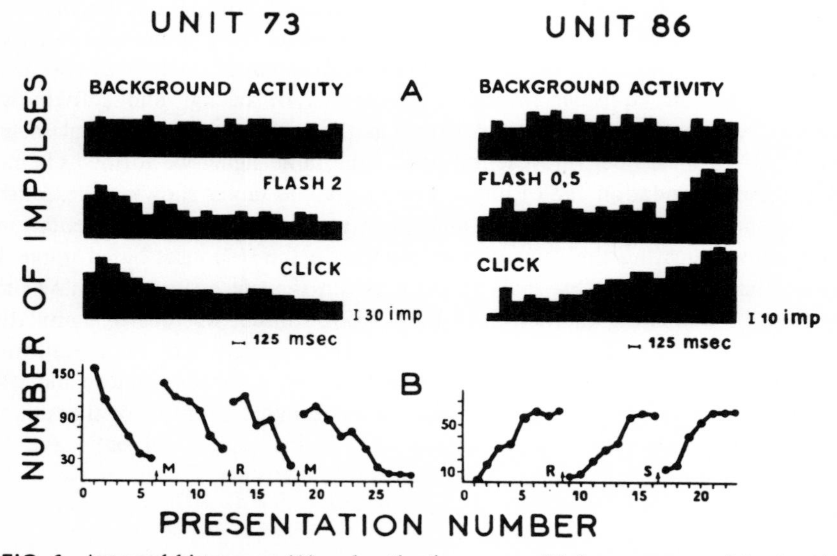

FIG. 1 Averaged histograms (A) and extinction curves (B) from neurons of the A-and I-types. The histograms present averaged data for 10 applications of the signals. Graphs: ordinate, total number of discharges during the interval between two successive signal presentations; abscissa, time. The form of responses to stimuli of different modalities for neurons of each type, as well as the mirror character of the reactions and of the dynamics of extinction are shown.

However, the responses of these neurons are not uniform, being divided in two distinct subgroups, an activatory subgroup (A) and an inhibitory one (I). In the case of A neurons, the response consists of a tonic 2–5-fold increase of the activity over the background. In the case of I neurons, the stimulus usually suppresses spontaneous activity, bringing it down to the zero level. All the temporal characteristics (latencies, duration of the response, number of presentations required for the extinction process) are very similar in both types of neurons, although some individual variations are seen. The averaged PST histograms present mirror images of each other (Fig. 1A).

But in the case of hippocampal neurons, the real nature of the responses is not reflected in the averaged histograms, due to their dynamic character, since the initial maximal A and I responses are summated in them with the final states which are at the level of background activity. The symmetrical character of the dynamics of A and I neuronal responses is clearly seen in the graphs of the total number of impulses during the period between each two successive stimuli (Fig. 1B). In the case of A neurons, these graphs represent typical extinction curves. The steep slope of the curves indicates the fast rate of extinction; spontaneous activation (judged by the animal's movements), pauses, extrastimuli, etc. lead to the recovery of the initial response; however the recovery begins at a somewhat lower level on successive occasions (Fig. 1B, U 73). In the case of I neurons, the extinction curves are similar to the graphic representation of the process of learning, i.e., they increase gradually and reach an asymptote. In this case, too, the activating stimuli interrupt the above-mentioned process, returning it back to a level which is close to the initial one, but which gradually deviates from the latter in the course of the repeated signal presentations (Fig. 1B, U 86). The slopes of the extinction curves for the A and I neurons are approximately equal, i.e., the rate development of the processes is similar.

The dynamics of the responses of A and I neurons can be seen in greater detail in Fig. 2, which contains graphs showing the density of the spikes during the action of a consecutive series of stimuli for neurons of both types. Here the density of the spikes is plotted for periods of 125 msec during the action of 24 stimuli (clicks). The graphs clearly show a gradual extinction of the activatory and inhibitory responses, which is expressed in their gradual reduction to complete disappearance; a recovery of responses, almost to the initial level, occurs as a result of rest intervals and presentation of different stimuli. In this case, too, the parameters of the processes for both types of neurons are similar.

Thus, wherever the activity of the A neurons is maximal, the activity of the I neurons proves to be minimal. In the case of A neurons, the initially heightened frequency of the discharges declines to the background level, while in the case of I neurons, the frequency of the discharges gradually increases reaching the background (or somewhat higher) level. The same final result, disappearance of the initial responses and the restoration of the background activity, is achieved in directly opposite ways.

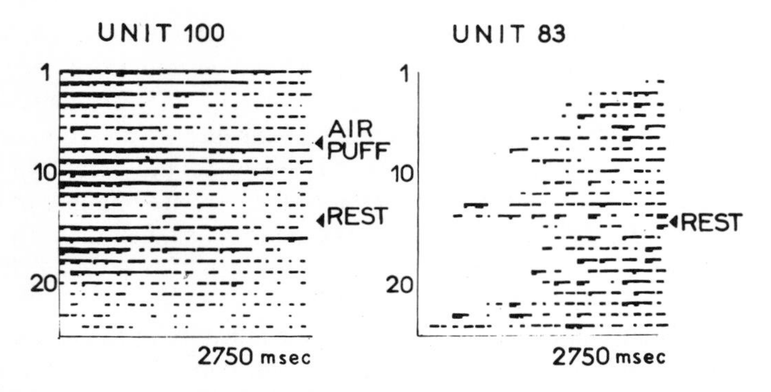

FIG. 2 Graphs showing the density of spikes for intervals of 125 msec during successive individual applications of a click. Unit 100 is a neuron of the A-type; unit 83, a neuron of the I-type. These graphs reveal the development of extinction as a gradual shortening of the response (activation, Unit 100; or inhibition, Unit 83) and interruption of the extinction process during the presentation of a different stimulus (air puff) and best intervals in the series of the signals. Along the vertical line, 24 successive applications of the signal; along the horizontal line, density of the spikes per 125 msec for a period of 2750 msec between successive applications of the stimulus. The number of points is proportional to the number of spikes for the period (125 msec). Stimulus onset coincides with the beginning of each horizontal line.

The coincidence of the temporal parameters of the A and I neurons is by no means accidental. These coordinated but opposite responses may be explained in the following two ways:

1. The A and I neurons function in parallel. In this case, there must exist a single source exerting influence on them; the concrete result of its action is determined by the properties of the postsynaptic element. An integration of the output signals of the A and I neurons would be senseless, in view of the symmetric character of their parameters; consequently, it must be assumed that they are addressed to different elements (or different structures), exerting a parallel double action—excitatory and inhibitory. Analogues of such a double action of the hippocampus can be found in investigations of the EEG and evoked potentials.
2. The A and I neurons are connected in series. In this case the A neurons might be regarded as the input system of the hippocampus, such that they exert an inhibitory action on the I neurons which constitute the basic efferent system of the hippocampus. Then, a gradual decline of the activity of the A neurons (according to the mechanism of "collateral" or "parallel" inhibition, Eccles, 1965a; Sokolov, 1965) would lead to an increase of the activity of the I neurons. The increasing activity of the I neurons may exert a regulatory influence on the lower executive mechanisms of the orienting reflex (reticular formation). This influence may be regarded as bearing a direct inhibitory character.

The impossibility of precisely determing the correlations of the latencies of the A and I neurons precludes a definite answer to this question.

On the whole, the following should be taken into consideration:

1. With the first applications of new signals, the activity of the majority of reactive neurons of the hippocampus (54%) declines (I neurons).
2. When the signals are repeated, the total activity of the hippocampal neurons increases.
3. A change in the activity of all the reactive neurons leads to a sharp decrease of the signal-to-noise ratio in the output of the hippocampus, which is due not only to the weakening of the signal (A neurons), but also to the intensification of a high-frequency noise (I neurons).
4. The activity generated by hippocampal neurons bears the character of a massive, summated efferent signal. When the stimulus is repeated, this gradually developing type of activity persists in the background; the resultant background activity may serve as a tonic modulator of the activity of the regulated systems.
5. The temporal parameters of responses of the hippocampal neurons and of their modifications are close to the corresponding parameters of the orienting reflex at the macrolevel.

B. Adequate Stimuli for Hippocampal Neurons

As stated above, the input of the hippocampal neurons is as a rule multimodal, whereas the responses are uniform during the action of any stimuli. In this connection, it might be said that selectivity with respect to the characteristics of the stimulus is by no means characteristic of the hippocampal pyramids. However, such a proposition will prove wrong, if we take into account that the properties of the stimulus for the organism are not limited to its physical parameters. Any change in the series of identical stimuli leads to recovery of the habituated neuronal response. In this respect the system of hippocampal neurons proves to possess maximal selectivity. The selectivity of the responses ensures the detection of changes in any parameters of the signal relative to the initial fixed level. Experimental data show that the possibilities of such selection may approximate the differential threshold in all spheres of sensitivity.

In order to illustrate the abilities of hippocampal neurons to detect changes in the signals, we shall present a number of methods which lead to an immediate recovery of the extinguished neuronal reactions (besides the gross modality changes mentioned above). They reveal the significance of the factor of "novelty" of the signal, in combination with some additional influences.

1. Changes in the intensity of a signal of a single modality. Repeated application of a signal with a fixed intensity results in a suppression of the responses of both the A and I neurons (see above). A critical test which reveals the significance of the novelty factor consists of a sudden decrease of the signal intensity without changing any of its other physical parameters. Naturally, such a recovery of the response cannot be accounted for by a change in the threshold, by fatigue of the peripheral organ, etc.

However, though the significance of the novelty factor appears here with particular clarity, there exists a complex interaction between the physical

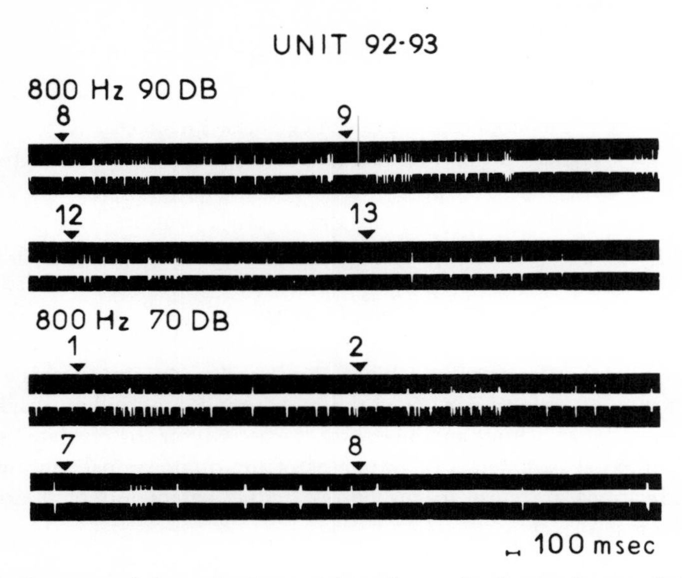

FIG. 3 Recovery of the activation reaction of two simultaneously recorded neurons–Nos. 92, 93–during a change from a strong sound (90 dB), the response to which was extinguished by the 13th presentation of the signal, to a sound of moderate intensity (70 dB). Numbers indicate the serial number of the stimulus presentation (at the arrows).

strength and the novelty of the stimulus. In an extended form, this is showing the example of cells No. 92–93 which were recorded with one electrode in field CA_2 and which are typical representatives of the group of A neurons (Fig. 3). Possessing similar types and dynamics of tonic activatory reactions, these cells also exhibit certain differences. Thus, cell 92 (spikes of a smaller amplitude) was characterized by a higher spontaneous activity, a shorter latency, a slower extinction process, and a lower absolute threshold; the extinction process in this cell was accompanied not only by a decrease in the frequency of the discharge, but also by their collecting into "doublets," which, according to some authors, indicates profound inactivation. The initial application of a tone of 800 Hz and 90 dB evoked a marked activation in both cells, which became weakened as a result of repeated presentations of the signal. A transition to the application of the same tone with an intensity of 70 dB again resulted in an increase of the activity. Subsequently, such change in the intensity of the sound (in the direction of its increase or decrease) was accompanied by an initial rise in the frequency of discharges (Fig. 3).

The final effect, however, resulted from the action of two factors: the novelty (change) of the stimulus and its physical strength. This is clearly seen from a comparison of a series of averaged histograms for a consecutive application of an equal number of stimuli of different intensities (Fig. 4). Strong sound stimuli (90–100 dB) evoke a considerably greater activation than weak (50–60 dB) and moderate (70–80 dB) stimuli, in conformity with the "law of force." At the same time, the level of activation is a function of the number of the given series of stimuli; this is proved by a comparison of responses to a tone of 70 dB applied in the second and eighth positions during a stimulus series with those to 90 dB applied in the sixth and tenth places. Activation to the 90-dB tone applied in the tenth place persists, though it is weaker. But the response to a tone of 70 dB (in the eighth place) almost disappears. On an averaged histogram, the short period of initial activation becomes fully levelled out due to the rapid process of extinction. This process may be seen in greater detail on the series of extinction curves which are presented in the order used in this experiment (see Fig. 6).

These graphs show that extinction is slower for stimuli of higher intensities. However, following each change of stimulus intensity, there is initial recovery of the neuronal response. At the same time in the course of the experiment the general level of reactivity gradually declines, while the rate of the extinction accelerates.

These phenomena are similar to those observed during objective measurements of behavior, where the components of the orienting reflex are used as indicators for measuring the thresholds of hearing (Vinogradova & Sokolov, 1955; Sokolov, Vinogradova, & Paramonova, 1964). Objective indices show an increase in threshold due to a gradual extinction of the orienting reflex; this is first reflected in the action of weak stimuli. In the present case, the thresholds of the most sensitive cell (No. 92) increase from 40 to 70 dB (Fig. 5).

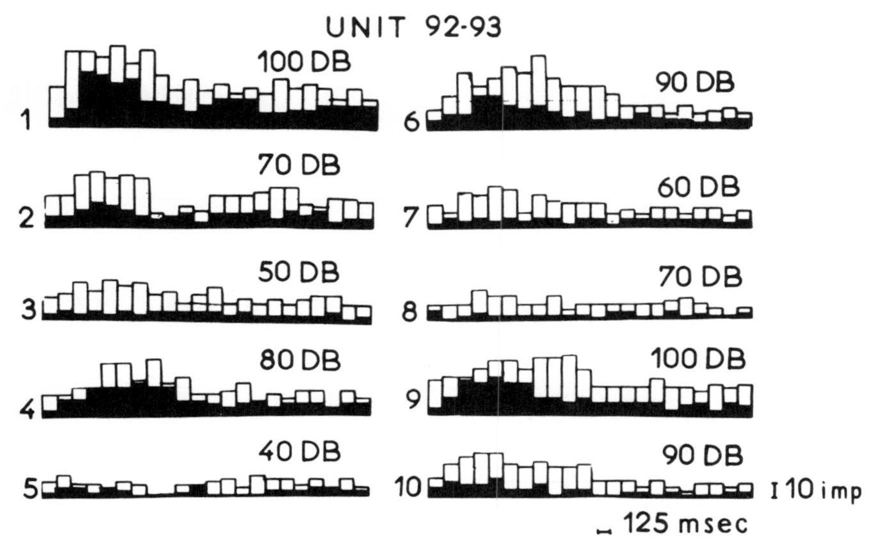

FIG. 4 A series of histograms of A responses from cells 92 and 93 to a tone of 800 Hz presented with different intensities. Explanations are given in the text.

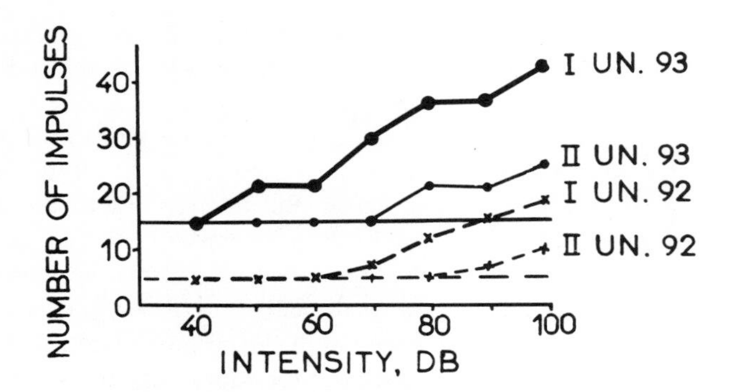

FIG. 5 Thresholds of firing of cells 92 and 93 at 800 Hz in the initial series of signal presentations (heavy lines, I) and in the final series (fine lines, II). The horizontal lines indicate the average background level for these cells. The threshold is determined as the average number of spikes exceeding the level of spontaneous activity. Intensities in decibels are indicated according to the scale of the oscillator: the threshold intensity (for a human subject) equaled 30 decibels.

Thus, when the intensity is changed, the action of the novelty factor is partially masked by the influence of stimulus intensity; at the same time the novelty factor may manifest itself in a very distinct form, during a transition from a strong stimulus to a weak one.

2. *Modification of the quality of the stimulus within a single modality.* As mentioned above, the cells of the dorsal hippocampus respond to tonal stimuli within the limits of an audible range of sounds, without any signs of preference for an "optimal" or "characteristic" frequency. A sudden recovery of the response as a result of a change in the frequency after preliminary extinction can be seen in Fig. 7 (neurons 92, 93). The response, which disappeared after 14 applications of an 800-Hz tone, recovers when this tone is replaced by a tone of

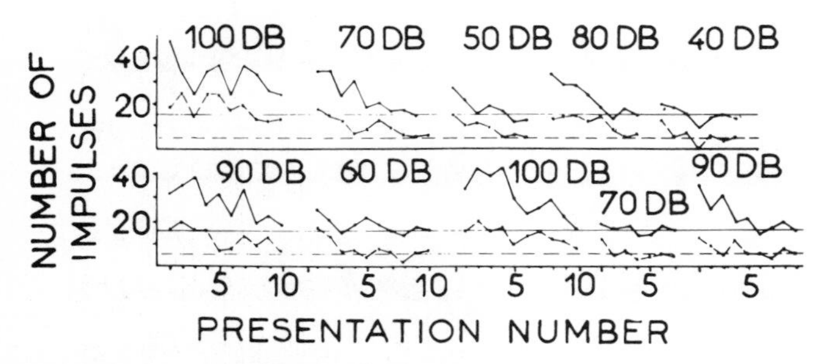

FIG. 6 Dynamics of reactions in the experiment with the application of a 800-Hz tone of different intensities in the same cells (Nos. 92 and 93), expressed in the form of consecutive extinction curves. Each change of intensity results in a recovery of response level in comparison with the directly preceding level, but in the course of the experiment generalized extinction of the effect occurs which influences the weak signals to a greater degree.

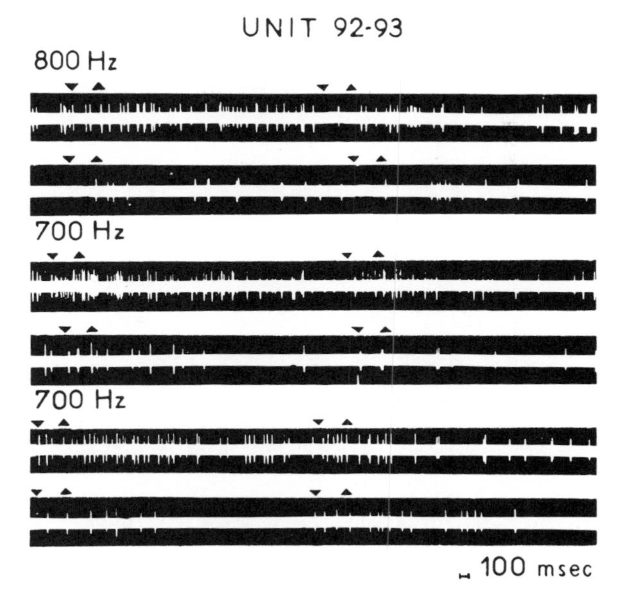

FIG. 7 Recovery of an extinguished response of cells 92 and 93 as a result of a change in the frequency of the tone.

700 Hz, and subsequently by a tone of 300 Hz. A similar phenomenon can be seen in other neurons of the A and I type (see Fig. 10).

A recovery of the initial response occurs as a result of any change in frequency. When the level of extinction is stable enough, the response recovers as a result of a change in the stimulus of only 50 Hz (in the middle-frequency range). Smaller amounts of frequency change were not tested, but it may be asserted that in this way it is possible to determine the differential threshold of signal discrimination at the neuronal level, just as at the level of the macrocomponents of the orienting reflex.

At the same time, the intensity of the recovering response depends on the degree to which the frequency of the new signal is similar to that of the initially extinguished stimulus. The intensity of the response is an increasing function of the differences between the stimuli. On the one hand, this fact may be regarded as a manifestation of the so-called generalization of extinction, and on the other, as a result of the growth of the "disconcordance signal" (see below) due to the application of markedly different tones. By many repeated presentations of one frequency, it is possible to obtain an artificial decline of neuronal reactivity to stimuli of this frequency as well as to the adjoining frequency band, while normal responses are preserved to more distant tones. This phenomenon is shown in Fig. 8, formed on the basis of responses of the cells described above (Nos. 92, 93). A tone of 800 Hz, which was used during a long period (investigation of the action of various intensities—see above), evokes a weakened

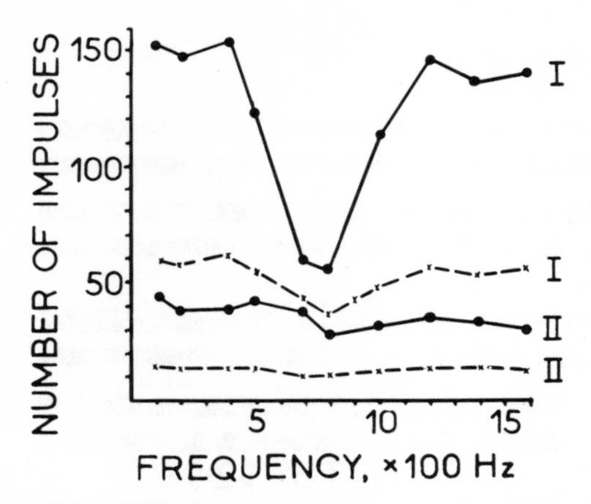

FIG. 8 Graph showing a selective decline of the reactivity of cells 92 and 93 during a prolonged extinction of one of the elements of the sound-frequency scale. The effect of generalization of the extinction is seen in (I). Similar curves, obtained after a prolonged application of different tonal stimuli are plotted below (II). Ordinate, total number of spikes; abscissa, range of the pure tones applied (one step equals 0.1 kHz). Solid circles indicate activity of the cell 92; crosses indicate that of the cell 93. Each point has been plotted according to the results of summation of the three first signals in each series. The horizontal lines indicate level of the background activity.

(due to extinction) and rapidly declining response. The same relates to a tone of 700 Hz, which was not previously applied, but is close in its frequency to the initial tone. The farther from this frequency band, the stronger become the responses. Thus, the cells begin to function as an inhibitory frequency filter in the band of the habituated frequencies (Fig. 8).

However, during application of long series of different frequencies this process becomes masked by another one: Just as in the case of a change in intensity, the level of the response recovery depends not only on the degree of the absolute difference of the signal, but also on the serial position of its application during the experiment. This is expressed in a gradual increase of the rate of the extinction process taking place with each introduction of a new series of tones, as well as in a general decline of the level of recovery of the response. The effect of repeated stimulus presentation may for some time be overpowered by the effect of the degree to which the signal differs; but in the case of a more or less protracted application of the signals, all the responses show a decline.

Thus, the recovery of the response depends on two factors—on the degree of the difference between the new signal and the preceding one (relative novelty), and on the serial number of the series in the sequence of the experiment (absolute novelty of the signal), since in the course of the experiment, in spite of a change in the quality of the signal, the very principle of such a change loses its novelty, and a general decline of the responses to all elements of the given class of stimuli takes place.

3. Modifications of the temporal parameters of the signal. Later on we shall return to an analysis of the ability of hippocampal neurons to register the duration of macroperiods of time (Section C). Here we shall approach another aspect of this question from and consider the possibility of evoking neuronal responses by changing the stable temporal characteristics of the signal.

After the extinction of a signal of a constant duration, the neuronal response can be recovered by means of a sudden increase of the duration (and to a lesser degree by its sudden decrease). Rapidly extinguishing responses of such a neuron to a tone of 3 sec become fully recovered when the duration of the stimulus is increased to 5 sec.

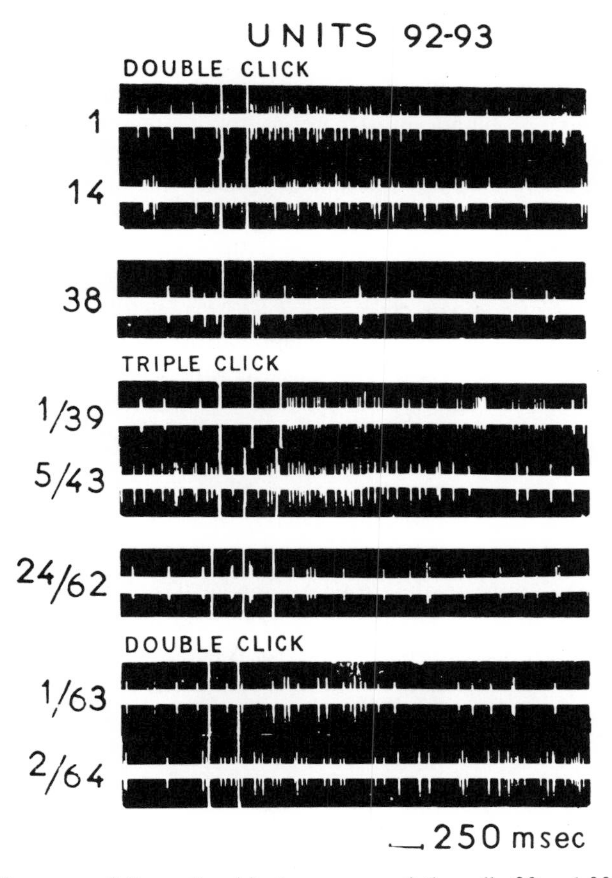

FIG. 9 Recovery of the extinguished responses of the cells 92 and 93 as a result of a transition from a double click to a triple one, and vice versa. The first number designates the serial number of the changed signal; the second number designates the total number of the applications of clicks and of their complexes. Vertical lines in the records denote click presentations. Details are given in the text.

A special form of time-dependent phenomena was observed during the application of short, low-frequency rhythmic stimuli, where it is the number of signals in the rhythmic series, and not the frequency itself, which constitutes the difference factor. This is shown in Fig. 9 for the same cells, Nos. 92 and 93. The application of a double click (with an interval of about 100 msec) evokes an increase in spike discharges, that becomes gradually extinguished. When, a triple click is then presented for the first time, the first two clicks in the complex do not evoke any changes in the reduced activity, coinciding with the habituated stimulus but after the third click an increase in discharges is first observed in the more labile cell 93, and subsequently, also in cell No. 92. During a second application of the triple click, the activation response arises after the first click of the complex. The same occurs after the extinction of the response to the triple click and return to the double one: Following an interval, which corresponds to the period before the application of the third click (in the place where this click is absent), an activation response of both cells reappears. The reappearance of a response of hippocampal neurons to the temporal position of an omitted regular signal, after a habituation to long series of stimuli with constant intervals, may be interpreted also as a time-dependent phenomenon.

Thus, these particular neurons can "measure" the duration of the signals, and the intervals between them, as well as "count" the number of signals in a group, thereby detecting the deviations from a fixed pattern.

4. Modifications of complex characteristics of the stimulus. Properly speaking, the above-described discrimination of double from triple clicks by a neuron presents an example of differentiation of successive signal complexes. This is related to simultaneous complexes of stimuli. The recovery of responses after the addition of an element to a stimulus complex cannot serve as proof of the action of the novelty factor, since it may be explained by the action of the mechanisms of summation. Therefore, the removal of a component from the complex is more convincing. It is possible to extinguish a response to a simultaneous complex—a click (strong component) plus a tone (weaker component). When the response disappears, the stronger component, i.e., the click, is removed from the complex; then the simplified and weakened signal begins to evoke the response again. This fact is shown in Fig. 10 for A and I neurons.

This means that when the effect of novelty is evaluated at the neuronal level, not only are the elementary physical parameters of single sensory stimuli considered, but also the coincidence in time of different signals acting as a single, integral stimulus.

5. Changes of background stimulation (tonic dishabituating influences). Phasic signals are evaluated relative to a certain level of tonic background stimulation (influences of the external and internal media), which remains rather constant during an experiment. Any change of this level, which is caused, for example, by the application of a long-lasting stimulus (e.g., change of background illumination, a constant sound) acts as a "novelty" and leads to a change in the effect of previously applied and extinguished phasic signals. This

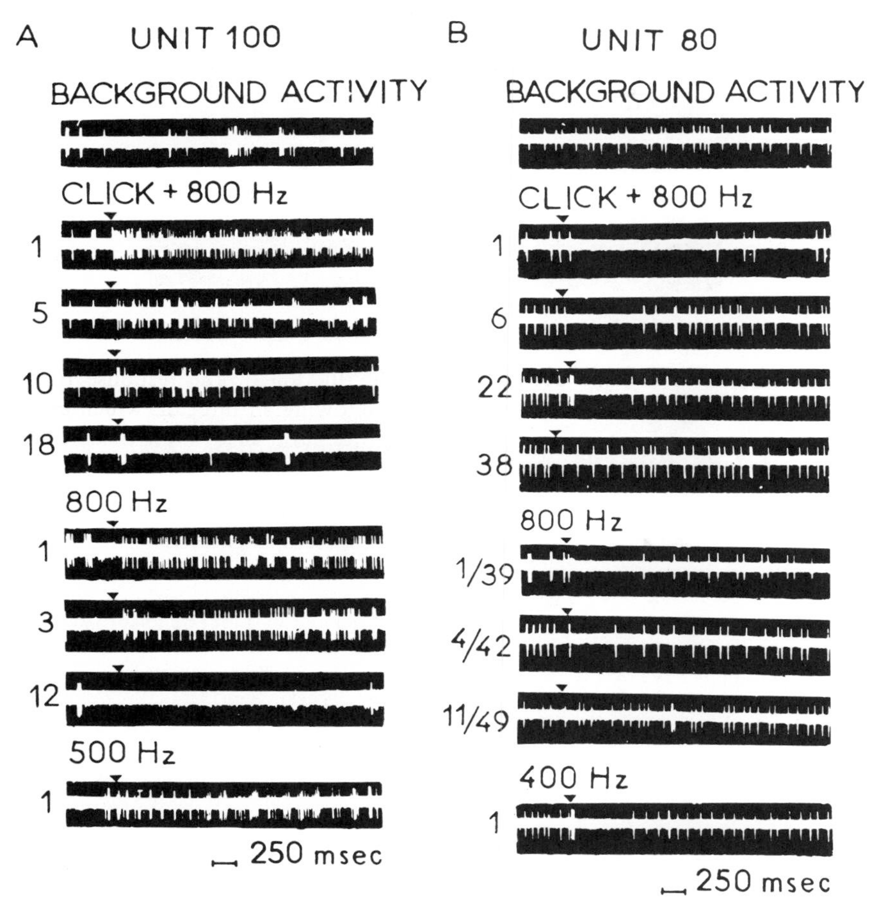

FIG. 10 Recovery of the response of A and I neurons as a result of subtraction of an element from a sound complex, the response to which has been extinguished (A and B). The recovery of the extinguished reaction after the change of the tone is shown below.

background change gradually loses its influence and becomes integrated as a new constant component of the environment. A return to the initial level in this case (discontinuance of the background stimulus which evoked the change) leads to a new readjustment ("subtraction from the complex") and to a new readaptation to the phasic signals acting against the modified background. Thus, the onset and end of a long-lasting background signal may themselves bring about a mobilization of the response to the novelty, and evoke a recovery of the responses to extinguished phasic signals. At the neuronal level these two effects may manifest themselves separately. The disinhibiting action of the background can be observed also when the background stimulus itself does not evoke any overt response of the cell. This can be seen in Fig. 11 for cell No. 48, which shows an

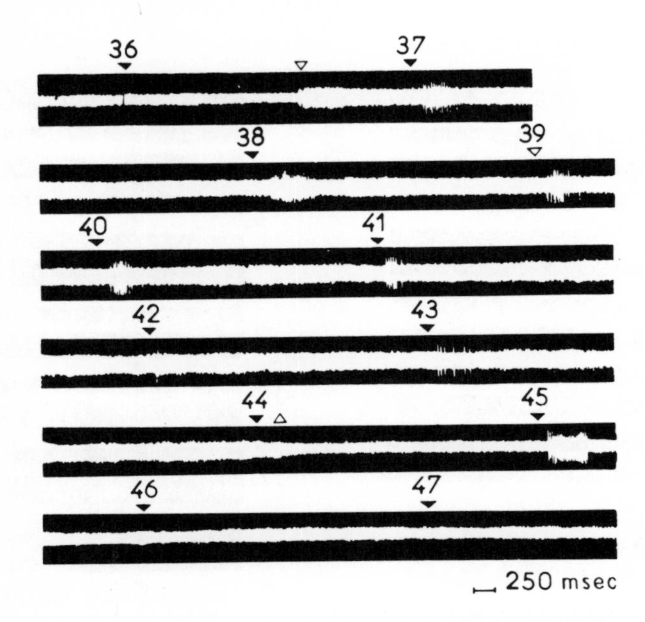

FIG. 11 The disinhibiting effect of the beginning and end of a 500-Hz long-lasting tone on the response of cell 48 to a flash of light. The flash is marked by solid arrows with the presentation number above, the tone by a noise on the line of the record, its beginning and end by down and upward open arrows respectively. Details in the text.

atypical phasic response to a light flash (and does not respond to a sound stimulus).* After complete extinction of the response to a flash of 5th intensity (in conventional units) the application of a long-lasting tone of 400 Hz, 80 dB does not itself evoke a response of the cell, but leads to a recovery of the response to a flash presented against its background (Fig. 11). After 8 applications, the flash ceases to evoke a response during the lasting action of the tone. When, after a full disappearance of the response to the flash, the tone is switched off, a new recovery of the response to the flash occurs. According to the general principle, the superimposition of the factor of physical intensity on the factor of novelty makes the action of the termination of the background signal a relatively less effective disinhibiting agent than at its beginning. This is likewise true of the cases when the long signal itself evokes a response. If the action of the signal is sufficiently long lasting, and the tonic A or I reaction declines against its background, removal of the tonic stimulus always results in a repeated recovery of the neuronal response (which is, however, less intensive than at the onset of

*Only three cells, lacking any spontaneous activity and having a definite phasic and modality specific response, were detected in the hippocampus. However, these cells may be related to all other hippocampal neurons because of the strongly pronounced dynamic changes of their responses.

the signal). This type of response must not be confused with responses of the "on-off" type manifested by the specific sensory neuronal systems. This effect itself becomes eliminated when the background signal is repeated, and the response to the "switched off" stimulus habituates first. The character of responses of this type is clearly seen in graphs which are based on an automatic calculation of the number of spikes for long-time intervals (with the aid of a "Volna" automatic pulse counter). These display a picture of tonic, modally nonspecific responses, which are characteristic of hippocampal neurons, and which consist of an increase or decrease of background activity during the action of sensory stimuli (Fig. 12). It can be seen that if the signal is long enough, the recovery of the level of activity begins during its action. The termination of the stimulus leads in these cases to the second change of the activity in the same direction. The disinhibiting influence of the background stimulation shows that, for the system of the hippocampal neurons, it is unessential whether the phasic signal changes relative to a constant sensory background, or whether the latter changes relative to a more or less standard phasic signal. This phenomenon may be regarded as a peculiar integration into a single complex of environmental (background) and phasic (foreground) factors.

It is easy to see that all the above-described ways of recovering extinguished responses of hippocampal neurons are none other than the well-known methods applied for evoking an orienting reaction (see Sokolov, 1959, 1964). The

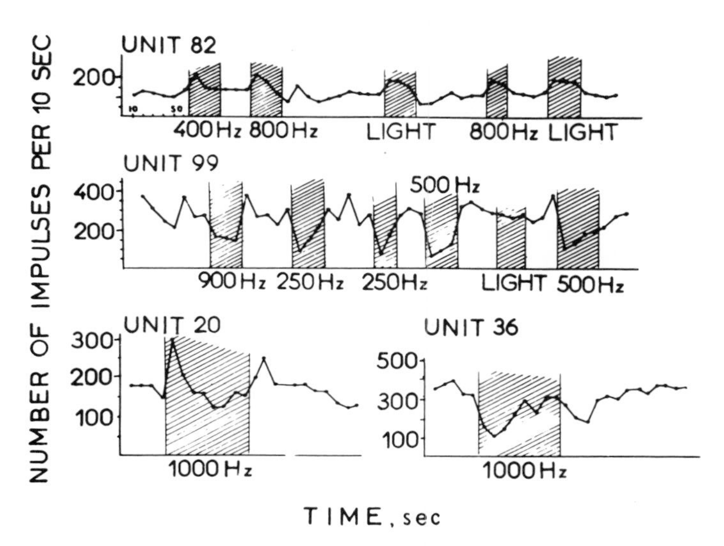

FIG. 12 Curves showing the responses of A and I neurons to tonic stimuli. The curves were obtained with the aid of a "Volna" automatic pulse counter: ordinate, total number of spikes per 10 sec; abscissa, time (the interval marked on the upper curve equals 10 sec). The A and I effect of the beginning and end of long-lasting signals (80 sec) is seen on the lower curve.

efficiency of these methods may be demonstrated on one and the same neuron (this has been partially done in the given section).

An analysis showed that the responses of hippocampal neurons are evoked not by a definite sensory modality of the signal or by any of its quantitative or qualitative characteristics, but by the special property of *novelty* of the stimulus. The hippocampal cells, which are markedly nonspecific with respect to the physical characteristics of the signal, manifest a high level of selectivity, detecting fine and complex changes in the applied signal or in the background against which the signal is presented. Thus, the responses of hippocampal pyramids arise as a result of any *noncoincidence* between the established trace system and the present stimulus or situation ("disconcordance signal").

C. Evaluation of Temporal Parameters by Hippocampal Neurons

A number of facts indicate that the hippocampus plays an important role in the evaluation of temporal parameters of signals. The particular structure of the hippocampus and its ability to exhibit long-lasting tonic responses testify to the possibility of a successive scanning of the signal in time. On the basis of an

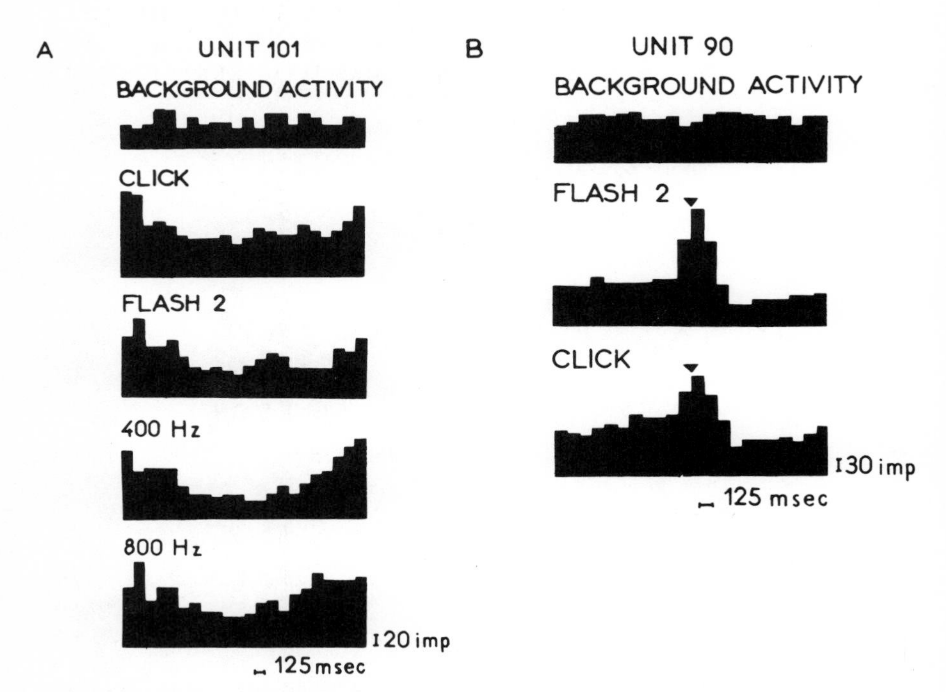

FIG. 13 Averaged histograms of the activity of neurons of the E type: (A) in the case of a usual structure of the PST histogram; (B) in the case of a reconstruction of the histogram in a such a way that the moment of action of the signal is in the center of the period of counting.

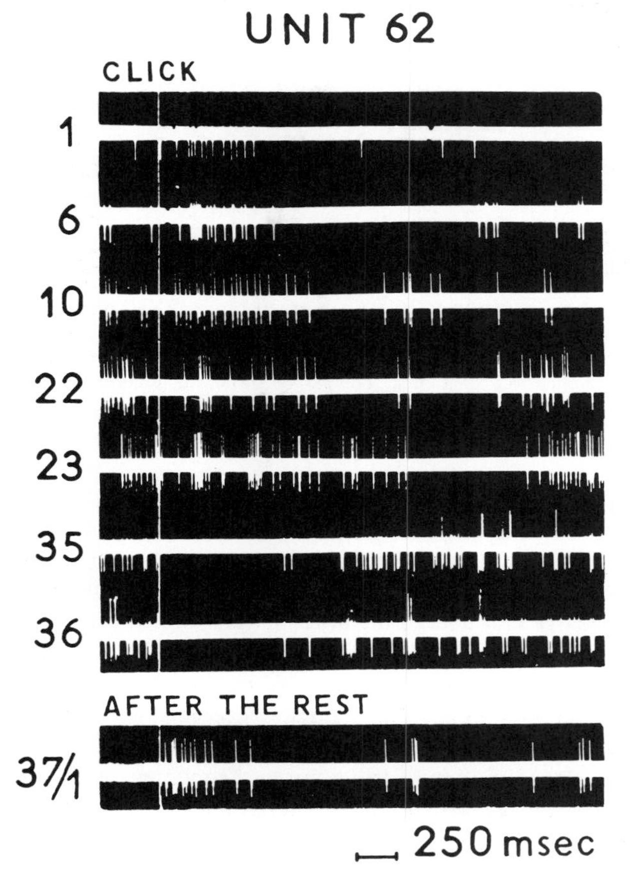

FIG. 14 Consecutive responses of a cell of the E type during rhythmic application of a click. Explanations given in the text.

analysis of the morphology of the hippocampus, McLardy (1959) advanced a theory according to which the hippocampus is regarded as a detector-coder for the temporal parameters of information.

The ability oı the hippocampal pyramids to evaluate macrointervals of time manifests itself in different ways. It is most pronounced in the group of "extrapolatory" neurons (E neurons). The individual properties of these neurons were variable, but their histograms and dynamics were characterized by some common features of essential significance. Their averaged PST-histograms are of a "saucerlike" (concave) form, which creates the impression that the cells have a double activation response separated by a period of a relatively low frequency of discharges (at background level), or even of some decline in activity (below the background) (Fig. 13A). The relative intensity and duration of the initial and

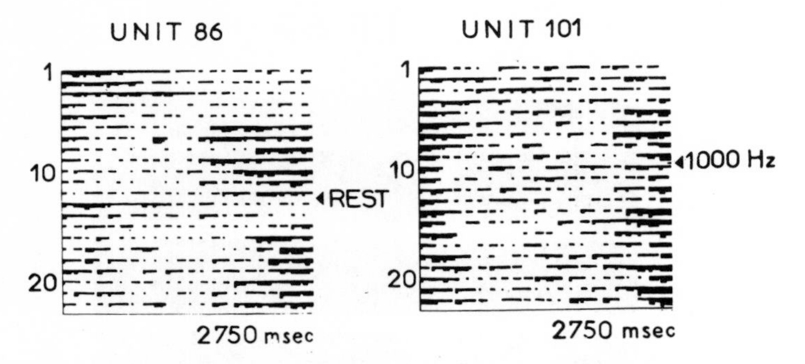

FIG. 15 Graphs of the distribution of the density of spikes for 2 neurons of the E type during successive applications of a click. The graphs show a gradual change in the response pattern (formation of the final discharge) and a change of this process caused by an interval in the series of stimuli (unit 86) and to a lesser degree by the action of an extrastimulus (unit 101).

final discharges of the histograms of these cells are different; in some cases there is a short, almost phasic acceleration (100–200 msec), in others, both discharges may possess tonic characteristics, being of a relatively equal duration, however. Finally, one discharge may exceed the other in duration. Of essential significance is the fact that analysis of the individual responses of the cells often does not reveal the form of responses which is presented by the averaged histograms–they result from complex modifications taking place in the course of repeated stimulus presentations. In a typical case, the cell initially responds with a diffuse discharge of a limited duration. Further, up to the moment of the next stimulus presentation, unorganized activity establishes itself at the prestimulus level or somewhat below it (up to complete inhibition). During rhythmic repetition of the stimulus, the response pattern begins to change at the end of the interstimulus interval and before presentation of the next signal there appears a condensation of the spikes. The initial discharge gradually weakens and shortens; simultaneously, the final discharge begins to grow and becomes longer. Owing to this, the period of reduced activity by which they are separated steadily shifts forward in time to the moment of a stimulus application. The response as a whole becomes shifted beyond the moment of stimulation (Fig. 14). The graphs showing the total numbers of impulses for the periods between stimuli do not reflect the major tendency of the dynamics, since often the same total numbers may be obtained in cases of an absolutely different pattern of responses (intensive initial and weak final discharge, and vice versa). Differential graphs for the first and second parts of the interstimulus period show the existence of reciprocal relationships between the final and initial discharges in most E neurons. An interval in the rhythmical applications of stimuli leads to a double effect: the initial discharge increases, while the final one becomes suppressed. Thus, on the whole, the response returns to its initial form.

An example of such dynamics is presented in a graph showing the density of spikes in cell 101 (Fig. 15A). In other, rarer cases, the formation of the final discharge is not accompanied by a disappearance of the initial response which preserves a relatively constant form (Fig. 15B).

The basic type of dynamics manifested by responses of E neurons might be interpreted as result of integration of the A and I responses which are typical for the hippocampus (a habituating A response + increase in number of discharges starting from the end of the interstimulus interval and peculiar to the extinction of the responses of I neurons). But apparently such an explanation is wrong. The time of emergence of the final discharge is determined by the duration of the interval between the stimuli—it arises as an anticipatory (extrapolatory) discharge before the moment of onset of the rhythmic stimulus. If we reconstruct the PST histogram in such a way that the moment of stimulus onset is in its middle (the final half of the period before the signal and the initial half after it), it will become quite clear that the signal is, so to speak, "embraced" on two sides (directly before the onset of the signal and immediately after it) by a zone of heightened activity (Fig. 13B). Individual histograms reconstructed in a similar way reveal a gradual shift of the activity to the zone before stimulus presentation. These findings support the conclusion that here we have an elaboration of an anticipatory response to the onset of stimulus presentation.

The ability to evaluate time parameters, which is highly characteristic of E-neurons, is also found in other cells of the hippocampus. A thorough analysis of the histograms of A neurons reveals a slight regular increase of the level of discharges before the end of the interval between stimuli. In some cases there was an increase in the frequency of discharges in the place of the absent stimulus, after a long series of signals presented with a constant interval. However, the high and irregular spontaneous activity of the hippocampus masks these effects to a considerable degree.

From the data described above, it is clear that in all the cases mentioned we deal with activatory extrapolatory responses. We failed to observe any shift of the onset of the inhibitory period in the I neurons forward, to the moment which precedes signal presentation. But the following fact arouses a certain suspicion. Although during the repetition of the signal the maximum level of the increasing activity usually falls during a period of 300 to 400 msec before presentation of the stimulus, the inhibitory reaction to this stimulus as a rule, begins at once, without any latency, as if "cutting off" the preceding activity. Such a sudden mobilization of inhibition can hardly be conceived if it is not prepared during a certain period before the onset of the stimulus.

Thus, hippocampal neurons are able to estimate macrointervals of time and to elaborate conditioned reactions to time (extrapolation).

DISCUSSION

A. Comparison of the Responses of Hippocampal Neurons with Those of Neurons of Other Brain Systems

The hippocampal neurons considered above form a distinct functional group. The following question arises: Are the aforesaid peculiarities inherent only in hippocampal neurons, or are they observed in other brain structures as well, and, consequently, is their significance more general? We shall briefly consider here only our own data relating to this question, since in this case all the investigations were carried out under the same conditions and with one set of stimuli, so that the differences observed cannot be explained on the basis of methodological variations, all the data being directly comparable.

The caudate nucleus, whose physiological characteristics are in some respects similar to those of the hippocampus, was the main object used for the purpose of comparison. This nucleus is a polysensory structure with pronounced inhibitory functions (see Vinogradova, Sokolov, p. 155 in this volume). Some limited material for comparison was obtained also from neurons of the motor cortex, since it was supposed that this polysensory system, serving as an output of excitation for all sensory areas of the cortex, may participate in the organization of reactions of the orienting reflex type. As an opposite example of a pronounced specific structure, we used some material obtained previously from investigations of the rabbit's visual cortex, in the course of which similar systems of stimuli and conditions of stimulation had been used (Vinogradova & Lindsley, 1963).

The comparison was done according to three basic criteria, which, in our opinion, made it possible to differentiate the systems responsible for specific transmission and processing of information from systems which are performing selective reception of the "novelty" of information. The comparative data are presented in Fig. 16. The comparison revealed the following:

1. Criterion for multimodality of the input. The system of the orienting reflex on a macrolevel comes into operation irrespective of the sensory modality of the signal. On the neuronal level, it is neurons with a multimodal input which may be regarded as a corresponding analogue. From Fig. 16A, it can be seen that the visual cortex possesses a high level of modal specificity (which is not surprising), and that almost the same level is characteristic of neurons of the caudate nucleus (which is somewhat unexpected, taking into account the high level of convergence in the structure as a whole). The level of multimodal convergence is high in the motor cortex, but proves to be the highest in the hippocampus (70%).

2. Criterion for nonspecificity of the output. The manifestations of the components of the orienting response are uniform under the action of different stimuli. The absence of specific coding of the incoming signal in the temporal distribution of spikes (pattern) may serve as an analogue of this phenomenon on the neuronal level. The majority of the visual cortex neurons manifest responses of a specific type; the patterns of these neurons depend on the physical characteristics of the stimulus. This is characteristic also of many neurons of the caudate nucleus, where even neurons with multimodal convergence may respond specifically to each of the stimulated modalities, preserving the coding of incoming information in the neuronal output. In the motor area diffuse reactions of a pronounced activating character predominate, but along with them

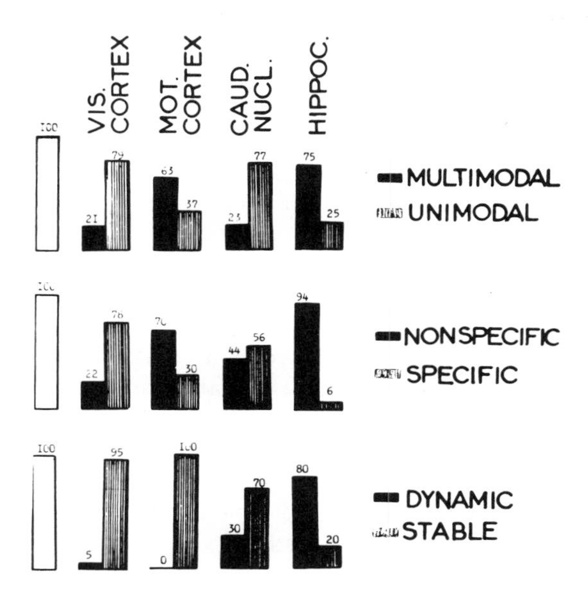

FIG. 16 Diagram showing the distribution of neurons from four brain structures in accordance with three criteria of the orienting reaction. Explanations are given in the text.

there is a large group of neurons with responses of a specific type. In the hippocampus such neurons comprise only 6% of the total number; all other responses consist of uniform diffuse tonic A and I effects.

3. Criterion for dynamic changes. The most prominent feature of the orienting reflex is its ability to habituate as a result of repeated presentations of a non-reinforced signal, and at the same time to preserve the possibility of an instantaneous recovery of the reactivity of the system in the case of a change in the signal. This process must undoubtedly have its correlate on the neuronal level; it cannot be effected by the neuronal structures which preserve response constancy during numerous applications of the stimulus. Of great importance in the given case is the coincidence between the temporal parameters of the dynamic processes at the micro- and macrolevel. Figure 16B shows that the number of neurons which display dynamic changes on the neocortical level is very low. According to our data, in the visual cortex even a group of multimodal neurons with a nonspecific type of response manifests a stable reaction during a repeated presentation of the signal. Elements of dynamic characteristics were found in a considerable group of neurons of the caudate nucleus (30%). But the modification of the responses in this structure displays some peculiarities which show that, in this case a secondary modulation of a stable specific response actually takes place as a result of diffuse activatory influences (see Vinogradova & Sokolov, p. 155 in this volume).

These features do not permit identification of the dynamic adaptation of the elements of the caudate nucleus with the extinction of the response to the novelty. Real extinction of this kind, which correlates the time of development with the extinction of the orienting reflex, is observed only in the neurons of the hippocampus (among all the investigated structures). Dynamic changes manifest themselves in the overwhelming majority of reactive hippocampal neurons (80%).

Thus, all the three basic criteria, which were proposed by us on the basis of analysis of the macrostructure of the orienting response, prove to coincide only in the hippocampus. An analyses of the recent data published in the literature concerning the existence of dynamic changes in neuronal responses and reactions to the "novelty" of a signal in different structures of the brain shows that in spite of some exceptions (Horn & Hill, 1966; Skrebitsky & Gapich, 1963), these phenomena are rare and atypical for the specific sensory relays and for the primary cortical areas. In these structures an "obligatory transmission of information" takes place, which is to a considerable degree independent of the state of the organism (sleep, wakefulness, emotions) as well as of its past experience (influence of preceding signals), and which allows the organism to obtain objective data concerning the external environment. Other structures effect a "dynamic transmission of information," where factors such as the initial state of the organism, the influence of past experience, the emotional equality of the stimulus, its significance, etc. are of prime importance. Available data indicate that such neuronal structures belong mainly to two large interconnected systems of the brain: (1) the nonspecific structures of the brain stem and thalamus (Bell et al., 1963, 1964; Horn & Hill, 1966; Jasper, 1964) and (2) the limbic system. According to our data, these phenomena are observed in the hippocampus (dorsal, Vinogradova, 1966; ventral, Semyonova & Vinogradova, unpublished data) as well as in the mammilary bodies (Konovalov & Vinogradova, unpublished data). Judging by published data, maximal similarity to the responses of the hippocampal neurons is shown by the cells of the amygdaloid nuclear complex (Creutzfeldt *et al.*, 1963; Machne & Segundo, 1956; Sawa & Delgado, 1963; Gloor, 1961). This similarity is by no means accidental, since these two structures enter a single functional complex, where the specific function of the amygdaloid nucleus is determined by its participation in organization of emotions, which is absent in the hyppocampus. It is assumed that in this complex the amygdaloid nucleus is concerned with the function of information processing depending on the presence and quality of reinforcement and on the significance of the signal for the organism, while the hippocampus is concerned with the function of processing and registration of information depending on its novelty (Douglas & Pribram, 1966). It serves as a filter which lets the new signals through and filters those elements of information which are already stored in memory. Consequently, these two structures together provide a heightened action of the signal if it possesses one of two properties: novelty and/or significance.

B. The Principle of Action of the Hippocampus as a Comparator

Various facts indicate that the hippocampus obtains information from other brain structures after its complex processing and integration. The hippocampus has no direct connections with the specific sensory systems. The signals reach the hippocampus through the septohippocampal pathways after processing in the nonspecific structures of the midbrain and diencephalon, and through the temporo-ammonic pathways from the enthorhinal cortex, to which information from the non-primary areas of the neocortex may converge. On the way to the hippocampal pyramids, the signals become additionally relayed to (and, consequently, also processed in) the septal nuclei and in the granular cells of the dentate fascia.

It may be assumed that the specific quality of sensory information in this area is almost, if not fully, lost (Gloor, 1961). Nevertheless, stimuli evoke long-lasting tonic responses in the hippocampus, which are apparently of great significance for the final action of the signal on the organism. In the first study of hippocampal neurons (Renshaw, Forbes, & Morison, 1940) it was stated that the sustained effects in the hippocampal pyramids are probably due to the reverberation of excitation. The hippocampus has a rich structural basis for maintaining long-lasting processes started by a short stimulus, which may be explained by the circulation of the excitatory process along closed neuronal circuits. A powerful system of recurrent collaterals of the pyramidal axons—both within a single field and between different fields, and connections between the hippocampi and the entorhinal areas of both sides, the classical "Papez Circuit" which passes through the non-specific structures of the diencephalon and midbrain, as well as an inversely directed circular connection—all present multiple chains that are completed through the hippocampus. These lengthy and massive processes in the hippocampus may be of double significance: (1) as a repeated passage of the same sensory information for additional processing and recording; (2) as formation of a powerful and sustained controlling signal which is directed to the efferent mechanisms, with respect to the given structure (reticular formation and cortex).

The hippocampal pyramids possess oriented basal and apical dendrites on which synapses from different afferent pathways terminate in a strict order. As a result, these afferent inputs have different weights, depending on the proximity of the terminal synapses to the soma of the pyramidal cell. Thus, the temporo-ammonic pathways terminate on more distal parts of the apical dendrites than the septohippocampal pathway; this may necessitate a longer summation of the excitation of dendrites for generating a propagating action potential. It is true that some authors question the assumption that afferents from both main inputs of the hippocampus may terminate on the same elements. Their doubts are based on histological data (predominant termination of the cortical afferents in field CA_1 and of the septal afferents in field CA_{3-4}, as well as on electrophysiological data, von Euler, 1960). Nevertheless, there exist some opposite data as well (Spencer & Kandel, 1965; Andersen *et al.*, 1964a); in addition the interaction of both inputs may be achieved through the internal associative systems of the hippocampus.

The organization of the neuronal structure of the hippocampus and the pattern of its responses to sensory stimuli allow us to regard the hippocampal pyramids as a system of comparators of signals which come from two sources. It may be assumed that the hippocampus receives integrative information concerning a given stimulus from sensory centers of a lower level through the ascending septofimbrial pathway. The hippocampus also receives information through the temporoammonic system, which is formed in areas of the neocortex (depending on the presence or absence of a corresponding trace). Different forms of interaction resulting from these two sources of information may create different states in neurons of the hippocampus. Under the action of a new signal (i.e., if the signal coming from the lower sensory centers does not find any equivalent in the cortical signals), the A neurons respond with a powerful tonic increase of their discharges, thus marking the entry of the cortical signals. Due to the reactions of the I elements, during the first applications of the stimuli; the level of total output of the hippocampal cells markedly declines. This must lead to the elimination of the tonic influence of the hippocampus on the reticular formation. Thus, favorable conditions are created for the evocation of an orienting reflex, as well as for the passage of new signals for registration in memory. At this time a primary analysis of the stimulus takes place, which is intensified by ascending influences of the non-specific system and the efferent realization of an orienting reaction which, in its final form, is apparently integrated in the reticular formation of the brain stem.

It is known, that the theta rhythm predominates in the electrical activity of the hippocampus at this time; in accordance with the initial concepts of its functional significance (Grastyan, Lissak, Madarasz, & Donhoffer, 1959; Grastyan, 1959) it may be regarded as an expression of the general inhibitory state of the hippocampus. These concepts are confirmed by data obtained as a result of intracellular recording from the pyramidal neurons (von Euler & Green, 1960; Spencer & Kandel, 1965; Andersen *et al.*, 1964a, b). The authors mentioned before emphasized the dominance of the inactivation process and of the IPSP during orthodromic stimulation of the pyramids; these authors regarded theta-activity in the hippocampus as a result of summary IPSP during recurrent inhibition of the pyramids by the system of basket cells.

It is possible that long-lasting excitation, which is maintained in the hippocampus due to cyclic processes (in the A neurons), contributes to the initial retention of the trace and forms a basis for the gradual and parallel development of a stable fixation of the trace ("neuronal model of the stimulus," Sokolov, 1960) in some other structures (presumably, in the nonprimary areas of the neocortex).

The formation of such a model leads to the emergence of a "coincidence" of signals coming from two different sources. At this period, first a limitation of the inhibitory responses of the I pyramids develops, and then a generation of high-frequency discharges by them. This process may be regarded as one of the probable mechanisms which determines the gradually developing active blocking of the passage of sensory signals. A high-frequency "noise" generated by the I-neurons blocks the executive mechanisms of the reticular formation, which discontinues the realization of the orienting reaction and stops the further passage of the signal for analysis and recording. A change in the signal, as a result of which no corresponding trace is detected, leads again to a noncoincidence of the incoming excitations, at the same time the activity of the I neurons and the tonic suppression by the hippocampus of the orienting reflex manifest a decline.

Thus, the real cause of excitation of the hippocampal neuronal reactions may be regarded as the divergence between the influence of the actual afferent signal and its trace fixed in the system of memory ("disconcordance signal"). This system possesses high selectivity with respect to the noncoincidence of the signals, and is fully nonspecific as regards any of their other characteristics. The suppression of the activity of the I neurons during the action of the disconcordance signal ("novelty") is then a necessary condition for the admission of the signal to the activating structures and for the realization of the orienting reflex. The maximum increase of the activity of these elements during the formation of the trace, which coincides with the discontinuance of the responses of the A neurons, operates as a mechanism which actively blocks the passage of the repeating signal to the executive mechanisms.

C. The Orienting Reflex and Memory

As already pointed out in the introduction, recent reviews of the functions of the hippocampus (Vigoroux, 1959; Douglas, 1967; Meissner, 1966) state, on the basis of the complex of available data, that two functions are better proved than the others: memory and inhibition of the orienting reflex. It is commonly believed that these two functions are mutually exclusive. This paradox is eliminated to a considerable degree when we regard the extinction of the orienting reflex as "negative learning" (Sokolov, 1965), i.e., actually as a manifestation, in a negative way, of the formation of a memory trace. Essentially, the orienting reflex is a reaction which is determined by the absence of an acting signal in the stored experience. The presence of the "new" is impossible without the existence of the "old." The extinction of the orienting reflex is a process of transformation of the new into the old, i.e., of some trace formation. The persistence of an unextinguishable orienting reflex in pathological conditions is the result of disappearance of the stored traces, which are necessary for a comparison with the new signal, or of a derangement of the process of registration of new traces. The formation of a "model of the stimulus" during the extinction of the orienting reflex is regarded by Sokolov (1960) as a process of formation of a parametric trace in the nervous system.

The hippocampus inhibits the manifestations of the orienting reflex in a gradual way, as the repeating signal becomes fixed in the nervous system. In a normally functioning brain these processes present are unified and are controlled by a common regulating mechanism. Owing to this, the hippocampus performs a single function–it allows the passage of information for registration, so long as it is new (memory), and blocks it when it loses its further informative property (extinction of the orienting reflex).

The following should, however, be added. As far back as 1893, Ramon-y-Cajal stated that the cellular elements of the hippocampus are of an obviously motor type. He compared the structure of the hippocampus, which was divided by him into regio superior (CA_{1-2}) and regio inferior (CA_{3-4}), with the 5th and 6th layers of the pyramids of the motor cortex extended as one continuous stratum. Brodal (1947) emphasized the purely efferent type of the hippocampal structure. The analogy with the motor cortex should not, perhaps, be exaggerated, but this type of big pyramidal cell with its long axons indicates that the vigorous diffuse reactions generated by them, in all probability, bear the character of an intensive controlling signal which switches on and off the structures as a whole, modulating their activity on the basis of changes in their own background level. This system proves to be well adjusted to regulating the orienting reflex. At the same time the participation of the hippocampus in the system of memory, apparently plays an essential and yet auxiliary role with respect to some other mechanisms. An analysis of the activity of the hippocampal cells shows that the preservation of any specificity of information in them is really impossible. Diffuse and generalized shifts of activity can hardly serve as a basis for recording discrete changes. Data available in the literature show that traces of long-term memory presumably are not stored in the hippocampus (Penfield & Milner, 1958, and many others) and that, perhaps, it is not involved in so-called "short-term memory" (Drachman & Arbit, 1966, and others). However, when it is damaged, the following occurs: (1) the transition of the traces from short-term memory to long-term memory becomes deranged, and (2) the negative influence of interference at the stage, which precedes the consolidation of the trace, increases. It is possible that the hippocampus, while prolonging the action of the given signal due to the formation of tonic neuronal activity operates as a mechanism intensifying the dynamic processes which precede registration and ensure the elimination of interfering signals by means of the previously described mechanism of blocking the orienting reflex.

12

ACTIVATION AND HABITUATION IN NEURONS OF THE CAUDATE NUCLEUS

O. S. Vinogradova
E. N. Sokolov

Moscow State University
and
USSR Academy of Sciences

Investigations of hippocampal neurons have revealed characteristic features of their response dynamics which may be regarded as correlates of the process of extinction of the orienting reflex at a macrolevel (see Vinogradova, p. 128 of this volume). This structure was chosen for research purposes on the basis of existing concepts of its inhibitory influence on the activating reticular formation.

However, neurophysiological data available in the literature show that structures other than the hippocampus (the neocortex, the nonspecific thalamus, and the synchronizing mechanisms of the lower brain stem) are characterized by an antagonistic interaction with the activating system. Theoretically, each of these structures may contribute to the suppression of the arousal reaction (see Vinogradova, 1961).

According to some authors, the function of inhibiting the activation effects, both in the behavior and in the EEG, is to a considerable degree inherent in the caudate nucleus. This was comprehensively demonstrated in a series of investigations carried out by Buchwald *et al.* (1961a, b, c), Heuser *et al.* (1961), Horvarth *et al.* (1964), and other researchers. Stimulation of the caudate nucleus leads to the development of sleep (Spiegel & Szekely, 1961; Heath & Hodes, 1952), as well as to suppression of ongoing activity and conditioned responses (Buchwald *et al.*, 1961b; Buser, Rougeul, & Perret, 1964). Low-frequency stimulation of the caudate nucleus evokes the emergence in the EEG of "caudate spindles," inhibition of secondary evoked potentials in response to sensory stimulations of different modalities (Krauthamer & Albe-Fessard, 1965; Demetrescu & Demetrescu, 1962; Butkhusi, 1965), and suppression of spontaneous neuronal activity in different brain structures (in the thalamus, Horvarth & Buser, 1965; Krauthamer & Feltz, 1965; in the cortex, Spehlman, Creutzfeldt, & Jung, 1960; Lehman, Koukkou, & Spehlman, 1962).

In the caudate nucleus itself, an extensive multisensory convergence was revealed as a result of investigations of evoked potentials (Krauthamer & Albe-

Fessard, 1964), as well as the responses of single neurons (Albe-Fessard *et al.*, 1960; Sedgwick & Williams, 1967). So far, the functions of this polysensory structure remain uncertain (see the reviews of Jung & Hassler, 1960; Laursen, 1963; Cherkes, 1966). Yet, the inhibitory properties of the caudate nucleus provided the basis for some authors regarding it as an antagonist of the ascending reticular activating formation, and even suggesting its connection with the selective regulation of attention. In a chapter devoted to the extrapyramidal system published in the *"Handbook of Neurophysiology,"* Jung and Hassler (1960) stated that the caudate nucleus "may play an important role in the selective inhibition of impulses from other sensory systems or areas of activity which do not belong to the most significant (at the given moment) system of excitation [p. 867]." This statement directly implies the participation of the caudate nucleus in the selective processes of attention.

The above-mentioned facts served as a basis for a series of experiments aimed at disclosing the properties and dynamic characteristics of the neurons of the caudate nucleus. We considered such an investigation of particular importance, since, the data obtained from it could permit us to judge with greater confidence the functions performed by the hippocampus in the system which controls the orienting reflex. The specific role of the hippocampus would be evident if caudate neurons behaved differently; if the data proved to be similar, it would be classed with other antagonists of the reticular formation.

Proceeding from this, we investigated neurons of the caudate nucleus as one of the structures which participate in the suppression of responses when the stimulus loses its novelty.

METHODS

The experiments were carried out on unanaesthetized rabbits. Recording was done from the head of the caudate nucleus. The methods of recording and stimulation are described in detail in our article on the hippocampus (see p. 129 in this present volume). Neuronal responses were analyzed by averaged post stimuli histograms based on repeated stimulus presentations. In order to evaluate dynamic changes, we computed serial histograms for blocks of five stimulus presentations. This method, which had not been applied to responses of hippocampal neurons, was dictated by the peculiarities of the caudate nucleus neurons. We shall substantiate the use of serial histograms later on, when describing the experimental results.

RESULTS

The Specific Character of the Responses

The overwhelming majority of the 72 investigated neurons of the caudate nucleus possessed a very low level of spontaneous activity (3–4 impulses per sec

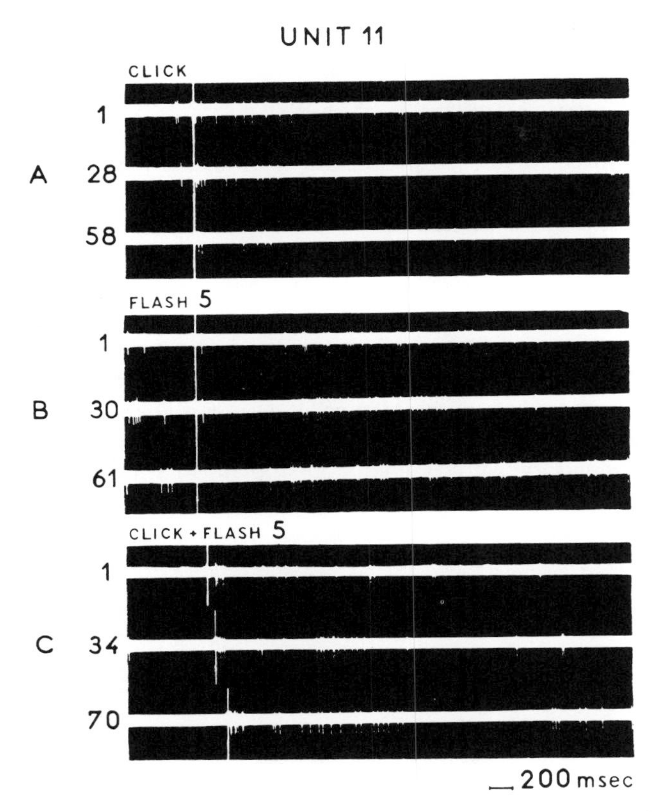

FIG. 1 Different responses of cell 11 from the head of the caudate nucleus to: (A) a click; (B) a flash (intensity–5 conversion units); (C) simultaneous presentation of flash 5 and click. Stimulus onset at vertical bars. Serial number of stimulus presentation is given on the left side of the records.

on the average), especially when the animals were in a state of rest. In neurons responsive to stimulation (43%), spontaneous activity was more often represented by rare single spikes. Spontaneous activity was more commonly in the U form of compact spike bursts (20–40 spikes in each) following one another at considerable intervals (often lasting 30–40 sec) during which the spikes were absent. Neurons with this type of spontaneous activity constituted a large part of the group of cells which did not respond to sensory stimulation.

In the group of responsive neurons, responses to sensory stimulation stood out in bold relief against the background described above. During the presentation of brief stimuli (flash, click) responses were of a pronounced phasic character. Their latencies, for most of the reactive cells, equaled 16–25 msec during the first presentations of the stimuli. The maximal latencies observed reached 170–220 msec. A considerable number of the investigated elements responded only to signals of one of the sensory modalities applied (23% of reactive neurons only to light stimuli and 37% only to acoustic stimuli). The remaining 40% of the reactive neurons responded to both light and acoustic signals.

However, the most essential feature of the responses, no matter whether the latency was short or long, or whether the input of the cell was unimodal or multimodal, proved to be a specific pattern of the spikes which depended on the quality of the sensory signal.

This can be illustrated in the example of cell 11 (Fig. 1). During the presentation of a click (Fig. 1A), the cell responds with a short primary discharge which passes into a diffuse secondary discharge almost without any pause. During flash stimulation (Fig. 1B), the primary discharge is less pronounced and in a number of cases is even fully absent; however, an inhibitory pause manifests itself, being followed by a vigorous secondary discharge. When the complex click plus flash (Fig. 1C) is presented, a different type of response arises. A strong primary discharge (stronger than during the action of a click) is accompanied by a short, but distinct inhibitory pause (about 300–400 msec) which is followed by a secondary discharge (more compact than during the action of a flash). These structural peculiarities of the responses are clearly reflected in the averaged histograms (Fig. 2).

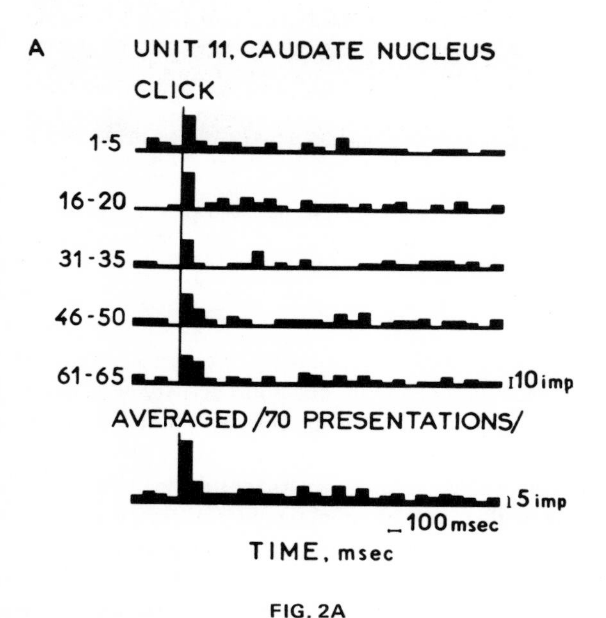

FIG. 2A

FIG. 2 Serial histograms for responses of cell 11 during the presentation of: (A) a click; (B) a flash (4 different intensities were applied—the histograms are in the serial order of presentation, not according to increasing flash intensity; (C) 'flash 5 plus click.' Averaged histograms for 70 applications of the signal are presented below. The counting period (bin width) was 100 msec. Stimulus onset was at the beginning of the histogram.

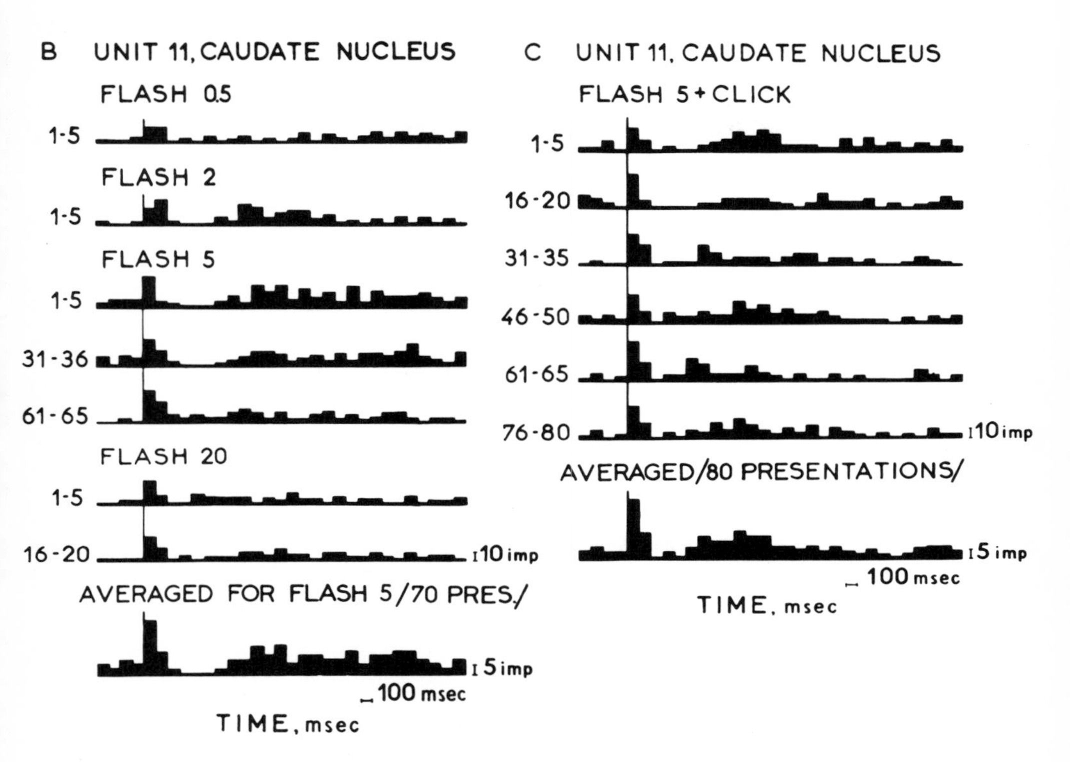

FIG. 2B and C

In other cases, the responses to light and acoustic stimuli bear an obviously antagonistic character. This is shown in Fig. 3 for unit 18: the flash evokes a short activation response in the shape of a primary discharge with a duration of 80 msec, while the click evokes a suppression of the spontaneous activity during an approximately similar period of time.

The specificity of the observed responses and their dependence on the quality of the stimulus are seen also within one sensory modality. On the histograms mentioned above of the responses of cell 11 during the action of a flash (Fig. 2B), it can be seen that the pattern of the secondary discharge differs during the presentation of signals of different intensities. It is fully absent during the action of stimuli of low intensities (flash 0.5 in conventional units), as well as of strong stimuli (flash 20). The secondary discharge arises in the shape of a distinct spike burst only during the presentation of flash 2 and reaches the maximum degree of expressiveness during the presentation of a stronger flash 5, which proves to be the optimal stimulus for this cell.

The selectivity and peculiar character of the responses are particularly apparent during the presentation of pure tones. As distinct from hippocampal

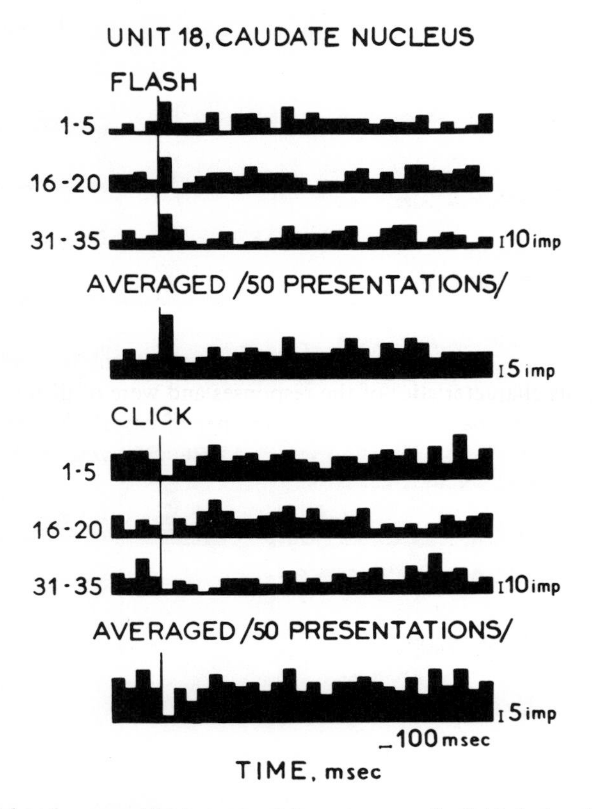

FIG. 3 Serial and averaged histograms of the responses of cell 18 during the presentation of a flash and a click. The opposite character of the responses is clearly seen.

neurons, the cells of the caudate nucleus are by no means indifferent to the frequency characteristics of the signals applied. Many of them detect the "optimal" frequency band. The tuning of these cells to frequency is not as narrow and precise as in the case of neurons of the specific auditory system; however, they respond to a restricted band of frequencies exhibiting a diminished response to both higher and lower frequencies. Different cells possess optima in various parts of the sound-frequency scale (4–8 kHz, 1–2 kHz, 0.5–0.8 kHz). Cells with an optimum in the region of high frequencies predominate, which correlates with the peculiarities of the rabbit's auditory function.

Responses to tones of different frequencies are distinguished not only by the degree of increase or decrease of discharges, but also by the discharge pattern. This is shown in Fig. 4 where cell 56 (which had a high spontaneous activity, seldom encountered in the caudate) responds to the first application of a 1000-Hz tone with very weak diffuse inhibition and to the first application of a 3000-Hz tone with a pronounced protracted suppression of activity; but to a tone of 5000-Hz it reacts with a distinct specific response–inhibition with a latency of about 80 msec during the presentation of the tone, a well-defined off-discharge, and a long secondary period of inhibition.

Thus, even in the case of convergence of different signals on one neuron of the caudate nucleus, the specificity of the information concerning the physical properties of the signal is preserved.

Dynamic Components of Activity

In the course of the application of long-lasting series of a signal, the neuronal responses showed a number of changes which indicated a modification of the state of the system as a function of monotonous stimulation. These changes related to various characteristics of the responses and were of different types.

1. Stabilization of the response. In a number of cases the repetition of the signal stabilized ("emphasized") the pattern of the response. In the case of the previously mentioned example (cell 11, Fig. 1B, 2B), it was expressed in a decrease of the latency (62–12 msec), preservation of the primary discharge, more pronounced inhibitory pause, and increase in the density of the secondary discharge. The same phenomenon is observed in the response of cell 56 to a tone of 5 kHz presented in Fig. 4. Such changes of the response ("emphasis") are

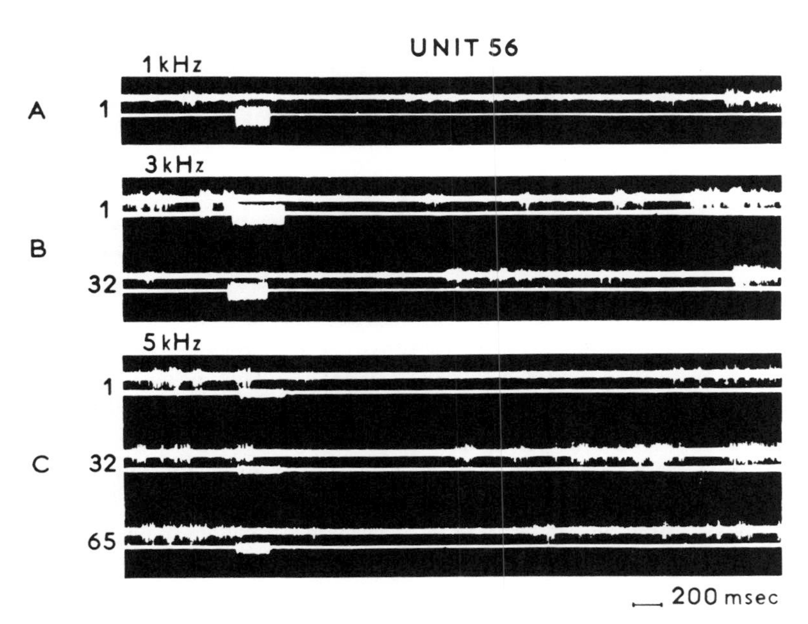

FIG. 4 Different types of responses of cell 56 during the presentation of pure tones of different frequencies, and different dynamic changes of these responses as a result of repeated presentations of the tones. Explanations are given in the text.

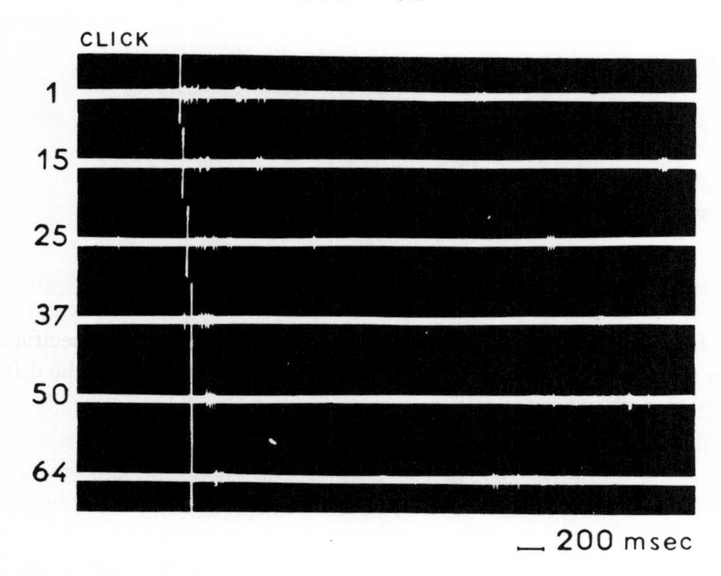

FIG. 5 A change in the pattern of the response of cell 42 as a result of numerous applications of a click.

characteristic of some specific structures (superior colliculus, Dubrovinskaya, 1967; lateral geniculate body, Beteleva, 1967).

2. Weakening of the response. In 30% of the neurons of the caudate nucleus, a partial habituation of the response was observed when the sensory signal was repeated.

This effect was expressed in longer response latencies, which gradually increased severalfold. In Fig. 5 (cell 42) it can be seen that the response initially arises with a minimal latency, equalling 24 msec. As a result of subsequent numerous repetitions of the signal, the latency increases to 75, 112, and 164 msec. Simultaneously, the response weakens. The secondary discharge undergoes the most effective and relatively rapid changes; it becomes eliminated with stimulus repetition. This suppression of the secondary discharge is seen in Fig. 5 (where it disappears by the 25th application of the click), as well as in Fig. 1A, where the secondary discharge does not fully disappear, but becomes considerably reduced.

The primary discharge is affected to a lesser degree. Sometimes it manifests only the effect of stabilization which takes place simultaneously with the suppression of the secondary discharge (Fig. 1A). In other cases, the primary discharge is somewhat reduced, but only to a specific level, which remains constant for a given cell; after reaching some definite magnitude (5–6 spikes for cell 42 in Fig. 5), it preserves a constant form, irrespective of the duration of the repeated signal. In some cases, as a result of protracted repetitions of the signal, the primary response, which at first consisted of 3–5 spikes, decreased to 1–2 discharges, but it never disappeared (see Fig. 8). The observed reduction of the response ("habituation") assumed a pronounced character by the 50th–70th repetition of the signal, and reached a final, stable form by the 100th–120th repetition (Fig. 6). In not a single case could we observe a full disappearance of the specific response, even when more than 200 signals were applied.

3. *The action of optimal and nonoptimal signals.* The above-described dynamic changes relate to the presentation of stimuli which evoke an optimal specific effect, i.e., the most pronounced, stable response characteristic of the cell. If, in the case of a selective specific reactivity of the cell to a definite acoustic frequency band, there exists a boundary zone with a less specific action, just as in the previously described case of the action of pure tones, the difference in the pattern of the responses is expressed not only in their type, but also in their dynamics. This is clearly seen in Fig. 7 which presents a series of successive histograms for cell 56. The weak diffuse inhibitory response to a tone of 2000

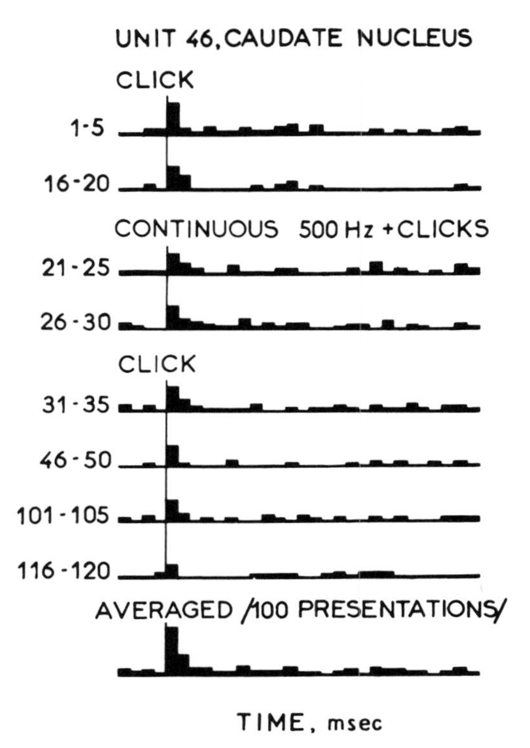

FIG. 6 Serial histograms of the responses of cell 46 during numerous applications of a click. The 3rd and 4th blocks are given against the background of a continuously presented tone of 500 Hz. An averaged histogram is given at the bottom.

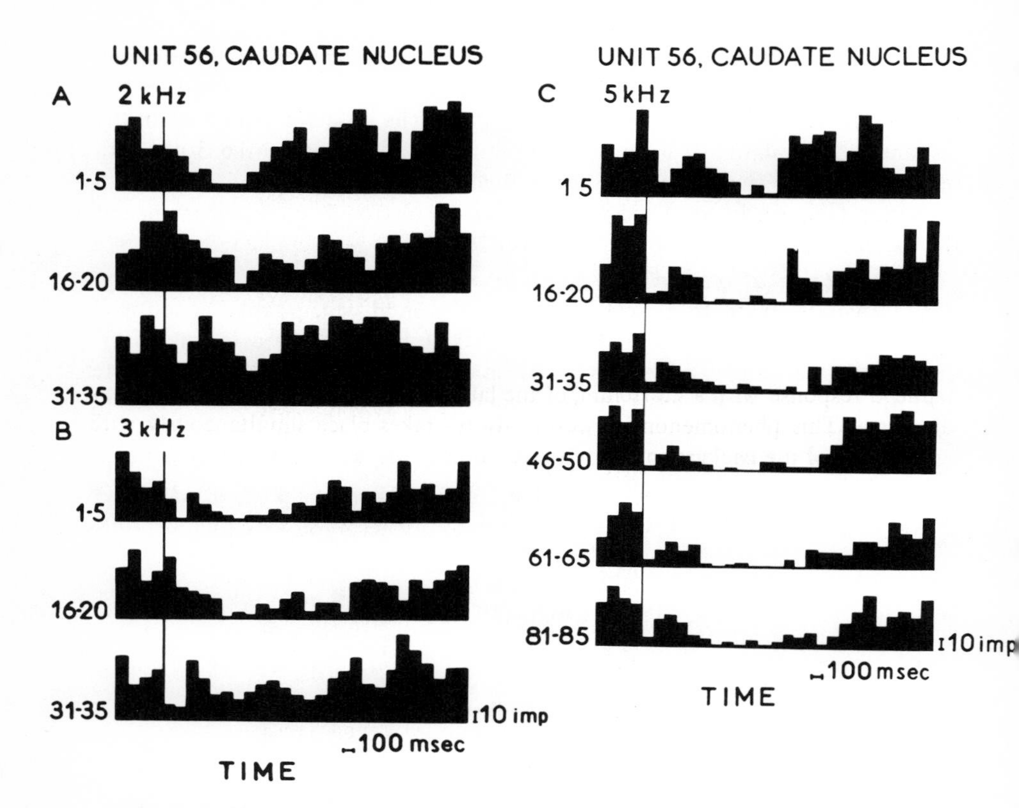

FIG. 7 Serial histograms of the responses of cell 56 to tones of different intensities, showing different types of response dynamics to presentation of the optimal frequency of 5000 Hz (C) and for a diffuse inhibitory response to tones which are distant from the optimal frequency to different degrees, 2000 (A) and 3000 (B) Hz.

Hz is masked by the background activity after the 31st–35th presentation of the signal (Fig. 7A). The more pronounced response to a tone of 3000 Hz is still evident after 31–35 presentations (Fig. 7B). The response to a tone of 5000 Hz (initial inhibition an off discharge and protracted secondary inhibition) is not extinguished at all, but becomes stabilized, reaching its maximum level of expressiveness after 60–80 applications of the tone (Fig. 7C). Thus, it is the degree of similarity of the tone to the "preferable" frequency (and not the total number of applications) which determines both the intensity of the effect and its stability.

This phenomenon shows that the responses of neurons of the caudate nucleus present a complex of specific and nonspecific effects which converge on one element. In a number of cases seemingly one and the same component of a reaction may be of different origins, judging by its stability. Thus Fig. 1A and B shows that the first applications of a flash, as well as of a click, evoke diffuse secondary discharges which differ only in their temporal characteristics. However, subsequent applications of the signals reveal their different nature whereas click repetition causes a decrease in the duration of the secondary discharge, repetition of a flash results in a stabilization of the secondary discharge.

4. The action of extra stimuli on the evoked and background activity. The assumption that specific and nonspecific influences converge on neurons of the caudate nucleus is confirmed by the effect of presentation of strong, long-lasting additional stimuli. The introduction of an additional stimulus (which does not by itself evoke neuronal responses except for an increase of the level of the background activity) either during the interstimulus interval or against the background of such a long-lasting signal leads to the "disinhibition" of the reduced response with a shortening of the latency and recovery of the secondary discharge. This phenomenon, however, always takes place simultaneously with an increase of the background activity. As soon as the activation influence of the extra stimulus on the background ceases, the disinhibition of the response disappears as well. This phenomenon can be seen if Fig. 6 is throughly analysed. This figure shows parallel changes of the evoked response and of the low background activity during the repetition of the signal and against the background of a disinhibiting tone of 500 Hz. The same can be seen in Fig. 8 which contains a record of the activity of this cell. This figure shows a neuronal response to a click characteristic of the caudate nucleus (the peculiar feature of

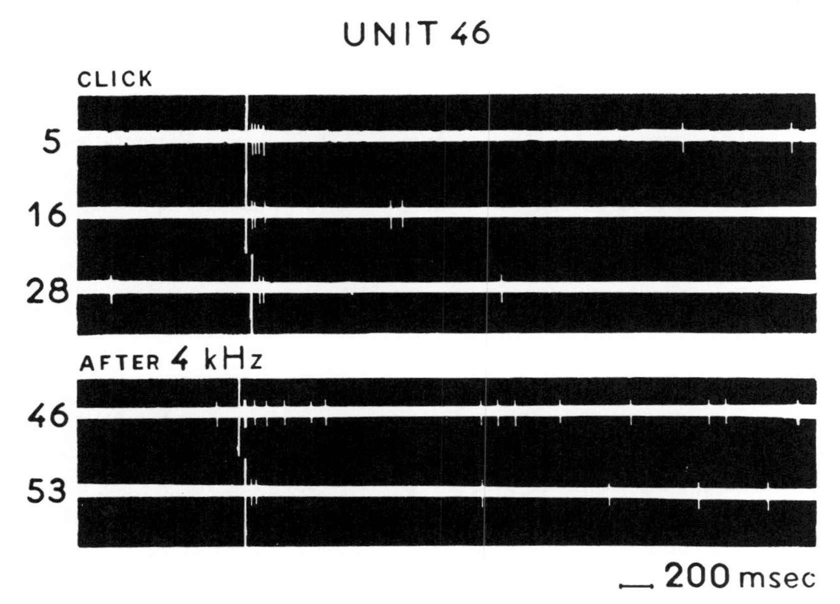

FIG. 8 Weakening of the response of cell 46 as a result of repeated applications of a click and disinhibiting action of a 4000-Hz tone on the spontaneous and evoked activity of the cell.

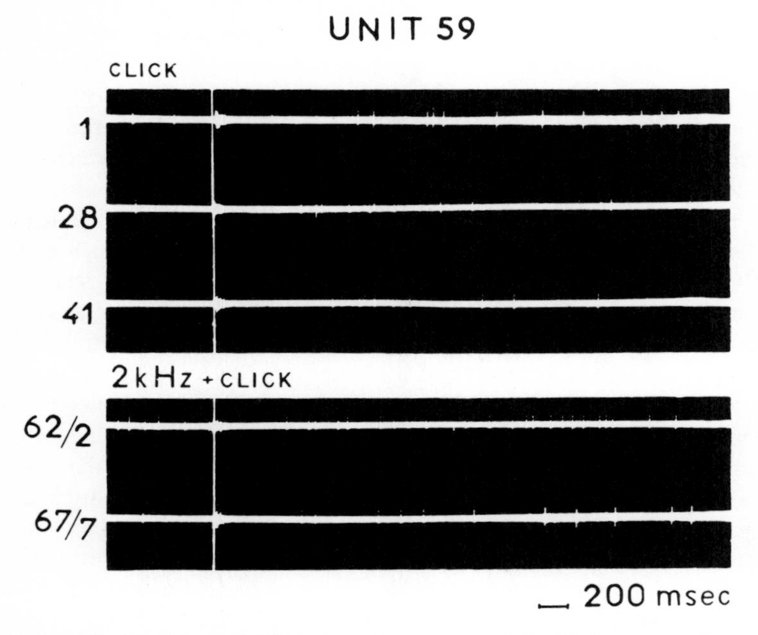

FIG. 9 Cell 59 with an inhibitory response to a click. Click presentation against a background of a 2000-Hz tone is accompanied by an increase of the background activity without appreciable changes in the inhibitory response.

this case is the absence of a secondary discharge); the response was reduced, the latency increased from 37 to 50 msec, and the number of spikes decreased from 4 to 2. (The response becomes stabilized in this form.) Presentation of a 4000-Hz tone during the interstimulus interval leads to a reduction of the latency to 25 msec, to an increase of the number of spikes in the discharge, and to a rise in the background frequency of the discharges (Fig. 8, No. 46). As a result of further repetition, the initial background level and the stable form of the response rapidly recover (Fig. 8, No. 53).

The same type of influence is exerted by an extrastimulus (a tone) on the background activity even when the main acoustic stimulus–a click–evokes a specific response in the shape of primary inhibition of the spontaneous discharges. This is shown in Fig. 9 (cell 59), where the neuron responds to the click with a constant inhibitory reaction that varies in its duration, probably with a slight rebound. A continuous tone of 2 kHz, presented for several minutes, raises the level of the background activity, and slightly decreases the duration of the inhibitory pause, without changing its basic characteristics.

A thorough analysis of the histograms shows that the simultaneous reduction of the background and of the response appears also when blocks of successive histograms are compared. It can be seen from Fig. 6 that an initially higher level of the response is always observed against the background of heightened spontaneous activity. When the signal is repeated, spontaneous activity shows a regular decline during 15–30 applications, after which the response undergoes certain changes in the direction of stabilization. In five cases the neuron did not react to the brief (e.g., click) stimuli which were presented in the experiment, but responded with an increase in background activity to strong and long-lasting stimuli. During repeated presentations of the signals this effect extinguished after 5–15 applications, just as was observed in the case of hippocampal neurons.

All these facts demonstrate that the responses of most neurons of the caudate nucleus result from a convergence of two types of influences–specific and nonspecific; the degree of participation of these two influences may be dissimilar for different neurons and during the presentation of different stimuli on the same neuron.

DISCUSSION

When analysing the dynamic changes taking place in neurons of various brain structures, it is necessary to use caution in the interpretation of the effects observed, especially in identifying them with the phenomena of emergence and extinction of the orienting reflex.

First, it is necessary to distinguish between the tonic level of wakefulness and the phasic effects of activation which may manifest themselves against this background (Sokolov, 1958). As is known, these two phenomena are closely related and correlate with each other. A protracted repetition of a monotonous stimulus results not only in the disappearance of the phasic orienting reaction, but also in a gradual and regular decline of the level of wakefulness. However, rapid extinction of phasic activation to a particular stimulus may not be accompanied by any appreciable parallel decline of the general activity level. At the same time, the working state of various brain structures depends undoubtedly on the general level of wakefulness, although it does not necessarily reflect the phasic effects of the recovery and extinction of the orienting reflex. In view of this, many dynamic changes, which are observed in neurons and which are similar to the transformation of the orienting reaction (gradual weakening of the effect), can be only indirectly or partially connected with it, being directly dependent on a more general and stable factor, the level of wakefulness.

Besides these two different types of influences (orienting reaction and tonic changes of the level of wakefulness), some slower changes take place which are characteristic of the specific systems.

These phenomena, which were originally described on the basis of modifications of the evoked potentials in the specific sensory relays (Galambos, 1956; Hernandez-Peon, Jouvet, & Scherrer, 1957; and others), apparently present protracted adaptational changes. Their development requires numerous repetitions of the stimulus for many hours. Analogues of such reactions manifesting a very slow adaptation were observed also in neurons of the specific sensory systems. These phenomena cannot be regarded as connected with the orienting reflex either; the general tendency towards a decrease of the effect produced by stimulus repetition is insufficient for equating processes which markedly differ in their temporal parameters.

An analysis of the responses of neurons of the caudate nucleus (short latencies, phasic character, presence of a definite temporal distribution of the spikes, dependence of the character of the response on the concrete physical properties of the stimulus), shows that they bear a pronounced specific character. In a considerable number of cases unimodality, or selectivity with respect to definite qualities of the stimulus, proves to be their characteristic feature. But even in the case of sensory convergence ("modal nonspecificity of the input"), the neurons respond with a phasic, structured reaction, which varies for different signals, retaining information about the quality of the acting stimulus in the temporal distribution of the spikes ("modality specificity of the output").

In accordance with numerous data obtained by different authors on the specific structures (see for example the present volume), these responses prove to be of a highly stable character, and manifest slowly directed changes whose final formation requires scores of repetitions of the signals. The dynamics of these responses are revealed during analysis of data in blocks of five stimuli, with the omission of 10 signals between the neighbouring blocks subjected to analysis. This would be impossible in the case of hippocampal neurons in which the entire dynamics, even during the very first applications of the signals, develop in the course of 5–20 repetitions of the stimulus.

An analysis of the data shows that responses consist of relatively stable initial components, as well as of later effects which are subject to a more or less rapid transformation. Such a division of the responses is often possible only when analysing the effect of long-term stimulus repetition and the dynamics of recovery of the initial response due to the action of an extra stimulus.

We suppose that the dynamics of the neuronal responses in the caudate nucleus are formed of at least the following two processes, whose mechanisms and origins are different: (1) a change of the general level of activation ("level of wakefulness"), which is produced by the activating structures of the brain stem and which declines in the course of a profound extinction of the orienting reaction; (2) slow processes of habituation (inactivation) in the system of specific transmission of information which may be expressed in an additional reduction of the response, as well as in its stabilization.

The first process is connected with the regulation of the working level of the system as a whole, which is reflected in a parallel change of the spontaneous and evoked activity of the neuron; the second process determines the change of the specific response of the neuron during protracted stimulus repetition. This duality of the neuronal responses of the caudate nucleus, which depends on the convergence of specific and nonspecific influences, is confirmed by the character of its afferent connections (on the one hand, fibers from the specific nuclei of the thalamus, and collaterals from the internal capsule, and on the other from the medial and intralaminar nuclei of the thalamus and midbrain tegmentum).

As is known, during an orienting reflex practically all efferent systems of the organism are involved. This extended effector manifestation must be integrated in definite systems of the brain, and in all probability in the reticular formation of the brain stem. Its ascending activating influences spread widely to rostral structures, which in the system of the orienting reflex are, strictly speaking, also of an effector character with respect to these influences. In view of this, when evaluating the activation and the dynamic changes in separate nervous elements, it is necessary to distinguish direct effects connected with the organization and integration of the activation reaction from secondary effects which reflect the "efferent" change of the working level (i.e., level of arousal) caused by remote activating influences.

A comparison of the responses of hippocampal neurons (see the present volume) and those of neurons of the caudate nucleus shows that here we have two types of participation in the system of the orienting reflex and in tonic activation. Whereas in the first case, we may admit a direct participation of the system in the regulation of the orienting reflex, in the second case we deal with a secondary modulation of the stable activity of the system by activating influences converging on it.

13

RESPONSES OF THE RABBIT'S LATERAL GENICULATE BODY TO SOUND STIMULI AND TO ELECTRICAL STIMULATION OF THE RETICULAR FORMATION OF THE BRAIN STEM

T. G. Beteleva

Moscow State University

In order to study the effect of weak stimulation on the activity of the nervous elements of the visual system, most authors use single conditioning electrical stimulations of the reticular formation before the presentation of light test stimuli (Buser & Segundo, 1959; Long, 1959; Dumont & Dell, 1960; Okuda, 1962; Tyc-Dumont, 1964). These investigations established that the evoked potentials and the neuronal responses to a flash become intensified, if the latter is applied 100–200 msec after reticular stimulation. Such an intensification of the response by a preceding reticular stimulation is observed in all neurons and does not depend on whether they do or do not respond to isolated stimulation of the reticular formation. According to Buser and Segundo (1959) only 50% of neurons in the cat's lateral geniculate body (LGB) do so.

The responses of the visual system to inappropriate stimuli usually have been studied by means of acoustic and somatosensory stimulation. These responses were recorded in the optic nerve (Spinelli, Pribram, & Weingarten, 1965), in the LGB (Meulders, 1965) and in the visual cortex (Skrebitsky & Voronin, 1966; Skrebitsky, 1966; Skrebitsky & Gapich, 1963). These authors established the fact of a parallel emergence of the response of the visual system and of a generalized orienting reflex.

The purpose of the present work was to investigate the neuronal responses in the lateral geniculate body of a rabbit to acoustic stimuli and to single electrical stimulation of the reticular formation. By applying such single stimulations, we intended to reveal those responses in the LGB that depend on the direct reticulothalamic connections described by Scheibel and Scheibel (1966), and also to reduce the general influence on the electrical activity which would be caused by repetitive reticular stimulation.

METHODS

The activity of the LGB units was recorded in unanaesthetized rabbits fixed in an experimental stand. The recording was done by a tungsten microelectrode which had a resistance of 6–10 megohms. The microelectrode was inserted with the aid of a hydraulic manipulator, whose base was mounted on the skull by means of cement. Before this, a trephination opening with a diameter of 5 mm had been made in the skull. The surface of the brain was covered with agar.

The LGB neurons which responded to a flash were subsequently investigated with the application of acoustic and reticular stimulation.

Clicks were produced by a loudspeaker driven by 40-msec pulses. Series of clicks with a frequency of 0.6 Hz were used in the experiments. The responses to repeated acoustic stimuli were investigated in 26 neurons.

The electrical stimulation of the brain was performed through implanted bipolar electrodes and with the help of a "Biofizpribor" pulse stimulator. Single electric stimuli of 0.5–25 V, with a duration of 0.05 msec, were applied every 4 sec. The responses to reticular formation stimulation were investigated in 12 neurons recorded from two rabbits. Histological control showed that in one of the rabbits the electrodes were located in the area of the nucleus gigantocellu-

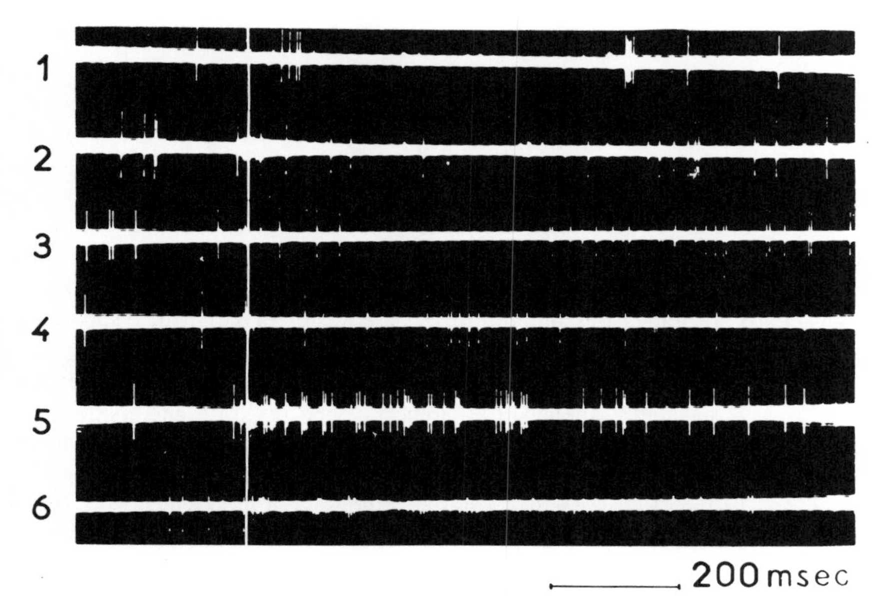

FIG. 1 Response of an LGB neuron to stimuli of different modalities (cell 221, rabbit 35): (1) flash; (2) click; (3–6) electrical stimulation of the reticular formation by single impulses of 4, 8, 9.6, and 12.8 V in intensity, respectively, and of 0.05 msec in duration. Stimulus onsets coincide with the vertical line.

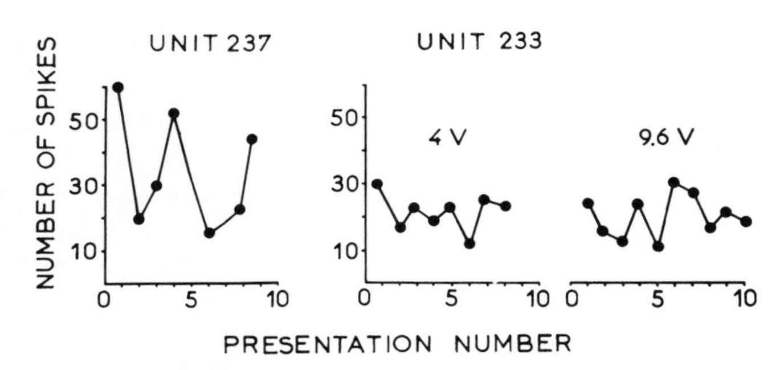

FIG. 2 Fluctuations of the number of spikes during 1500-msec periods during successive applications of inadequate stimuli. Left: responses to single sound stimuli (cell 237, rabbit 36); Right: Responses to threshold (0.05 msec, 4 V) and maximal (0.05 msec, 9.6 V) electrical stimulations of the reticular formation (cell 233, rabbit 36).

laris of the midbrain, and in the other rabbit, in the central gray matter. The responses of 4 LGB units to electrical stimulation of the visual cortex, in the area where the amplitude of the evoked potential to a light stimulus is highest (Polyansky, 1963), were recorded in the course of a control experiment.

RESULTS

Seventeen neurons (out of 26 investigated with acoustic stimuli) exhibited a modification of their spontaneous activity which lasted 1000–1500 msec after the presentation of the stimulus. In 9 of these neurons the response consisted of an initial decline of discharge frequency, which began within a period of 100–550 msec, and in its subsequent increase, within a period of 250–1000 msec. In the other 8 neurons the response to clicks began with an acceleration of spike activity. In some cases an inhibitory phase was observed in the responses of these neurons.

Most neurons which reacted to clicks exhibited variable responses. When the stimulus was repeatedly applied, the latency of the response and its duration changed. The response was small; separate responses could hardly be distinguished in the spontaneous activity. The difference between the response to a sound, and to the discharge which arises in response to a flash which has a stable pattern, is shown in Fig. 1. In 4 neurons (out of 17 which responded to the sound) this reaction was characterized by the appearance of spike bursts persistently repeating their pattern from one response to another.

A calculation of the number of spikes for a period of 1500 msec showed that within 10 successively applied stimuli no substantial changes arose in the responses to the sound (Fig. 2, unit 237).

Responses to electrical stimulation of the reticular formation were observed in 8 out of the 12 recorded neurons. The pattern of the responses and their parameters depended on the voltage of the stimulating current.

The thresholds of the responses of individual neurons to the stimulation of the reticular formation varied from 0.8 V to 6.4 V, but in most cases they equalled about 4 V. The responses to threshold and weak suprathreshold stimuli were similar to the neuronal responses of single sound stimuli. They consisted of an alternation of periods of decreased and increased activity. The similarity of the responses to sound and reticular stimulation is illustrated in Fig. 3, which presents poststimulus histograms of the responses of the same neuron to both kinds of weak stimuli.

Owing to the considerable variability of the responses to reticular stimulation of threshold and suprathreshold intensities, it was difficult to distinguish them in the spontaneous activity and to establish their temporal parameters. From the histograms, which were formed on the basis of averaging the responses to 10 successively applied stimuli, it may be seen that most responses to stimulation of

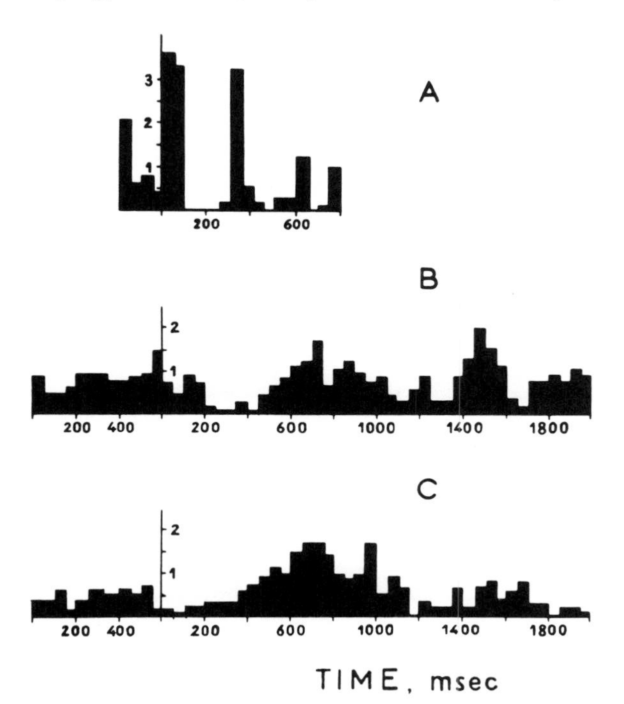

FIG. 3 Poststimulus histograms of neuronal responses to stimuli of different modalities (cell 237, rabbit 36). (A) flash; (B) click, (C) electrical stimulation of the reticular formation (0.05 msec, 4 V). The responses to 10 successive applications of the stimulus have been averaged. Ordinate, average numbers of spikes per response. The moment of the stimulus presentation coincides with the ordinate.

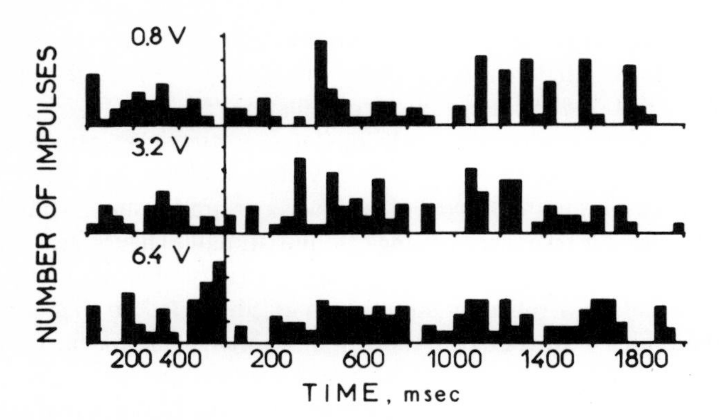

FIG. 4 Poststimulus histograms of neuronal responses (cell 220, rabbit 35) to electrical stimulation of the reticular formation by single impulses (0.05 msec) of different intensity. The intensity of stimuli is indicated by figures at the left-hand side of the histograms.

low and moderate intensities begin with a period of suppressed activity which in 200–500 msec is followed by a period of intensified activity. The duration of the neuronal reactions to a reticular stimulation equals 1000–1500 msec.

An increase in the intensity of the stimulating current changed the pattern of the response. In most neurons it led to a more rapid replacement of the inhibitory phase by intensified firing (Figs. 1 and 4), while in others, on the contrary, it increased the duration of the inhibitory phase (Fig. 5).

It is important that most neurons respond to increased reticular stimulation only up to a definite limit. In some neurons the action of a current of such

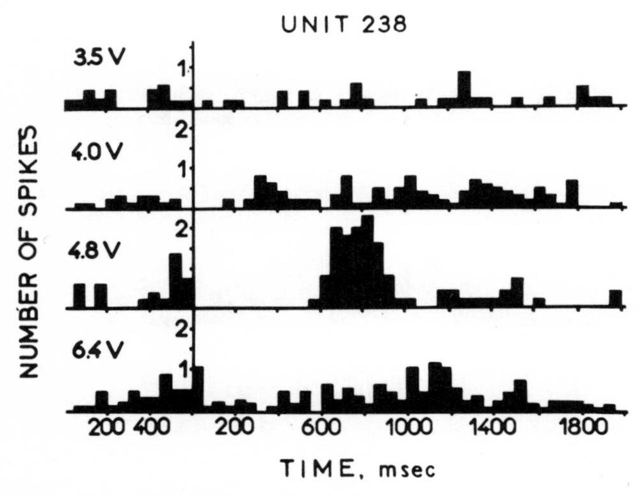

FIG. 5 Poststimulus histograms of neuronal responses (cell 238, rabbit 36) to electrical stimulation of the reticular formation by single impulses (0.05 msec) of different intensity. Designations are the same as in the preceding figures.

maximum voltage evokes a high-frequency spike discharge of a declining amplitude. A discharge of this kind is shown in Fig. 1 (5). After the discharge the activity of the neuron may recover at once, as shown in the figure, or some time later. In a number of cases the neuronal discharges ceased for a very long time and their recovery was not observed (Fig. 1, 6). A reaction of this type was regarded by us as reflecting excessive neuronal depolarization. In the case of another type of reaction, an increase of stimulus voltage led to deterioration or disappearance of discharges (Fig. 5).

A calculation of the number of spikes for a period of 1500 msec showed that 10 repetitions of the stimulus did not evoke substantial modifications of the response at any intensity of the electrical stimulation applied (Fig. 2B).

For a comparison with responses to reticular formation stimulation, we also recorded the responses of four neurons to stimulation of the visual cortex. In all cases, single electrical stimulations of the cortex resulted in an intensification of neuronal activity, which began with a latency of 100–300 msec and lasted up to 800–1500 msec after stimulus presentation. The inhibitory phase, which is characteristic of neuronal responses to sound and reticular stimulation, was not observed during stimulation of the cortex. An increase of the voltage of the stimulating current from 4 to 40 V and a prolongation of the action of the stimulus from 0.05 to 50 msec led to a more pronounced acceleration of activity. In the case of such parameters of the stimuli, no depolarization of the neuron was observed. A tentanization of the cortex produced quite a different action (Fig. 6). Stimulation of the cortex with a frequency of 100 Hz brought about a depolarization of the neuron at very low intensities of the stimulating current. When stimulation was started, there was a gradual acceleration of activity and then the appearance of spike bursts. The grouping of the spikes was accompanied by a decline of spike amplitudes. When the tetanus was discon-

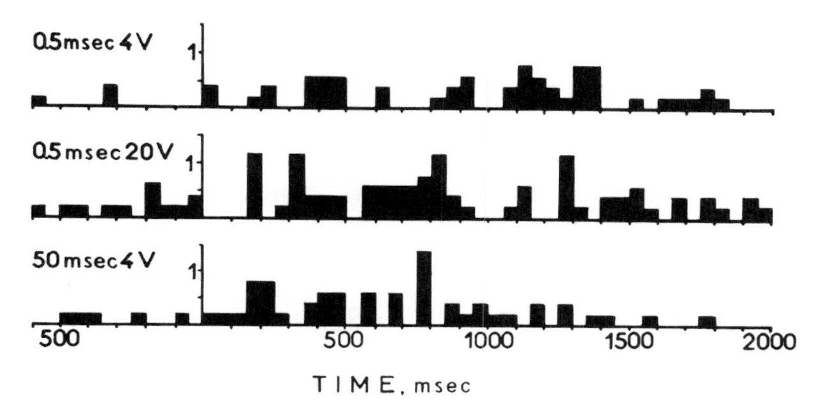

FIG. 6 Poststimulus histograms of neuronal responses to electrical stimulations of the cortex (cell 247, rabbit 37). The duration and intensity of the stimuli are indicated at the left-hand side of histograms.

tinued, this type of activity persisted, but the frequency of the impulses within the bursts showed a gradual decrease. The frequency and amplitude of the spikes recovered a few minutes after the discontinuance of the tetanus.

Thus, neurons of the LGB, which respond to flashes of light, also respond to reticular and cortical stimulation. Sound and reticular stimuli evoke a similar two-phase reaction which is characterized by similar latencies and durations. This similarity proves to be particularly great in the case of weak stimulations of the reticular formation. The difference between the responses to these two kinds of inadequate stimuli is that the responses to stimulation of the reticular formation are more consistent and vary in individual neurons to a lesser degree than the responses to sound stimuli. The responses of the LGB neurons to electrical stimulation of the visual cortex are characterized by higher threshold and, unlike the responses to reticular stimulation, possess only one phase.

DISCUSSION

The disclosed similarity of the responses to sound and reticular stimulation may be regarded as proof that the influence of inadequate stimuli on the visual system pass through the reticular formation of the midbrain. This is confirmed by a number of studies published in the literature. Thus, in the case of the visual neurons, a high correlation was established between responses to painful and sound stimulation (Murata, Cramer, & Bach-y-Rita, 1965; Skrebitsky, 1966). Spinelli *et al.* (1965) showed that the responses of the optic nerve to sound and somatic stimulations depend on the level of wakefulness.

During electrical stimulation of the visual cortex, the character of the responses and the parameters of the stimuli which evoked them differ considerably. This leads to the assumption that nonspecific stimulations are transmitted to the LGB along direct reticulothalamic pathways, without cortical involvement, as described in the work of Scheibel and Scheibel (1966).

It was found that most responses to sound and reticular stimulation consist of periods during which the activity decelerates and accelerates. This coincides with the data of Wilson, Pecci-Saavedra, and Doty (1965) concerning the two-phase character of the action of reticular stimulation upon the specific responses of the visual system. According to these authors, the inhibitory effect possesses a lower threshold and is recorded with a shorter latency than the excitatory one.

It is noteworthy that not all neurons of the LGB respond to stimulation of the reticular formation. This confirms the data obtained by Buser and Segundo (1959) from an investigation of the cat's LGB. A still greater part of the neurons does not respond to single sound stimulations; this disproves the assertion of Meulders (1965) concerning the diffuse influence of weak stimuli on all neurons of the LGB. This discrepancy is apparently explained by the fact that Meulders recorded evoked potentials, and not responses of single neurons.

Unlike Skrebitsky (Skrebitsky, 1966; Skrebitsky & Gapich, 1963), we did not observe any decrease of the responses to sound and reticular stimulations during 10 applications of the stimuli. This difference in the results is probably due to the kinds of the applied stimuli: Skrebitsky used tones which lasted more than 1 sec and which were analogues of a tetanic stimulation of the reticular formation, leading to the emergence of a generalized orienting reflex. The responses recorded in our experiments were similar to those obtained by Jung (1961) for neurons of the cat's visual cortex during single stimulation of the nonspecific nuclei of the thalamus. According to the assumption of Jung, these nonspecific reactions bear a relation to different specialized functions of the reticular formation which are connected with the regulation of eye movements, as well as with the maintenance of perceptual constancy.

14

NEURONAL MECHANISMS OF SYNCHRONIZATION AND DESYNCHRONIZATION OF ELECTRICAL ACTIVITY OF THE BRAIN

N. N. Danilova

Moscow State University

As is already known, frequent repetition of the same stimulus leads to the extinction of the orienting reflex and to the development of sleep. Repetition of the signal not only ceases to evoke orienting responses, but exerts an inactivating, inhibitory influence; this is expressed by the intensification of slow synchronous waves in the EEG (Danilova, 1961). Thus, we may speak of the existence not only of phasic or tonic inhibition of the orienting reflex, but also of a special inactivation reaction (Sokolov, 1963d).

This concept concurs with available data concerning the existence, in the nonspecific thalamic system of an inactivating, inhibitory system (Monnier, Hölsi, & Krupp, 1963), the stimulation of which provokes both behavioral and electrocortical sleep.

The following question arises: What is the neuronal substratum of the development of EEG synchronized rhythms, as an inactivation reaction arising during the extinction of the orienting reflex? This question is connected with another, more general question concerning the neuronal mechanisms of synchronization of the electrical activity of the brain. Electrophysiological investigations concerned with this topic thus far have failed to yield definitive results.

Some research has established the absence of a correlation between the electrical activity of single neurons and the slow waves which are recorded from the brain with the aid of microelectrodes (Li & Jasper, 1953; Li *et al.*, 1956; Mountcastle, Davis, & Berman, 1957; Purpura, 1959). Other investigators emphasize that the correlation of the EEG slow waves and the spike activity of single neurons increases when the animal is falling asleep, or when anaesthesia becomes more intense, as a result of which the slow waves assume a particularly pronounced character (Verzeano, 1955, 1956; Verzeano & Calma, 1954; Creutzfeldt, 1961; Amassian, 1961).

Nor is the mechanism of synchronization of the slow waves clear. Whereas some authors state that the impulses circulate along neuronal chains (Laufer & Verzeano, 1967; Verzeano, 1955; Verzeano & Negishi, 1960); others (Andersen & Sears, 1964; Andersen *et al.*, 1964) emphasize the role of recurrent inhibition.

The available data lead to the assumption that neurons of different types are differently connected with the mechanism of synchronization. Of particular importance from this point of view is the search for neurons which would bear a direct relation to the phenomenon of synchronization of the brain slow electrical activity which develops as a result of extinction of the orienting reflex.

The present work was devoted to investigating the role of neurons of the non-specific thalamus in the generation of synchronized slow rhythms of the rabbit's brain. For this purpose, the electrical activity of single neurons of the nonspecific thalamus was recorded simultaneously with the gross electrical activity of the hippocampus, reticular formation of the midbrain, visual cortex, and nonspecific thalamus of the opposite hemisphere. The gross recording was performed using implanted macroelectrodes.

The selection of the nonspecific thalamus for investigation was determined by its double function. On the one hand, it is an inactivating system and participates in the development of synchronized rhythms, in particular, rhythms of sleep. On the other hand, the nonspecific system of the thalamus (diffuse-projection system, in the terminology of Jasper) ensures a selective nonspecific activation of the cortex, which depends on the modality of the acting stimulus.

Taking into account these functions of the nonspecific thalamus, it may be assumed that research into the electrical activity of its neurons will make it possible to approach the comprehension of the neuronal mechanism of synchronization and desynchronization of the electrical activity of the brain.

The investigation consisted of two parts: the first part was devoted to studying the relationship between the spike activity of a single neuron of the nonspecific thalamus and the slow waves which are recorded by means of implanted electrodes from various structures of the brain; the second part was devoted to studying the correlation of the slow wave and spike activities which were recorded from the nonspecific thalamus with the aid of a single microelectrode.

METHODS

The investigation was carried out on 24 unanaesthetized rabbits which were fixed in a special stand during the experiments.

Neuronal activity was recorded extracellularly by a tungsten electrode. On one of the channels, we recorded the spike component of the neuronal electrical activity after its passage through a band-pass filter which passed all frequencies

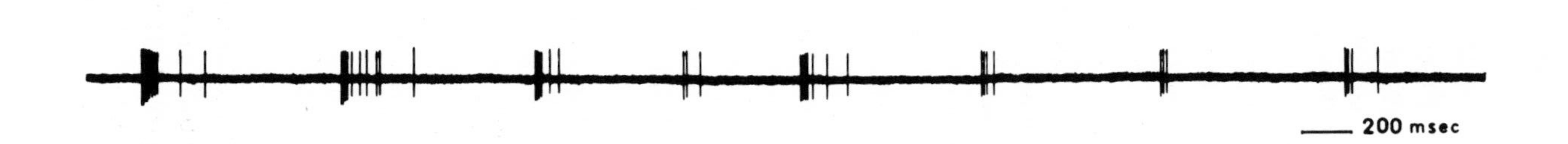

FIG. 1 Continuous recording of the background activity of a neuron with spike grouping (cell 50). It can be seen that along with single spike discharges, the neuron generates groups of such discharges.

from 300 to 8000 Hz, and on the other channel, the neuronal electrical activity which included both the spike and slow components (1–8000 Hz).

The microelectrode was inserted remotely with the aid of a Hubel micromanipulator. During the experiment, the piston of the micromanipulator with the electrode was mounted on a plug which had previously been fixed on the surface of the skull over a trephined hole. Fixation of the plug was accomplished by means of a stereotaxic instrument. Position and depth of the microelectrode insertion were determined according to the atlas of Fifkova and Marsala (1962).

The electrical activity of the occipital cortex, hippocampus, midbrain reticular formation, and nonspecific thalamus was recorded unipolarly on an 8-channel electroencephalograph. The indifferent electrode was placed in the nasal bone. The position of the micro- and macroelectrodes in the brain was controlled histologically.

During an experiment we studied the neuronal responses to repeated applications of the same sensory stimulus at intervals of 15 sec, each stimulation lasting 5 sec. Usually, during the investigation of each neuron, 20–60 stimuli were applied. The duration of each experiment was 5–6 hr.

In order to disinhibit the orienting reflex, the repeatedly-applied stimulus was replaced by another signal; this was followed by the presentation of the previous sensory stimulus. In the course of the experiment the new disinhibiting signals were usually presented several times.

Rhythmic clicks (at the rate of 10 per sec), with an intensity of 80 dB above audibility threshold, delivered from a photophonostimulator (FFS-02), were used as repeatedly applied stimuli. Clicks at frequencies of 5 and 15 per sec, as well as air puffs on the surface of the head and hind paw, were used as disinhibiting stimuli.

In order to establish the relationship between the slow synchronous rhythms arising in the EEG of the rabbit's visual cortex and the groups of spike discharges of single thalamic neurons, we determined the time of emergence of the slow waves in the EEG of each group, and the sequences of groups of neuronal spike discharges. We then formed correlation fields characterizing the degree of relation of these two processes with respect to the time of their emergence and disappearance. The moment of their emergence was determined relative to the beginning of the interval between the stimuli presentations.

To characterize groups of EEG slow waves, histograms of the distribution of the wave periods were made. Neuronal spike activity was evaluated by a histogram of the distribution of intervals between the spike bursts, the indispensable condition being the absence of single spikes in the intervals between these bursts; in the opposite case, such intervals were not taken into account.

When slow waves connected with neuronal spike bursts were also recorded with the help of a microelectrode, the amplitudes and durations of separate phases of these waves were measured and the dependence of these parameters of the slow waves on the number of spikes in the group was investigated.

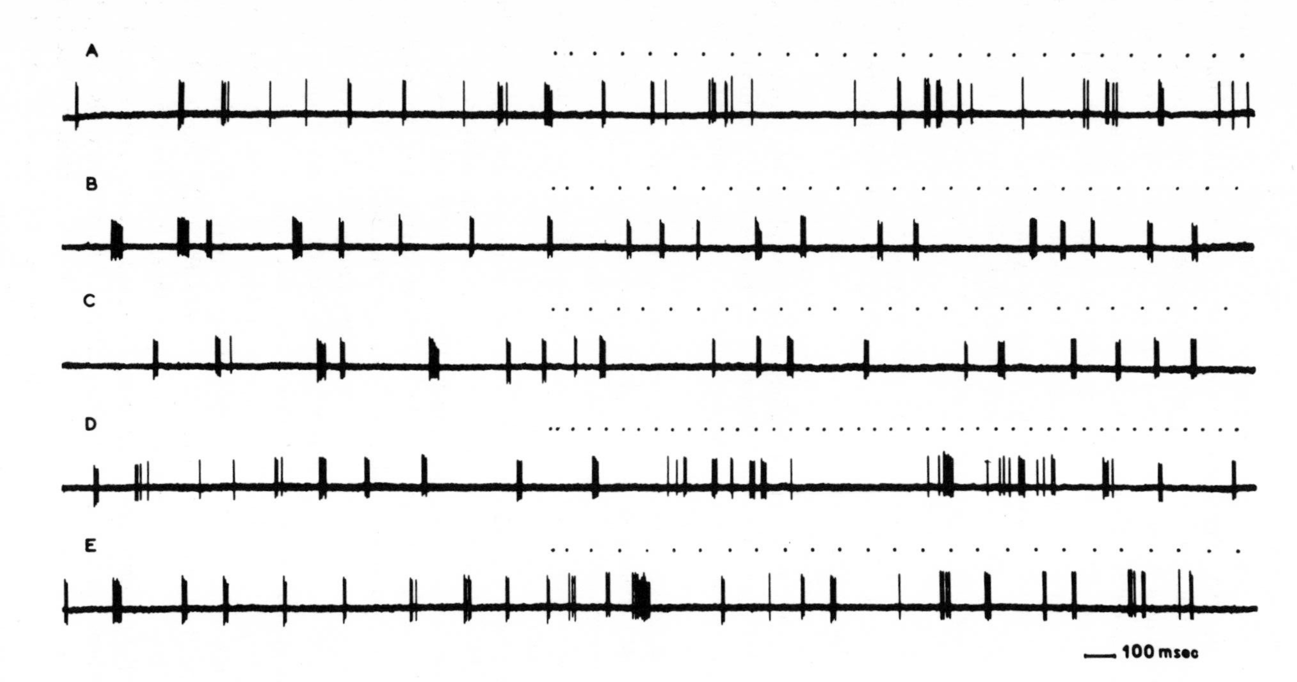

FIG. 2 Extinction and disinhibition of a neuronal response (cell 165) to clicks (10/sec). (A, B, C, E) lst, 8th, 11th and 13th applications of the stimuli. (D) extra stimulus–clicks at a frequency of 15 per sec. The clicks are designated by dots.

RESULTS

Neurons with Spike Grouping

A study of the spike activity of single neurons in different structures of the thalamus (nucleua reticularis, nucleus ventralis lateralis, nucleus medialis dorsalis, nucleus lateralis posterior) revealed the existence of a large group of neurons with spike grouping; they comprised 50% of all 130 cells investigated.

The main feature of these neurons was a grouping of the spikes into bursts. Each group of spikes presented a high-frequency discharge consisting of 2–12 spikes. The frequency of the neuronal discharges within the burst depended on their number. In a group of 2–6 spikes the latter followed with a frequency of about 300 per sec. In groups containing up to 12 spikes, the neuron discharged at a lower frequency.

Usually the spontaneous activity of neurons of a given type consisted of single spike discharges, as well as discharges of spike groups (Fig. 1). Often spike groups followed one another, forming sequences of spike bursts, which alternated with periods when the neuron discharged only with single spike potentials. Thus, the formation of sequences of spike groups constituted the second characteristic feature of these neurons.

Neurons with spike grouping exhibited a certain relationship to the system which is responsible for the orienting reflex. A sensory stimulation which is new to the animal (the first applications of a repeated signal, as well as extra stimuli) destroyed the spike grouping in these neurons, replacing the bursts by irregular single spike potentials (Fig. 2A). However, with habituation to the repeatedly applied stimuli (clicks at a frequency of 10 per sec), such stimuli lost their ability to destroy the spike groups characterizing the neuronal background activity. From Fig. 2B it can be seen that the response is weakened by the 8th application of the stimulus; by the 11th application it has disappeared. During click stimulation, the neuron discharges with the same spike bursts as before the application of this stimulus. An extra stimulation (15 per sec clicks) destroys the spike groups which are then replaced by single spike discharges (Fig. 2D), after which the original response recovers to subsequent presentations of 10 per sec clicks (Fig. 2E).

This process of extinction of the neuronal response could be observed simultaneously with the constant background activity of the neuron. In a number of cases, however, along with a weakening of the response, an intensification of spike grouping occurred in the neuronal background activity. In the intervals between the stimuli, separate groups were replaced by sequences of spike bursts; their total number, as well as their duration, was steadily growing.

Such intensification of the spike grouping was due to the extinction of the orienting reflex, since any alteration in the presentation of rhythmic stimulation

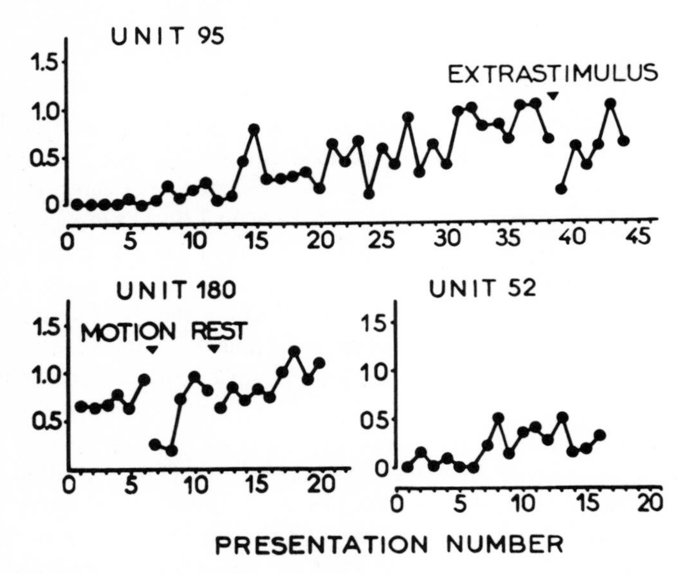

FIG. 3 Growth of the number of spike bursts in the background activity of three neurons (cells 95, 180, 52) in the course of experiments with repeated presentations of clicks. The duration of the acoustic stimulation was 5 sec; the frequency of the clicks was 10 per sec. It was delivered with a constant interval of 15 sec. Ordinate, average frequency of the spike bursts, characterizing the neuronal background activity before each click. Abscissa, index of the record of the neuronal background activity.

weakened this grouping. Any change in the signal parameters (introduction of an extra stimulus, modification of the usual interval at which the stimuli were presented, as well as intervals in their presentation) led to the emergence of single spikes in the background activity of the neuron.

Figure 3 shows a change of the frequency of group discharges in the background activity of several neurons observed during the experiment. A repetition of the same signals intensifies the spike grouping. In cell 95, spike bursts did not appear at once, but only after several stimulus presentations. In cells 180 and 52, spike bursts were in evidence from the very onset of recording. A spontaneously occurring movement of the animal, an added interval in the stimulus presentation, or an extra stimulus disrupts the process.

Usually the process of spike grouping developed simultaneously in a number of neurons, although the phases of emergence of their spike bursts did not coincide. Likewise, the response of these neurons to sensory stimulation was simultaneous. The grouping of spike discharges proved to be disrupted in the entire ensemble of the neurons. Figure 4 presents sections of simultaneous records of the spontaneous activity of two neurons recorded with a single microelectrode.

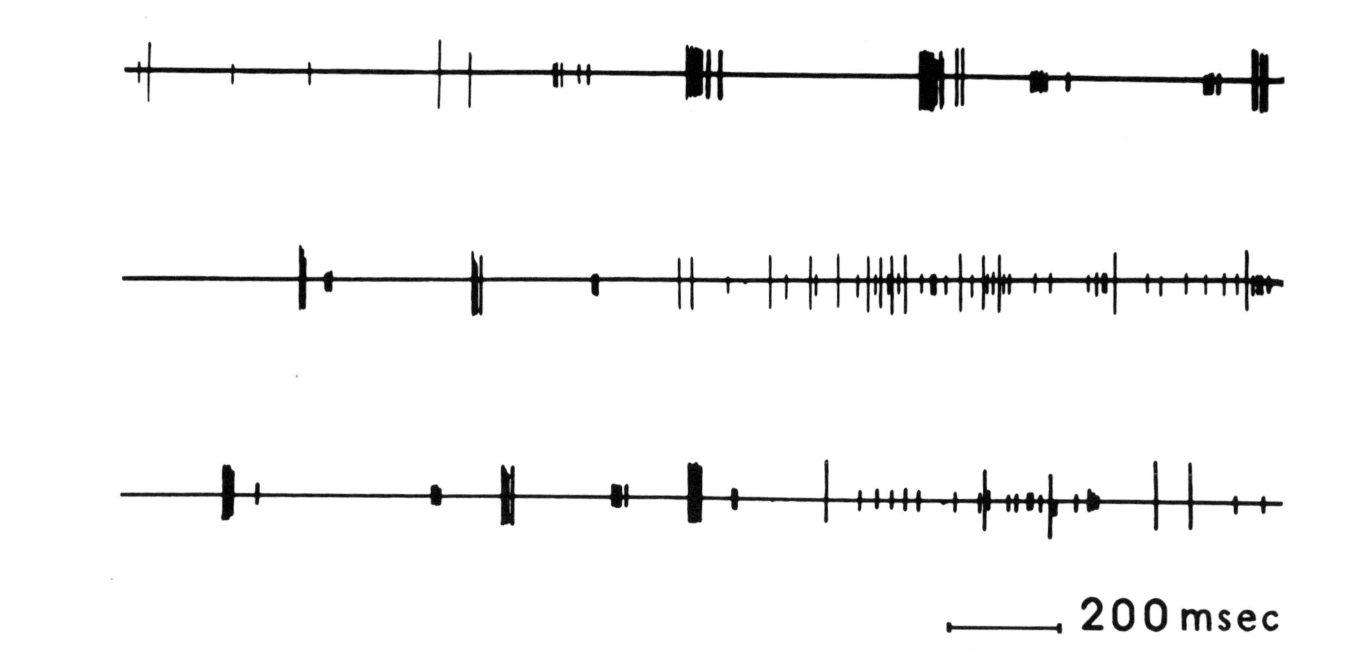

FIG. 4 The functional relation of spike activity of two neurons simultaneously recorded by means of one microelectrode. It can be seen that the spike grouping arises simultaneously in both neurons.

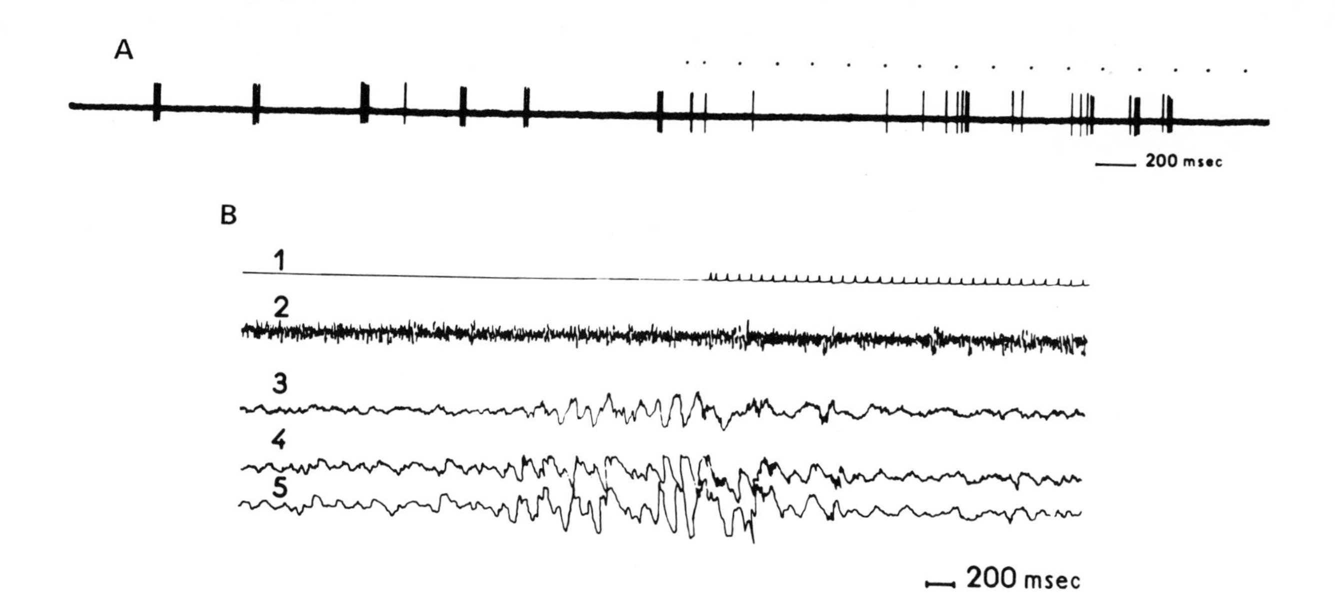

FIG. 5 (A) Record of the spike activity of a single thalamic neuron (cell 116). (B) Electrical activity of the hippocampus (3), visual cortex (4), reticular nucleus of the thalamus (5): 1. mark of clicks at a frequency of 10/sec; 2. EMG of extensor brachii et carpi; this figure shows that before the application of the stimulus there occurs slow synchronous waves in the electrical activity as well as a sequence of spike bursts of the thalamic neuron.

It can be seen that the periods of spike grouping in both neurons coincide. The factors which caused the emergence of single spikes in the background activity of one neuron proved to be responsible for the emergence of single spike discharges in the other neuron as well. However, full synchrony is not observed in the discharges of both neurons.

Correlation between the Sequence of Spike Bursts in Neurons of the Nonspecific Thalamus and Groups of EEG Synchronized Waves

Experiments with repeated presentations of the same signals showed that along with intensified spike grouping in the thalamic neurons, there were observed groups of slow waves in the electrical activity which were recorded by means of implanted macroelectrodes from the visual cortex, hippocampus, lateral thalamus, and midbrain reticular formation. In the course of experimentation, the groups of slow waves increased in amplitude and duration, gradually filling the entire interval between the stimuli. In all the brain structures where the electrodes had been implanted, groups of slow waves appeared simultaneously. Series of slow waves in the electrical activity of the investigated brain structures were abolished by any new stimulus or by any change in the experimental conditions. Thus, a change of the usual stereotype of presenting clicks at intervals of 15 sec eliminated the groups of slow waves which had previously emerged.

In other words, groups of slow waves of electrical activity behave in the same way as spike groups of thalamic neurons. They become intensified with the extinction of the orienting reflex and are eliminated as a result of its elicitation. In addition, slow waves appear during periods when the discharges of thalamic neurons consist only of spike groups.

Figure 5 shows that a sequence of spike groups of a thalamic neuron (5A) corresponds to bursts of slow synchronous waves in the electrical activity of the hippocampus (3), visual cortex (4), and lateral thalamus (5) (5B) which appeared at the end of an interval, i.e., before the next (32nd) presentation of 10 per sec clicks.

Sensory stimuli (clicks) destroy both the slow synchronous waves and the spike grouping. During stimulation only single spike discharges are recorded, while slow waves are absent in the electrical activity.

Below we present the results of a correlation of the spike activity of thalamic neurons with the slow synchronous waves of the EEG of the rabbit's visual cortex.

Figure 5 contains data characterizing the correlation of two sequences which were formed by spike groups of a single thalamic neuron, and by slow waves of the EEG of the visual cortex. The abscissa in Fig. 6A indicates the moments when sequences of spike groups appeared in the neuronal background activity (the latter was recorded during the intervals between the stimuli). The moment

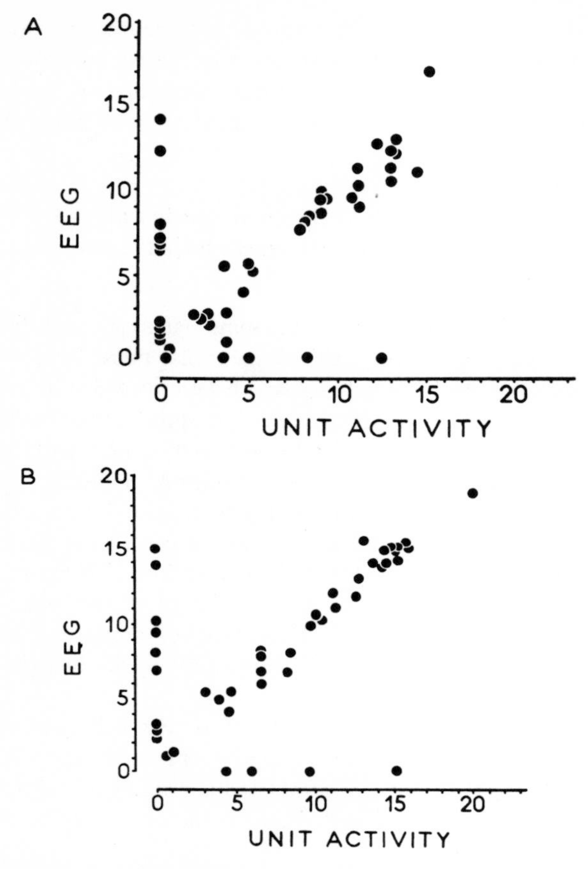

FIG. 6 Correlation between the moments when sequences of neuronal spike bursts (cell 116) emerge (A) and disappear (B), and the groups of slow synchronous waves in the EEG of the rabbit's occipital cortex.

of appearance of each such sequence was measured in seconds from the onset of the corresponding background recording. The ordinate indicates the moments when groups of slow waves (measured in the same way) appeared in the background EEG of the rabbit's visual cortex. The graphs show that a positive correlation exists between the times of emergence of these two kinds of sequences. The points of the correlation field are disposed along one line.

A similar picture is observed in the graph which presents the scatter diagram of the moments when sequences of spike groups in the background activity of the same neuron end, as well as of the moments when groups of slow waves in the EEG disappear (Fig. 6B). However, the same figures demonstrate certain cases when the appearances of these two kinds of sequences do not coincide; the points are disposed along either the abscissa, or the ordinate. Apparently, the emergence in the EEG of groups of synchronized slow waves, in the absence of spike bursts in one of the thalamic neurons, indicates only that these slow waves are determined by group discharges of other neurons whose spike activity was not recorded by us. On the other hand, the emergence of spike groups (which form a sequence) in only one neuron is insufficient for the development of a group of synchronized slow waves in the electrical activity of the cortex. In some cases, therefore, the discharges of the thalamic neuron, in the shape of spike bursts, are not accompanied by synchronization of the slow activity of the rabbit's visual cortex.

The relationships described above are also reflected by the fact that the duration of a sequence consisting of a series of slow EEG synchronous waves, as a rule, exceeds the duration of a sequence of spike bursts recorded from a single neuron. The average period, within which slow waves arise in a group of synchronized EEG oscillations, is smaller than the average length of an interval between groups of spike discharges in a single neuron.

Figure 7 presents a histogram showing the distribution of the periods within which group discharges, following one another consecutively, appear in a single neuron, as well as a histogram showing the distribution of slow waves forming groups of synchronized oscillations in the EEG. Those histograms are based on

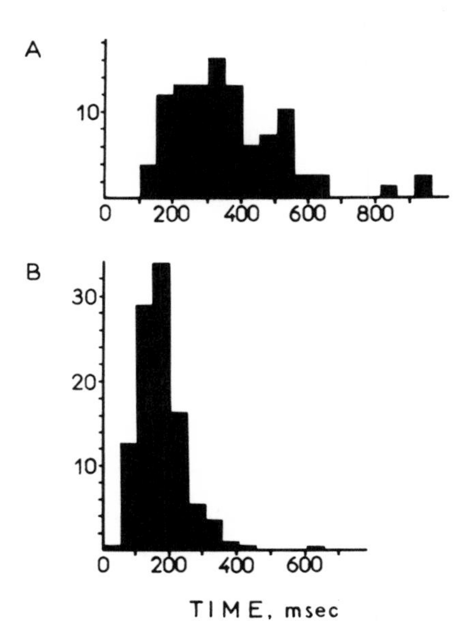

FIG. 7 (A) Histogram of the distribution of intervals at which spike bursts arise inside the sequences formed by them. Cell 116, total number of the intervals measured is 77. (B) Histogram showing the distribution of the periods of slow waves forming synchronized rhythm of the EEG of the rabbit's visual cortex which were recorded simultaneously with sequences of spike bursts of neuron No. 116. Total number of periods measured is 449.

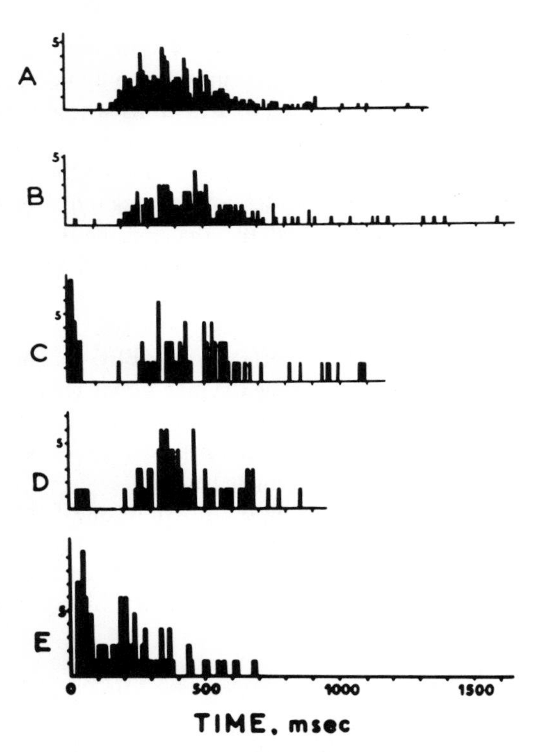

FIG. 8 Histograms of the distribution of intervals at which spike bursts follow one another, for 4 neurons: (A) No. 133; (B) No. 160; (C) No. 132 M; (D) No. 132 B. Each histogram is formed of 429, 170, 67, and 65 intervals, respectively. (E) Histogram of the distribution of intervals formed by spike bursts from two simultaneously recorded neurons (cells 132 B and 132 M). Number of intervals of which the histogram has been formed is 86.

an analysis of the cases in which slow waves emerged in the EEG simultaneously with sequences of spike bursts of the thalamic neuron. It can be seen that the histogram which presents the frequency of emergence of spike bursts in the background activity of the thalamic neuron (Fig. 7A) is more extended and has a maximum at a lower frequency than does the histogram which characterizes the period of slow waves forming groups of EEG synchronized potentials (Fig. 7B).

The characteristics of the periods within which spike bursts emerge in a single thalamic neuron, when the animal is in a state of rest and when groups of slow waves are recorded in its EEG, are shown in Fig. 8. The figure contains histograms indicating the distribution of these periods for four different neurons. Two of them, cell 132 M and cell 132 B (Fig. 8C, D), were recorded simultaneously, and neurons 133 (Fig. 8A) and 160 (Fig. 8B) were recorded in other experiments. The period within which spike bursts arise in the sequences ranges from 200 to 700 msec. Thus, these neurons can determine synchronized rhythms at a frequency ranging from 5 oscillations to fractions of oscillations per second. However, in some neurons the frequency of group discharges may be considerably higher, as is shown in Fig. 7B for a cell, the majority of whose interspike intervals was less than 200 msec.

The difference between the histogram which shows the distribution of the periods of slow synchronous EEG waves and the histogram of intervals between the group spike discharges decreases appreciably, if the latter characterizes intervals formed by spike bursts not of one, but at least two neurons. Such a histogram for two neurons, recorded simultaneously by means of one microelectrode, is shown in Fig. 8D (cell 132 B and cell 132 M).

Correlation of Slow Wave and Spike Activity of Thalamic Neurons Recorded with a Single Microelectrode

Further investigation of the neuronal mechanism of synchronized slow waves, which are recorded in the EEG during the extinction of the orienting reflex, was performed by studying the specific features of the electrical activity of thalamic neurons. Particular attention was paid to the correlation of the spike and slow components of the electrical activity recorded from the thalamic structure by a single microelectrode. In some cases we recorded biphasic slow waves which usually emerge simultaneously with a spike burst of the neuron.

Figure 9 presents a record of the electrical activity of a single neuron with spike grouping. Each spike burst arises during the initial phase of the slow wave, after which a more protracted oscillation with an opposite sign develops.

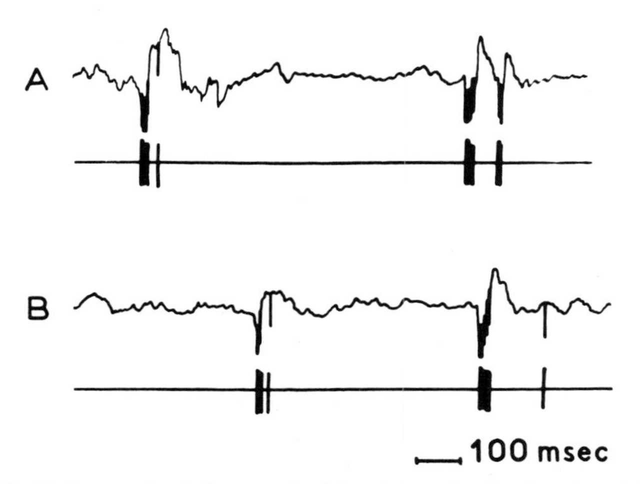

FIG. 9 (A, B) Fragments of the record of the slow and spike electric activity of a single thalamic neuron (cell 132 B) from a single microelectrode. Upper trace, slow wave and spike activity; lower trace, spike activity extracted from the upper record. Spike bursts correlate with slow waves. Single spikes do not bring about slow wave generation (B).

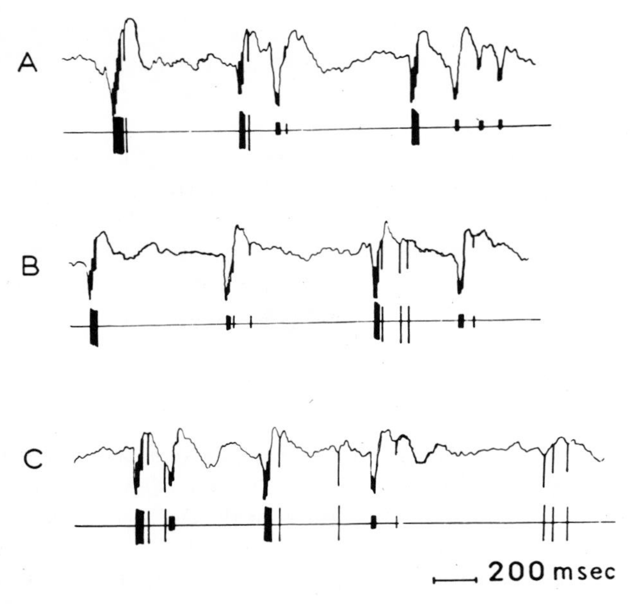

FIG. 10 Simultaneous recording of the slow and spike background activity of two neurons with one microelectrode (cells 132 B and 132 M).

Upper trace, total electric activity; lower trace, activity of two neurons differing in their spike heights: record of spike activity was obtained by filtering the total activity. (A, B, C) Fragments of records differing in relationship between the activities of the two neurons.

Consequently, the slow component contains biphasic slow waves, each having a definite duration. The development of the first phase is closely bound up with the emergence of a neuronal spike burst. The slow activity is formed of biphasic slow waves which appear as a result of spike bursts of each neuron of the given type.

Figure 10 contains a record of the background electrical activity of two neurons with spike grouping, recorded by means of one microelectrode. The spike burst of each neuron appears during the first phase of the slow wave, which is followed by an oscillation of the opposite sign. The interference of these biphasic waves of each neuron with an average duration of 100 msec produces the picture of slow synchronous waves of the thalamic structure. The intensification of spike grouping in the neuronal ensembles leads to an increase of the number of slow waves. New sensory stimulations eliminate both the spike bursts and the slow biphasic waves connected with them. They exert a similar action upon the neurons of the group described above, although various neurons may manifest different reactivity. Figure 11 shows a response of two neurons—simultaneously recorded by means of one microelectrode—to clicks (10 per sec). As can be seen, the spike grouping in both neurons is progressively eliminated during the stimulation. A stronger reaction is shown by the neuron whose

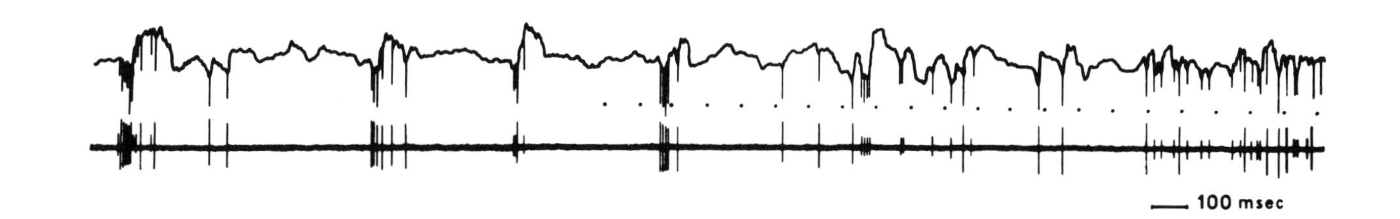

FIG. 11 Responses of two neurons (cells 132 B and 132 M) to sensory stimulation (clicks at a frequency of 10 per sec), recorded with one microelectrode.

Top trace, electrical activity of the neurons (including both the slow and spike components); Bottom trace, the spike component of the neuronal electrical activity. The clicks are designated by dots. It can be seen that the biphasic slow waves and the spike bursts connected with them disappear during click stimulation. A more intense response is given by the neuron with higher spike amplitude.

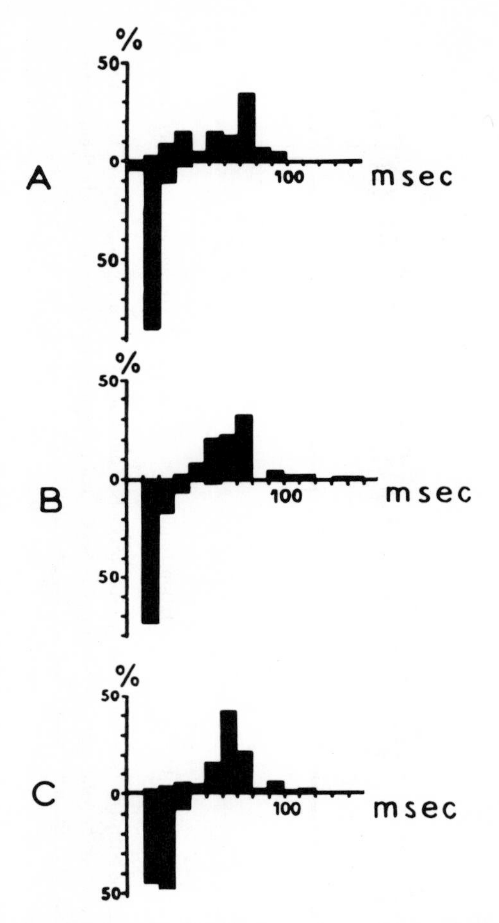

FIG. 12 Histograms showing the distribution of the duration of the first and second phases of the slow wave; the data are summarized for 3 cells (132 B, 132 M, 160) for cases when the slow biphasic wave correlates with a group of 2 spikes (A), 3 spikes (B), and 4 spikes (C). Abscissa, duration of the phases in msec. Ordinate, percentage of phases with different durations. The values of the first phase are plotted downward; those of the second phase are plotted upward. Each histogram is formed of 48, 106, and 112 values, respectively.

recording yields a higher amplitude of spike activity. During stimulation, all the biphasic slow waves, which are characteristic of a more restful state of the neuron, disappear, as do a part of the slow waves arising during group spike discharges of the neuron with low amplitude spike record (Fig. 11).

Neurons with spike grouping which are located close to one another, do not always discharge simultaneously, although they show a general dependence on the animal's functional state. In particular, either an intensification or weakening of the spike grouping takes place simultaneously in the entire neuronal ensemble. The intensity of the grouping, however, varies for different neurons.

Characteristics of a Biphasic Slow Wave Connected with the Emergence of a Neuronal Spike Burst

An analysis of a biphasic slow wave recorded with the aid of a microelectrode showed that its parameters (duration and amplitude) are stable. The duration of the initial phase is shorter than that of the second phase. Another characteristic is a definite increase in the duration of the first phase of the slow waves, with an increase in the number of spikes in the group.

Statistical characteristics of the duration of the phases of a slow wave depending on the number of spikes in the group are contained in the histograms of Fig. 12 (A, B, C), where summary results of three cells are presented. One can see that the duration of the first phases increases with the growth of the number of spikes in the group. In the cases when the slow wave correlates with a group containing 2 or 3 spikes, the predominant duration of the first phase is 20 msec (in 70–80% of the total number of the measurements performed–Fig. 12A, B). But when the group consists of 4 spikes, the duration of the first phase increases to 30 msec in almost 50% of all the cases analyzed. It is noteworthy that the increase of the duration of the first phase is accompanied by a shortening of the second phase; thus, the period of the biphasic slow wave on the average remains constant, equaling about 100 msec. At the same time, the stability of the period of the biphasic wave increases and its variability declines.

The stability of the period of the slow wave proved to be different for different cells. Thus, a period of 100 msec was particularly characteristic of cell 160 (46%) and was less often encountered in cell 132 B (23%). This is due, apparently, to the fact that cell 160 predominantly discharges with groups containing an appreciable number of spikes. Its groups usually consist of 4 spikes

TABLE 1
Percent of Groups Consisting of 2, 3 and 4 Spikes Recorded in the Background Activity of each of the 3 Investigated Neurons

	Number of groups (%)		
Number of spikes in the group	Cell 132 B	Cell 132 M	Cell 160
2	25	33	9
3	41	38	40
4	34	29	51
Total:	88	42	136

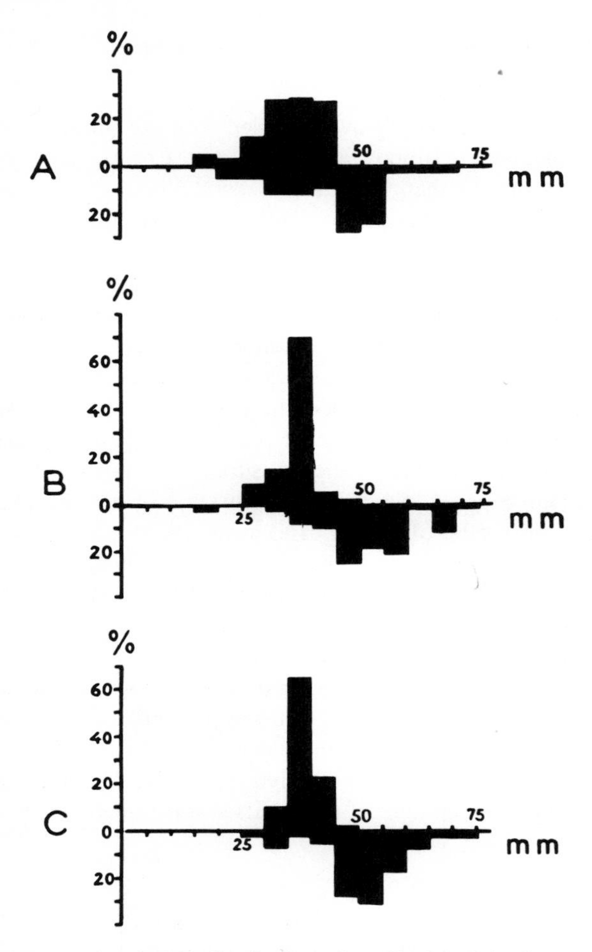

FIG. 13 Histograms showing the distribution of amplitudes of the first and second phases of the slow wave and their dependence on the number of spikes in the group.

(A, B, C) for slow waves which correlate with groups of 2, 3, 4 spikes. The data have been summarized for two cells (132 B, 132 M). It may be seen that with the growth of the number of spikes in the group, the amplitude of the first and second phases increases and the variability diminishes (the amplitude values are given in conventional units). Abscissa, value of the amplitude of the first and second phases. Ordinate, percentage of each amplitude (the values of the first phase are plotted downward; those of the second phase are plotted upward).

(see Table 1). And, as seen above, the stability of the period of a biphasic slow wave increases with the growth of the number of spikes in the group.

Not only the temporal characteristics but also the amplitudes of the first and second phases of the slow wave changed with the increase of the number of the spike discharges in the group. The growth of the number of spikes in the group leads to an increase in the amplitude of the phases of the slow wave. This dependence is shown in Fig. 13; the data are summarized for cells 132 M and 132 B. The slow wave, which correlates with groups of 4 spikes, has a higher amplitude than slow waves arising with groups consisting of 2 and 3 spikes. Both phases become more stable in amplitude; variability declines with the growth of the number of spikes in the group.

DISCUSSION

This investigation revealed, in unanaesthetized rabbits, a relationship between the emergence of a sequence of spike bursts of the thalamic neuron and the appearance of slow waves in the EEG of the visual cortex, with a high positive correlation between these bioelectrical phenomena. These properties are peculiar to a large group (50%) of the thalamic neurons.

Our data partially coincide with the results obtained by other investigators from anaesthetized animals (Andersen & Andersson, 1968; Negischi, Bravo, & Verzeano, 1963). These authors also failed to detect a high correlation between thalamic and cortical activity: in a number of cases the thalamic discharges, characterized by spike bursts, were of a shorter duration than the cortical slow waves, and the spike bursts emerged out of phase with the cortical slow waves. Our results also show that during extinction of the orienting reflex as a result of repeated applications of the same stimulus, the spike grouping of thalamic neurons becomes intensified in unanaesthetized rabbits. Along with the increase of spike grouping, there occurs an intensification of the slow waves in the electrical activity of the visual cortex, hippocampus, and lateral thalamus, which begin to form sequences, as do the spike groups of the thalamic neurons. During the presentation of additional sensory stimulation, the spike bursts, which characterize the neuronal background activity, disappear and are replaced by single spikes. Simultaneously, the slow waves disappear.

Like Andersen and Andersson (1968), we detected a correlation between spike bursts and a slow wave simultaneously recorded by means of a microelectrode from thalamic neurons. However, whereas according to our data, the duration of the biphasic wave was similar for different neurons and equalled about 100 msec, according to the results obtained by Andersen *et al.*, the period of the wave amounted to 120 msec; this longer period may be explained by the effect of the anaesthesia applied by the investigators in these previous experiments.

It may be assumed that the thalamus is the site of the pacemaking mechanism which determines the slow, synchronous rhythms of sleep recorded in the EEG of the visual cortex and other brain structures (Andersen & Andersson, 1968). From this viewpoint, neurons with spike grouping are neurons of the inhibitory, inactivating ("moderating") system which is localized in the thalamus (Tissot & Monnier, 1959; Monnier, Hösli, & Krupp, 1963). The intensified grouping of their discharges is a manifestation of this system, whose inhibitory influences, according to Monnier *et al.* (1963), are widely spread, and, in particular, regulate the excitability of the neurons of the motor cortex.

Most investigators regard the synchrony of the discharges in a neuronal group as an indispensable condition for the emergence of a slow wave. Thus, Laufer and Verzeano (1967) believe that a correlation between slow wave and spike discharges can be revealed only by simultaneous recording of the activity of several neurons.

According to our data, spike bursts associated with slow waves recorded simultaneously in some neurons by means of one microelectrode in an unanaesthetized rabbit are not necessarily synchronized. The high synchrony in the generation of spike bursts in a neuron ensemble is inherent for the anaesthetized animal with barbiturate spindles in the EEG.

In the course of experiments on unanaesthetized rabbits, we never recorded thalamic spindles formed by slow waves that regularly follow one another with a frequency of 8 to 12 per sec and which correlate with spike bursts, as demonstrated in the work of Andersen *et al.* (1964). Complexes in the shape of a biphasic wave with a spike burst emerging during the first phase, followed one another, and formed a sequence without merging and without producing a typical "spindle" pattern. The period of thalamic slow waves forming the sequence (recorded by microelectrode) as a rule was not constant and varied over a broad range. In our experiments EEG spindles were composed of slow waves with a period of 100–300 msec.

The absence of synchrony in the generation of spike bursts of these neurons during the emergence of slow waves in the electrical activity of the unanaesthetized rabbit indicates that the mechanism of their generation differs from that suggested by Andersen *et al.* (1964). These authors believe that the synchrony of discharges in an ensemble of neurons which exhibit spike grouping lies at the base of the emergence of rhythmical processes in the EEG. This synchrony is based on a simultaneous modulation of their excitability by internuncial inhibitory neurons, and the emergence of a postexcitatory phase characterized by a low threshold.

It is possible that the absence of spindles and synchrony in spike bursts is due to many influences which are more effective in the unanaesthetized than the anaesthetized rabbit.

The absence of synchrony in neuronal spike bursts may be explained in another way. Different neurons with spike grouping may be connected with various cortical areas, and consequently may be responsible for different cortical rhythms. Thus, numerous independent rhythms may simultaneously exist in the cortex (Bekhtereva, 1965).

Also, Andersen *et al.* (1967), demonstrated the controlling role played by groups of thalamic cells in the generation of electrical activity in small cortical areas to which these cells project. The spindlelike activity recorded within the cortical targets of these projection systems exhibited complete synchrony both in its emergence and disappearance, as well as in the frequency and phase of separate potentials within each spindle.

According to our data, synchronization of the discharges of each neuron characterized by a sequence of spike bursts becomes eliminated as result of sensory stimulation: The groups disappear and are replaced by single spikes. This effect, which may be designated as a reaction of desynchronization, possesses all the properties of a nonspecific activation reaction—one of the components of the orienting reflex.

It may be assumed that the reaction of desynchronization of spike activity is one of the neuronal mechanisms of the arousal response. It exists independently, along with the reactions of acceleration and deceleration of the neuronal spike activity, which are usually regarded as neuronal correlates of the activation reaction (Kogan, 1966; Skrebitsky, 1966; Sokolova, 1966; Vinogradova, 1965).

15

THE DESYNCHRONIZATION REACTION AND ITS EXTINCTION IN THALAMIC NEURONAL SPIKE ACTIVITY

N. N. Danilova

Moscow State University

Research into the electrical activity of individual neurons of the nonspecific thalamus is of particular interest from the point of view of the orienting reflex. According to Sharpless & Jasper (1956), the nonspecific system of the thalamus constitutes the morphological substratum which ensures the organization of selective attention.

So far the neuronal activity of the non-specific thalamus has not been studied in connection with the problem of attention. We know only of a short communication by Jasper (1964) that, during a surgical operation on the human brain, some neurons in medial thalamic nuclei exhibited habituation.

We shall consider here the results of studying both spontaneous activity and sensory-evoked responses in neurons of the rabbit's nonspecific thalamus in connection with the ability of these neurons to manifest the habituation effect.

METHODS

The investigation was carried out on 24 unanesthetized rabbits. During the experiments, the animals were fixed in a special stand. Recording of the neuronal activity was performed extracellularly with tungsten microelectrodes (a more detailed description of the methods applied is given in the preceding chapter of this volume).

Stereotaxic coordinates were determined according to the atlas of Fifkova and Marsala (1962) with the amendment of Meschersky and Chernyschewskaya (1959).

The position of the electrode in the brain was controlled histologically, by locating a small lesion made by the passage of current through the electrode. The following nuclei of the thalamus were investigated: n. reticularis, n. ventralis lateralis, n. ventralis medialis, n. medialis dorsalis, and n. lateralis posterior.

The aim of the experiments was to study the changes which take place in the neuronal responses to repeated applications of a 5-sec sensory stimulus. Acoustic clicks at a frequency of 10 per sec were most often used, although clicks of other frequencies (15 per sec or 4–5 per sec) were applied as extra stimuli. Two or three such disinhibiting stimuli were tested during each experiment.

In the course of some experiments, the EMG of the rabbit's fore paw, as well as the rabbit's respiration, was recorded on an 8-channel electroencephalograph. Special experiments were devoted to investigating the responses of the non-specific thalamic neurons to the protracted (30 sec) action of stimuli of different sensory modalities. For this purpose, the following kinds of stimuli were used: clicks and light flashes at frequencies of 4, 10, or 15 per sec from an FFS-02 photo-phonostimulator, as well as air puffs on the surface of the head and hind paw.

Finally, in five experiments the effect of Nembutal on the spontaneous activity of the neurons and on their responses to sensory stimulation was studied. Nembutal was injected intraperitoneally at a dose of 20 mg/kg.

RESULTS

Investigations of the neuronal activity of various thalamic nuclei made it possible to classify all the neurons studied into four groups on the basis of their spontaneous activity and particular responses to sensory stimulation (Danilova, 1966).

The first group consisted of neurons whose spontaneous activity was characterized by a high frequency of spike discharges, which emerged in a random sequence without forming any specific pattern. Acoustic stimulation led to an increase in the frequency of the discharges, whereas light stimulation usually proved ineffective.

The second group consisted of neurons with very rare spike discharges. Sensory stimulation usually exerted no influence on the activity of these neurons, or produced inhibition of whatever spike discharges were present.

The third group included neurons whose spontaneous activity consisted of groups of spikes and single spikes. These neurons were driven by rhythmic light and sound stimulation.

Neurons belonging to the fourth group were conventionally called by us "neurons with spike grouping." This group included 50% of the 130 neurons studied. We shall consider here the peculiar features of activity only of this last group of neurons.

The Background Activity of the Neurons and Its Relation to the Animal's Functional State

The neurons of this group showed well-defined spontaneous activity. The character of the spontaneous discharges depended on the animal's functional

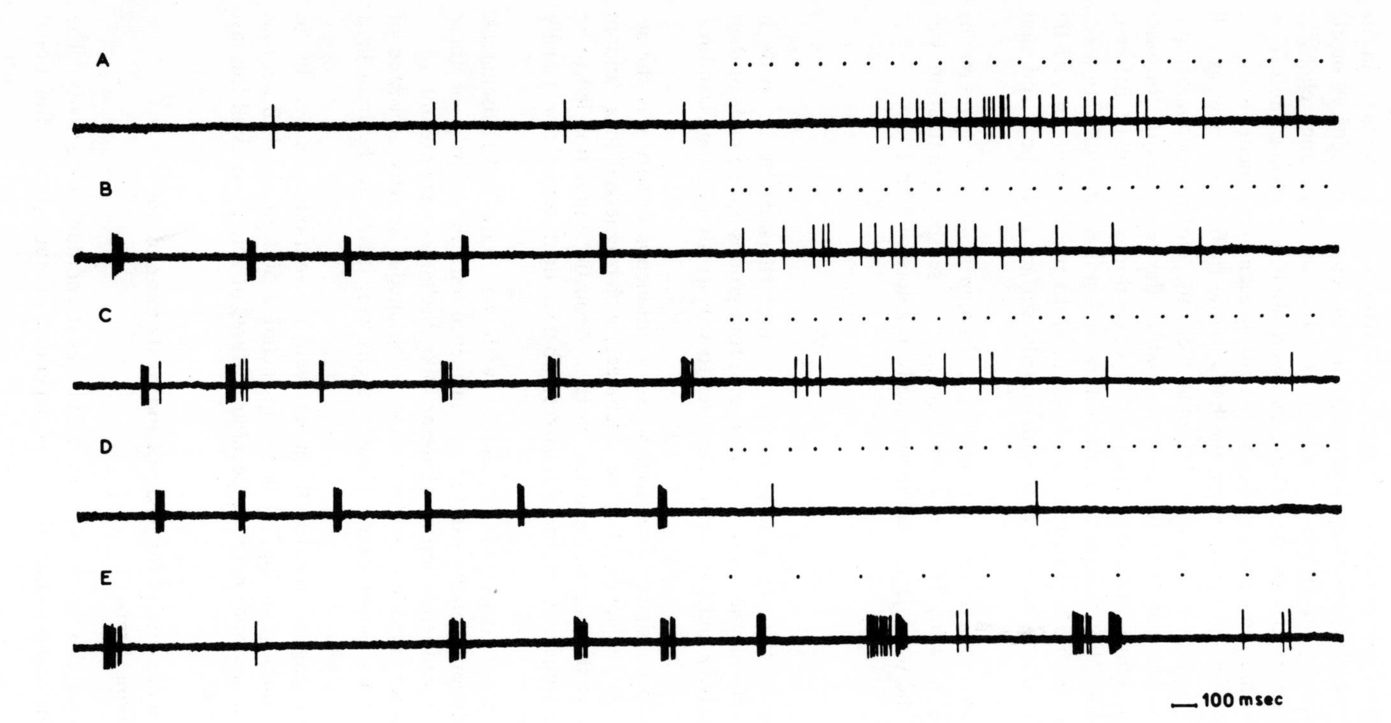

FIG. 1 Types of neuronal responses to sensory stimulation and their dependence on functional state of the animal: (A) during wakefulness, note absence of spike bursts (B–E) during sleep. During sleep, sensory stimulation generally eliminates spike grouping, replacing it with single spikes at high frequency (B), low frequency (C); or almost no spikes at all (D). Sometimes, bursts of spikes were not eliminated, but actually increased in their duration (E).

state. We could judge the rabbit's increasing excitation (arousal) by its more frequent spontaneous motor responses, which were reflected in the EMGs of the forepaw or even in motor artifacts. In these conditions, as a rule, the neuron discharged with irregular single spikes. The more intense the animal's spontaneous motor excitation, the higher the frequency of the single spike discharges of the neuron.

As the animal became calm and more accustomed to the experimental conditions (number of motor reactions showing a decrease), spike bursts began to appear in the neuronal spontaneous activity, along with single spikes.

Neuronal Responses to Sensory Stimulation Depending on the Animal's Functional State

Neurons with spike grouping exhibited a specific type of response to sensory stimulation; spike bursts characterizing the neuronal background activity usually disappeared and were replaced by irregular single discharges (Fig. 1B, C, D). The frequency of single spike discharges varied widely. Figure 1B shows when the stimulus led to the elimination of spike bursts and evoked a distinct increase in the frequency of single spike discharges. In other cases the same stimulus, which likewise eliminated the spike bursts, exerted an insignificant influence on the frequency of single spikes (Fig. 1C). When single discharges were absent in the background activity, the neuron often responded with a disappearance of the spike bursts (Fig. 1D) without an increase in single spikes.

It is noteworthy that the neuronal responses, which are presented in Fig. 1B, C, and D, and which differ only in the frequency of the emergence of single spikes during stimulation, were recorded from the same neuron in the course of an experiment with repeated stimulus presentations. These responses were recorded at the first, ninth, and twelfth applications of a 5-sec stimulus (10 per sec clicks). Sometimes the sensory stimulation does not fully inhibit the spike bursts in the neuronal background activity. The bursts persist, but undergo a certain transformation from compact bursts consisting of 2–5 spikes to more extended groups containing up to 12 spike discharges (Fig. 1E).

The emergence of the neuronal responses described above characterized by suppression of spike bursts, led to the natural conclusion that such responses may take place only in the presence of spike bursts in the background activity. But if they were absent in the spontaneous neuronal activity–which, as a rule characterized the animal's state of wakefulness–then the neuron responded to the sensory stimulation with an increase in the frequency of single spike discharges (Fig. 1A). Usually this response was not intense, and often was even absent.

Thus, the response of neurons of this group to sensory stimulation is formed of two components. The first is a suppression of the spike bursts; the second is an increase of the frequency of single spike discharges during sensory stimulation. Sometimes the response is expressed by the emergence of single spikes which were previously absent in the background activity. Very often both components of the neuronal response manifest themselves simultaneously. However, the elimination of spike bursts may occur without an increase in single spikes (Fig. 1D). Apparently, the suppression of spike bursts is an indispensable condition which makes possible the manifestation of the second component of the response.

The latency of the neuronal response was of a particularly distinct character when a large number of spike bursts were observed in the background activity of the neuron; they persisted for a period of 0.1–0.5 sec after the stimulus had been switched on.

Relationship of the Neuronal Response to the Modality of Sensory Stimulation

All 65 neurons of this group responded to acoustic stimulation (clicks), exhibiting the following characteristic feature: Spike bursts were replaced by single spike discharges. Similar responses were evoked by other kinds of stimulation. The application of air puffs proved particularly effective. The response to such stimulation was investigated in 34 of 65 neurons. In 23 neurons it was the same as in the case of acoustic stimulation; only 4 neurons responded to the air puffs with inhibition of single spike discharges while 6 proved to be nonreactive. In 26 neurons, a typical response emerged also during the animal's spontaneous movements. It must be borne in mind that in the course of some experiments we judged the presence of motor responses from the emergence of motor artifacts; this means that all cases of correlation between weak spontaneous motor and neuronal responses were not taken into account and were not included in the above number.

Thus, the character of the neuronal responses to acoustic, cutaneous, and proprioceptive stimulation was similar: Spike bursts were replaced by single neuronal discharges.

The effect of visual stimulation (flashes of varying frequencies) on the neuronal activity was somewhat different. The response to flashing light was investigated in 19 neurons. Eight of these were corresponsive. As to the reactive neurons, they formed two equal groups: One group (26%) responded to light in the same way as to stimulation of other modalities; the second group (26%) manifested an opposite effect. As a result of photic stimulation, the spike bursts in this group became intensified. The neurons which did not manifest any spike bursts in their background activity began to discharge such bursts during stimulation by flashes of light.

Consequently, most of the neurons proved not to be modality specific with respect to acoustic, cutaneous, and proprioceptive stimulation; some displayed the same type of response as to light stimulation. However, photic stimulation is usually ineffective, or it evokes the reaction of inactivation.

Most of the investigated cells were multisensory: 29 of 65 neurons responded to stimulation of two modalities and 14 to stimulation of three modalities. It

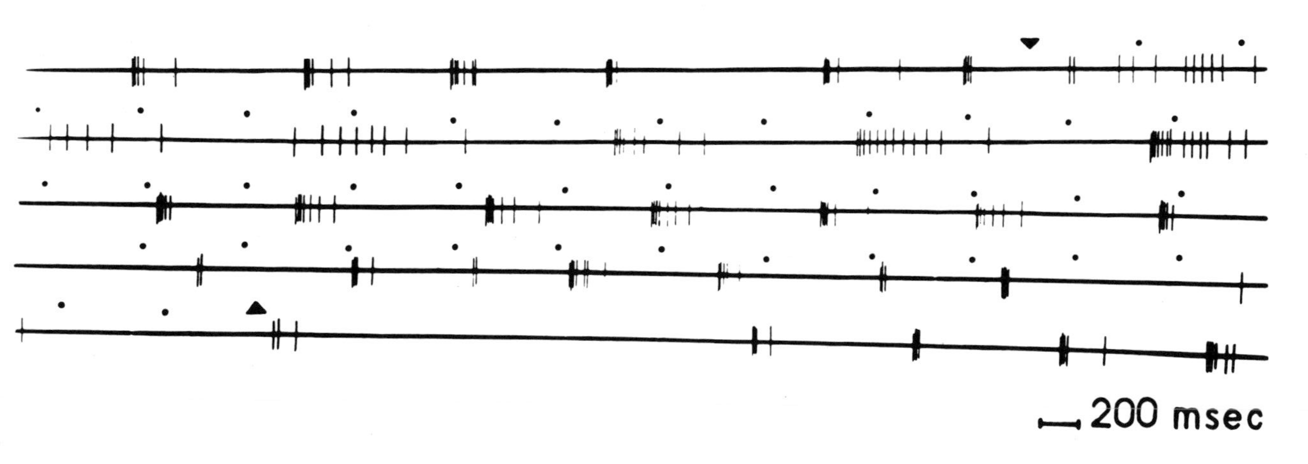

FIG. 2 Decrease of the neuronal response of cell 96 during continuous protracted (30 sec) stimulation. The arrows designate the beginning and end of click presentation at a frequency of 4 per sec (dots). The records are continuous from top to bottom.

must be taken into consideration that not every neuron was subjected to all kinds of stimulation.

Neuronal Responses to the Protracted Presentation of a Sensory Stimulus

The response of neurons with spike grouping to acoustic, cutaneous, and occasionally to photic stimuli usually appeared at the onset of stimulus presentation. If the response persisted during the entire period of stimulation, it proved more intense at the beginning than at the end. In other words, the phenomena of elimination of spike bursts and increased frequency of single spikes usually characterize only the initial period of stimulation; at the end of stimulation a decline in the frequency of single spike discharges and a recovery of the spike bursts are observed.

Sometimes this property became apparent immediately after the first presentation of the stimulus, particularly if it was of sufficiently long duration. Thus, in all cases when a stimulus with a duration of 30 sec was applied, there could be observed a weakening of the response during the action of the stimulus (Fig. 2). From this figure it can be seen that clicks at a frequency of 4 per sec eliminate the spike bursts. At the beginning of stimulation single spikes are randomly distributed, but within 1 sec they begin to join into extended groups. As the stimulation continues, they become more and more compact and the number of spikes per group decreases. Gradually the neuronal spike activity assumes the pattern of the background discharges of this neuron, although the acoustic stimulation persists. It is noteworthy that the weakening of the initial response and the subsequent recovery of the grouping, as a rule, pass through a stage of emergence of extended, protracted bursts containing up to 12 spikes.

Neuronal Responses to Repeated Stimulus Presentations

The emergence of a neuronal response only to the onset of stimulation manifested itself with particular clarity in experiments with repeated stimulus presentations. With the repetition of the stimulus, the initial response became shorter, owing to an earlier recovery of the spike bursts. Repeated application of the stimulus could lead to a full disappearance of the response. In this case, during the whole period of stimulation, the neuron discharged with only spike bursts. However, an extra stimulus could intensify or restore the response of abolition of spike bursts. A subsequent repetition of the stimulus again led to a weakening of the response.

Figure 3 shows a typical case of a decrease of the neuronal response as a result of repeated application of a 5-sec acoustic stimulus (clicks at a frequency of 10 per sec). The neuronal response was measured in millisec according to its first component, i.e., elimination of the spike bursts. The duration of this response was determined by the period from the onset of stimulation or from the preceding spike burst, which was recorded during the latency of the response to

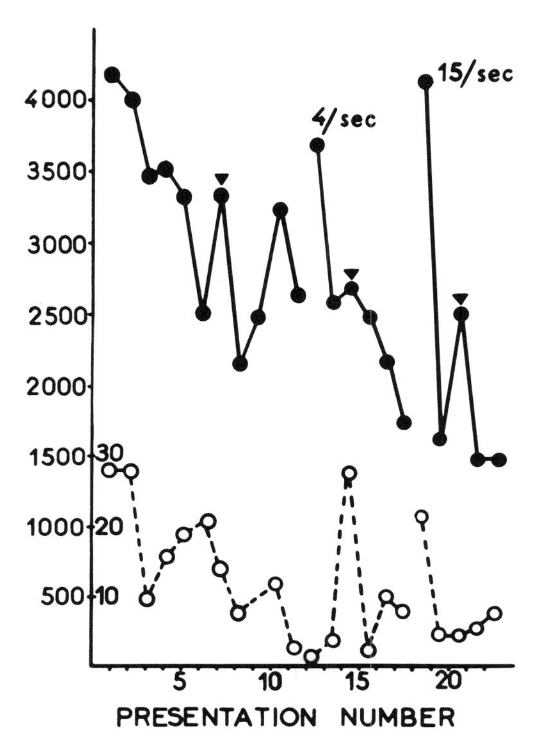

FIG. 3 Extinction and disinhibition of the neuronal response (cell 96) in an experiment with repeated presentations of an acoustic stimulus (clicks at a frequency of 10 per sec). Duration of the acoustic stimulation, 5 sec; duration of the interstimulus interval, 15 sec. The continuous line designates the duration of the first component of the response—elimination of spike bursts. The dashed line designates the total number of single spikes. Ordinate, duration of the first component of the response in msec, and number of single spikes in the second component; abscissa, serial number of stimulus presentation. The point at which clicks of one frequency are replaced by clicks of another frequency is indicated (4 per sec, 15 per sec). The arrows show the occurrence of strong motor artifacts.

the emergence of the first spike burst at the end of the stimulation. The magnitude of the second component (emergence of single spikes) was determined by the number of single spikes during the period when the spike bursts were absent. After 11 applications of the acoustic stimulus the duration of the first component of the response decreased from 4.1 to 2.6 sec. However, any changes in the stimulus (in particular, a shift in its frequency to 4 per sec or 15 per sec) resulted in an increase of the duration of the response. (A similar effect was produced by the animal's movements. In Fig. 3 the moments of movement are designated by arrows.) The presentation of clicks after the extra stimuli which increased the duration of the response, again led to its decline. The process of this decline developed more rapidly than at the beginning of the

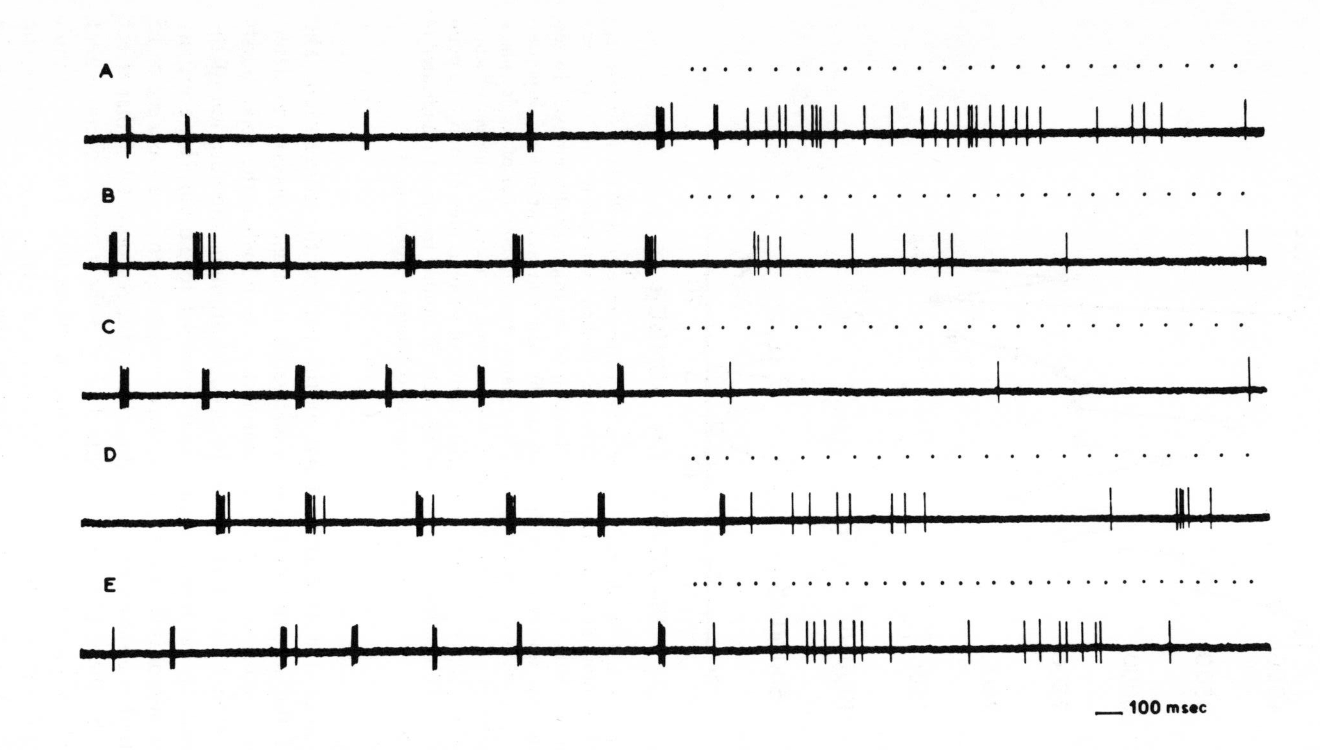

FIG. 4 Extinction and disinhibition of the neuronal response (cell 96) to clicks. (A, B, C, D) 2nd, 9th, 11th, and 17th presentations of clicks at a frequency of 10 per sec. (E) neuronal response to a different stimulus (clicks at a frequency of 15 per sec) applied instead of the 18th presentation of the previous clicks. The stimuli were presented at constant intervals of 15 sec. The dots designate click presentation.

experiment. The second component of the response, i.e., an increased frequency of single spikes emerging during the elimination of the spike bursts, also showed a decline with stimulus repetition. However, this component of the response decreased more rapidly than the elimination of spike bursts.

Thus, there is not complete correspondence between the changes of both components of the response to sensory stimulation. On the one hand, an increase in the duration of the period of spike burst absence does not necessarily signify a rise in the frequency of the single discharges. On the other hand, a growth of the number of single spikes may take place without any increase of the duration of the elimination of spike bursts. This creates the impression that although both components of the response can change in parallel fashion, exhibiting a decrease of stimulus repetition they are to some degree independent. The second component of the response, i.e., the increased frequency of single spike discharges, reaches a zero value more rapidly, while the first component of the response, i.e., the elimination of spike bursts, persists for a longer period of time.

Figure 4 presents a record of the spike activity of a neuron (cell 96) during the 1st (A), 9th (B), and 11th (C) applications of 10 per sec clicks. A gradual decrease of the response can be seen clearly. The number of single spikes, which replace the spike groups of the background activity, steadily declines until such spikes are no longer recorded, even though the stimulus continues to suppress the spike bursts. The 17th application of the 10 per sec clicks (Fig. 4D) leads to a spontaneous disinhibition of the second component of the response: Single spikes appear during stimulation. Presentation of a different stimulus (clicks at a frequency of 15 per sec) evokes a still stronger response (Fig. 4E).

The entire process of a decrease or recovery of the response of a given neuron develops against the background of its constant functional state; the frequency of spike bursts (about 2 bursts per sec) in the neuronal background activity remains constant during the whole experiment. Also invariable is the frequency of the emergence of single spike discharges, which ranges from 0.5 to 1.5 spikes per sec.

Protracted experimentation with repeated presentations of the same stimulus, however, leads to a change in the background activity of the neuron: The spike grouping becomes intensified and the frequency of spike bursts in the neuronal spontaneous activity increases. In these conditions, the extinction of the neuronal response to sensory stimulation continues to develop in parallel with the change in its spontaneous activity. The occurrence of this activity (spike grouping) depends on the orienting reflex; intensification of the orienting reflex, caused by a change in the habitual course of the experiment, leads to a decrease and elimination of spike bursts.

Effect of Nembutal Anesthesia on the Neuronal Responses

The dependence of the character and magnitude of neuronal responses on functional state was confirmed by experiments utilizing 20 mg/kg of Nembutal

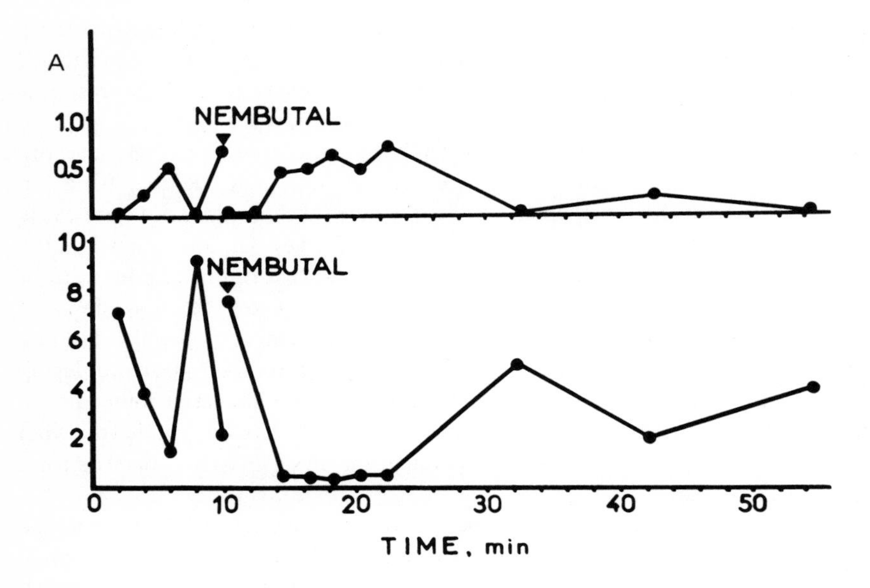

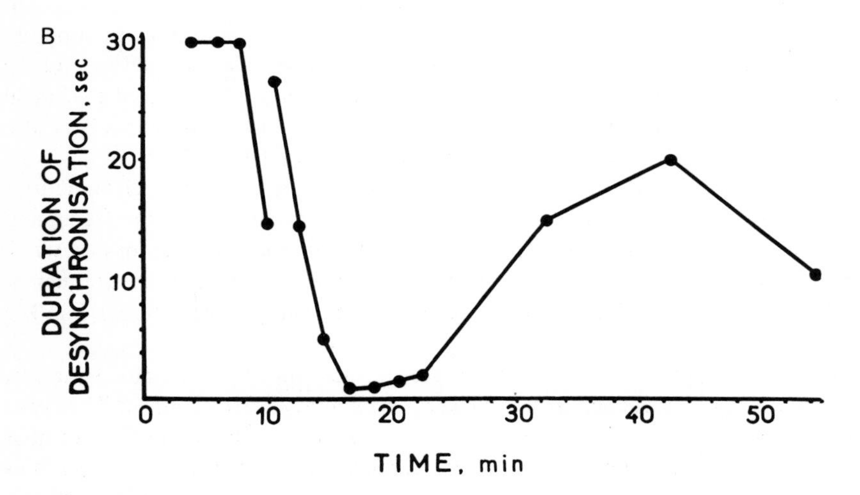

FIG. 5 (A) Change of background neuronal activity (cell 65) under the influence of Nembutal: ordinate, above, frequency of spike bursts, below, average number of single spikes per second; abscissa, time, in min. The injection of Nembutal is marked by an arrow. (B) Change of the value of the neuronal response (cell 65) to clicks at a frequency of 10 per sec (duration of the stimulus, 30 sec) under the influence of anaesthesia. Nembutal was injected at the•10th minute. Ordinate, duration of the first component of the response–elimination of spike bursts ("desynchronization"), in seconds; Abscissa–time, in min. The magnitude of the "desynchronization" response was measured as the sum total of intervals containing no spike bursts during each stimulus. Each of these intervals had to be at least 1 sec in duration in order to be counted.

injected intraperitoneally. During these experiments, the electrical activity of the neurons was recorded continuously—before, during, and after the administration of the drug.

Figure 5A shows a typical case of a change in the neuronal background activity under the influence of Nembutal. It can be seen that anaesthesia intensifies the spike grouping in the background activity of the neuron: The frequency of spike bursts increases, and simultaneously a decrease of the number of single spike discharges is observed. During approximately the fourth minute after the injection of Nembutal, the number of single spikes in the neuron's background activity dropped to the minimal level (about 0.5 spikes per sec). At the same time, the number of spike bursts in the background increased. The anaesthesia proved to be weak, however, and 18 min later a recovery of the initial background activity was observed.

During the period when the number of single spikes in the background activity of the neuron decreases, its response to sensory stimulation usually weakens (Fig. 5B). This figure shows a diminution of the neuronal response during the fourth minute after the administration of Nembutal, indicated as a decrease of the period within which the spike bursts are eliminated. Then, the neuronal response was practically absent for approximately 7 min. When the effect of Nembutal subsided, the initial background rhythm recovered, and at the same time the response reappeared for 10 sec or more.

Thus, Nembutal anaesthesia eliminates single spike discharges in the neuronal spontaneous activity, intensifies the spike grouping, and makes the spike bursts resistant to sensory stimulation.

DISCUSSION

A Two-Component Character of the Response

The response of the investigated neurons to sensory stimulation consists of two components. First, sensory stimulation eliminated neuronal spike bursts. This is an indispensable condition for the development of the second component, an increase in the frequency of single spikes.

Different methods of data processing were used for measuring the values of the two components of the response.

Thus, in order to establish the value of the first component, i.e., the elimination of spike bursts, the duration of this elimination was determined, while for evaluating the second component, i.e., the increase in the frequency of single spikes, it was necessary to determine their total number for the same period of time.

These changes in spike discharges are evident to some degree in histograms of interspike intervals.* The most typical change in the histogram, which arose

*Not shown in this chapter—(Ed.).

under the influence of sensory stimulation, was a shift of the interspike intervals toward a predominance of the longer ones. This reflected the very fact of elimination of the spike bursts, since the shortest interspike intervals were observed with the bursts. The increase of the number of longer interspike intervals and the decrease of the number of the shortest ones were often accompanied by a rise in the average frequency of the neuronal spike discharges. All this testifies to the existence of different influences on neurons of this type and to their complex interaction on this neuronal level.

Reaction of Desynchronization

Of particular importance for comprehending the peculiarities of the responses of neurons of this group is their first component. The very fact of the grouping of spike discharges and of their merging into sequences of several groups, as well as the correlation of these groups with series of slow waves, which are recorded in the cortex and in other brain structures (Danilova, 1966), apparently indicates that these neurons belong to a special synchronizing system. From this point of view, the elimination of groups may be designated as a reaction of desynchronization.

Similar data are contained in the works of Hori *et al.* (1963) and Hori and Yoshii (1965) who detected, in the encéphale isolé preparation of the cat, in thalamic nuclei (CM and VA), a special type of neuron which responds to electrical stimulation of the reticular formation of the midbrain with an elimination of spike bursts.

In a number of other investigations the phenomenon of grouping of the neuronal spike discharges was revealed in almost all the thalamic nuclei (Verzeano & Negischi, 1960; Negischi & Verzeano, 1961; Negischi, Bravo, & Verzeano, 1963; Andersen & Sears, 1964; Andersen, Brooks, Eccles, & Sears, 1964; Andersen, Eccles, & Sears, 1964). However, the desynchronization reaction was not detected. As a result of single electrical stimulations of the sensory nerves, or of the contralateral structures of the thalamus, the above-mentioned authors usually obtained an effect opposite to synchronization which was expressed in the emergence of a sequence of spike bursts. We observed the synchronization reaction in response to photic stimulation in only a small number of neurons. The divergence between our data and the data of others is apparently due to different effects produced by sensory and electrical stimulations.

It may be assumed that the extinction of the desynchronization reaction accompanied by an intensified spike grouping in the neuronal activity of the thalamus is connected with inhibition of the non-specific activating system of the midbrain, which develops during the extinction of the orienting reflex. A similar effect is observed also in the case of Nembutal anaesthesia. The blocking of the ascending activating influences contributes to the work of the synchronizing system of the thalamus, revealing some specific properties of its neuronal net.

A strong excitatory process, which is evoked by sensory stimulation, interrupts the work of the synchronizing mechanism, and calls forth the desynchronization reaction. Proceeding from the concepts of Andersen and others concerning the specific role of the internuncial, inhibitory neurons, which effect recurrent inhibition of neurons with spike grouping and thereby make possible their group discharges, it may be assumed that sensory stimulation suppresses precisely these neurons. The existence of such a mechanism was shown by Szentagothai (1967) for Renshaw cells.

The Nonspecific Character of the Neuronal Responses

The reactions of neurons of the given group exhibited all the attributes of nonspecific responses. They did not form any definite pattern in the sequence of spike discharges. During stimulation single spikes, which replace the spike bursts, appeared in a random order.

At least two types of stimulation converged on each neuron of this type—acoustic and cutaneous. In many cases certain movements of the animal, and sometimes photic stimulation, proved to be effective. In other words, the neurons of this group were multisensory.

Stimulation of different modalities evoked the same type of response, namely, a two-component response which included the elimination of spike bursts and an increase of the frequency of single spikes. The responses could differ only in the duration of the periods during which the spike bursts were absent, and in the magnitude of the increased frequency of single neuronal discharges.

Responses to visual stimulation constituted an exception. In some neurons, a response of the nonspecific type emerged, but with an opposite sign: synchronization of spike activity, and intensification of spike grouping. This response appeared with a long latency—from 0.1 to 0.5 sec.

The nonspecific character of the desynchronization reaction is also indicated by its predominant emergence at the moment of stimulus onset. At the end of stimulation, the intensity of the response (including both of its components) decreases. This is expressed by the fact that the highest frequency of single spikes is recorded at the beginning of stimulation. The decrease of the frequency of single spike discharges by the end of the stimulation coincides with the recovery of their grouping. The neuronal response to sensory stimulation showed other properties which are common with those of the orienting reflex. Thus, with the repetition of the stimulus, the response weakens, but can be restored by a different stimulus (in our experiments, by clicks of some other frequency). The process of extinction and disinhibition is most distinctly expressed in the elimination of spike bursts.

The detection of neurons with these properties in various nuclei of the thalamus apparently testifies to the existence of a single mechanism which ensures the reactions of activation and inactivation in the thalamus, as well as its particular role in the development of cortical nonspecific responses.

PART C
Overview and Hypothetical Schema

This section contains a single chapter by Sokolov. It is noteworthy both for its organization of diverse data and precise formulation of the circuitry and dynamics of brain events which are thought to underly elicitation and habituation of the orienting reflex. The schema (model) which he proposes here represents a large advance over the model which he first introduced to Western science in 1960.* At that time, it was hypothesized that the sensory cortex prevented elicitation of the OR as habituation progressed partly by presynaptic inhibition of sensory terminals projecting to the reticular formation. The present model features, among other structures, the hippocampus, and places it as an intermediary between the analyzers and the reticular activating formation. Thus the advance of data necessitates revision of theory, and Sokolov has never been one to hold a theory sacred. Characteristically, his present theorizing not only accounts for much known data, but points the way for future experimentation as good theory must. Sokolov remains one of the most perceptive and exciting brain-behavior theorists of the day.

*Sokolov, E. N. Neuronal models and the orienting reflex, in Brazier, M. A. B. (Ed.), *The central nervous system and behavior, Transactions of the Third Conference.* New York: Josiah Macy Jr. Foundation, 1960. Pp. 187–276.

16

THE NEURONAL MECHANISMS OF THE ORIENTING REFLEX

E. N. Sokolov

Moscow State University

1. Introduction

Polygraphic recording of somatic, vegetative, and electroencephalographic responses in man and animals have made it possible to differentiate the defensive, orienting, and adaptive reflexes by means of repeated applications of stimuli of different modalities and intensities. The adaptive reflex, being a local defensive reflex, is characterized by the following features: a relatively high threshold, specific type of response, local reflexigenic zone, different direction of changes taking place when the stimulus is switched on or switched off, and high stability when the stimulus is repeated. The defensive reflex, which is a reflex of the whole organism, differs from the adaptive one by an extensive generalization of the response, as well as by a still higher threshold and ability to arise as a result of stimulations of different modalities.

The orienting reflex is characterized by a low threshold, extensive generalization of the excitatory process, absence of any specific reflexogenic zone, uniform responses to the stimulus when it is switched on or switched off, and development of extinction which proves to be selective with respect to the parameters of the repeated stimulus.

In order to explain the selective extinction of an orienting reflex, with regard to the parameters of the signal, the concept of a "neuronal model of the stimulus" was introduced. This is a concept of a trace which registers the properties of the stimulus applied. The orienting response is thought to result from the emergence of a "disconcordance signal" ("mismatch") when the external signal does not coincide with the "neuronal model of the stimulus." Within certain limits the orienting response is more intensive the greater is the difference between the established "neuronal model of the stimulus" and the parameters of the acting stimulus (Sokolov, 1960).

Whereas the repeated signal is characterized by some stable properties, e.g., color, intensity, and localization in space, the "neuronal model of the stimulus" fixes all the parameters simultaneously. The magnitude of the orienting response increases proportionally to the number of the parameters of the signal which are simultaneously changed.

The "neuronal model of the stimulus" registers not only the elementary, but also the complex properties of the signal, such as coincidence or succession of several stimuli in time. This is proved by the emergence of an orienting response when one of the elements of the complex stimulus is excluded, or when the sequence of the elements in the complex is changed.

The selectivity of extinction relates also to temporal characteristics, to the duration and interval between the stimuli. A change of these characteristics also leads to the emergence of an orienting response.

The orienting response occurs not only under the action of indifferent, but also signalling stimuli. In the latter case the extinction of the orienting reflex is accompanied by a stabilization and consolidation of the conditioned reflex.

The extinction of the orienting reflex may take place against the background of a constant functional state which is characterized by the EEG spectrum. However, with the repetition of the stimulus, there develops, along with selective extinction, a generalized tonic inhibitory process which is reflected in a change of the EEG spectrum: slow waves appear and the high frequency components decrease. Such a change in the EEG, which is characteristic of the drowsy state, is sometimes accompanied by a disinhibition of the vegetative components of the orienting reflex. Consequently, when the stimulus is repeatedly applied, it is necessary to distinguish a selective (with regard to the parameters of the signal) extinction of the responses, which is not directly correlated with any change in the functional state, and a general change of the functional state which may disturb the selective extinction achieved in the active state.

However, investigations of the macroresponses do not allow a disclosure of the inner mechanisms of the orienting reflex. In view of this, we started an investigation of the responses of single neurons by methods similar to those which were applied for studying the manifestations of the orienting reflex on the level of macroresponses of the whole organism.

The experiments were carried out on unanaesthetized rabbits by means of extracellular microelectrode recording. Specific sensory systems (optic chiasm, superior and inferior colliculi, lateral geniculate body, visual cortex), as well as non-specific systems (hippocampus, midbrain reticular formation, and nonspecific nuclei of the thalamus) were subjected to investigation.

The electrical potentials, after amplification, were separated by a system of two filters. The first filter, which isolated signals within the range of 0.5–100 Hz, made it possible to record the slow components of the electrical activity. The second filter, which isolated signals within the range of 300–5000 Hz, allowed us to record the spike activity. The principal method used for studying the neuronal organization of the orienting reflex was a repeated application of the stimuli at specific intervals, which made it possible to ascertain the degree of stability of the responses of the recorded neuron. If an extinction of the reaction developed during the experiment, the degree of selectivity of the achieved extinction was studied by presentation of different test stimuli.

Attention was paid also to the processes reflecting changes in the functional state of the cerebral cortex and nonspecific thalamus.

The present work summarizes the data obtained under the guidance of the author by a group of researchers of the Department of Neuropsychology at the Faculty of Psychology of the Moscow State University.

2. The Mechanisms of Detection of Properties of Signals

All the investigated neurons may be divided into two large groups depending on the degree of development of the extinction process: neurons with extinguishing responses and neurons with stable responses. During repeated stimulation with light, the majority of neurons of the retina, superior and inferior colliculi, lateral geniculate body, and visual cortex manifested no phenomena of habituation comparable to the extinction of the components the orienting reflex. These neurons may be united into a group of specific afferent neurons. In a number of cases it proved possible to establish that these neurons are related to the detection of signal properties.

The general principle of detection of separate properties by neuronal nets is a convergence of neurons of the first order on a neuron of the second order, so that excitation of the last one reaches a critical level only when a definite pattern of the signal appears on the receptor surface. Detectors of edge, direction, and angle may serve as examples. In these cases, the property of a signal is coded by characteristic of the channel, i.e., of the output neuron in such a net.

An important mechanism of the detection of properties is lateral inhibition; it ensures the suppression of excitation in the neighboring neurons which with respect to detection of properties, are close to that neuron excited by the given property.

Detectors of intensity. A study of the orienting reflex on the level of macroresponses revealed selectivity of extinction with regard to the intensity of the signal. A decline in the intensity of the signal led to an orienting response, just as a change in the color of the stimulus or in its position in visual field. Hence, it may be concluded that at the neuronal level, intensity is coded according to the same principles as other properties of the signal. This was experimentally confirmed. Thus, in neurons of the lateral geniculate body, which respond to a light stimulus with short-latency discharges, the total number of spikes in the discharge varies from stimulation to stimulation. However, the number of spikes reproduced stably in the pattern of response is different, depending on the intensity. In some cells, the greatest number of stably reproduced spikes was observed in responses to stimuli of highest intensity, in other cells during the action of stimuli of moderate intensity, and in still others in responses to the lowest intensity applied in the course of experiment (T. G. Beteleva, Chapter 13 of this volume).

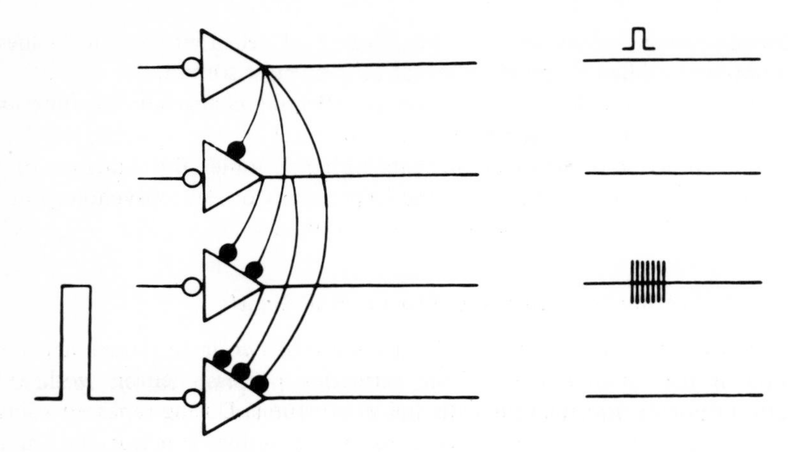

FIG. 1 Neuronal net for intensity detection. Neurons (triangles) receiving excitation, have different thresholds, which in the scheme increase from bottom to top. These neurons are connected by anisotropic lateral inhibitory links so that the neuron with the highest threshold inhibits all the neurons possessing lower thresholds. As a result, at the output of the net, only that neuron not affected by lateral inhibition is excited. This neuron is selectively "tuned" to the given intensity. Right-hand side: spike activity at the output of separate elements of the neuronal net; left-hand side, the signal presented. Solid circles, inhibitory synapses; open circles, excitatory synapses. Only one neuron which is selectively "tuned" to a definite intensity of the input signal reacts to it.

A study of the neuronal responses of the rabbit's visual cortex to flashes of different intensities also showed that there are neurons which are selective with respect to the intensities of the light stimuli (I. Chkhikvadze, Chapter 4). Of particular interest are neurons which manifest such responses only to weak stimuli. In this case the selectivity cannot be reduced merely to variations of the thresholds of excitation. A necessary condition is active inhibition of the responses to stimuli of high intensities.

The neuronal net, which provides detection of a definite level of signal intensity, can be schematically presented on the basis of anisotropic lateral inhibition, acting only in the direction of neurons with lower thresholds. In this case the intensity of the signal is coded as an index of the output neuron in the neuronal net, and by the frequency or distribution of the spikes in the discharge (Fig. 1).

Detectors of the speed of an object's motion. A group of neurons of the rabbit's superior colliculus (12% of all the neurons investigated in this structure) responded to a flash of light with a discharge consisting of 7–10 compact groups of spikes with a total duration of 150–200 msec. The duration of each group equalled 3–20 msec, with interspike intervals of 1–2 msec (N. V. Dubrovinskaya, Chapter 9).

The responses of these neurons did not depend on the intensity of the signal (except a certain change of the latency) and persisted during all the repetitions of the stimulus. It should be noted that various groups of a rhythmic discharge differ in their recovery cycles and appear independently of one another. The application of a rhythmic light stimulus evokes a frequency-specific response of the neuron. With the increase of the frequency of stimulation, the number of groups in the rhythmic discharge declines. Thus, four groups appear at the frequency of 10 per sec, two groups at the frequency of 25 per sec, and one group at the frequency of 30 per sec. Consequently, the critical frequency for separate groups proves to be different. It may be assumed that individual groups are generated independently of one another and summate on the neuron, passing along parallel channels.

With respect to some properties, these neurons seem to be similar to the neurons of the rabbit's retina which selectively respond to the speed of an object's motion in the visual field (Barlow *et al.*, 1964).

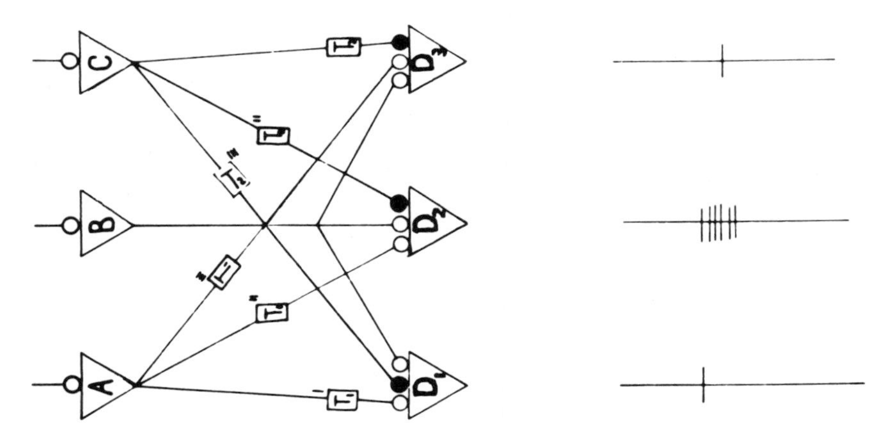

FIG. 2 Neuronal net for detection of the object's rate of motion (A, B, C) neurons receiving excitation from receptors located on one line. Integrating neurons D_1, D_2, D_3 receive direct excitation from the neuron B and additional excitation from the neuron A with different delays T_1', T_1'', T_1'''. In addition, the integrating neurons receive inhibitory influences from the neuron C with increasing delays T_2', T_2'', and T_2'''. Depending on the rate of the object's motion, only that integrative neuron (D_2) becomes excited, to which excitation comes from neuron A with the delay T'' equaling the time of the passage of the stimulus from the point represented by neuron A to the point represented by neuron B. As a result, the excitations from the neurons A and B are summated by neuron D_2. The rate of motion is such that there remains no time for the inhibition, which follows from neuron C to neuron D_2 with a delay T_2'', to suppress this reaction. If the object's rate of motion increases, neuron D_1 proves to be included, and if the speed decreases, neuron D_3 does. Neuron C ensures the selectivity of the reactions of the neuronal net with respect to the object's direction of motion and limits the time of excitation of neuron D. The direction of the object's motion is A–B–C. From left to right: two-layer neuronal net ensuring the coding of rate of movement by the index of the neuron at the output; reaction of neuron D_2 which is selectively "tuned" to the given rate of the object's motion. Designations are the same as in Fig. 1.

Schematically, the selective neuronal response to a definite speed of the object's motion may be presented as a result of convergence of excitatory influences (neurons A and B) and inhibitory influences (neuron C) coming from neurons of the first order on second order neuron D. A necessary condition for the excitation of neuron D coincides in time with the excitation of neurons A and B, as well as absence of inhibition from neuron C. This occurs only at a definite speed of the object's motion. The optimal speed to which the neuronal net reacts is determined by the time of delays T_1 and T_2. Each speed of the object's motion in the visual field may be presented as a result of excitation of one of a number of D neurons, which possess different delays T_1 and T_2 (Fig. 2). In this case the speed of the object's motion is coded as an index of the neuron.

Under the action of a flash of light illuminating the entire visual field, neuron D responds with a small spike burst which depends on the difference between the delays T_1 and T_2. The influences of several D neurons apparently converge on the neuron of the superior colliculus, which responds with a rhythmic discharge; each of the D neurons selectively react to a definite speed of motion.

Detectors of time intervals. During the application of stimuli with definite time intervals, the orienting response on the macrolevel habituates selectively with respect to the fixed interval between stimuli, and appears again when the interval is changed. If we apply the principle of coding signal properties by

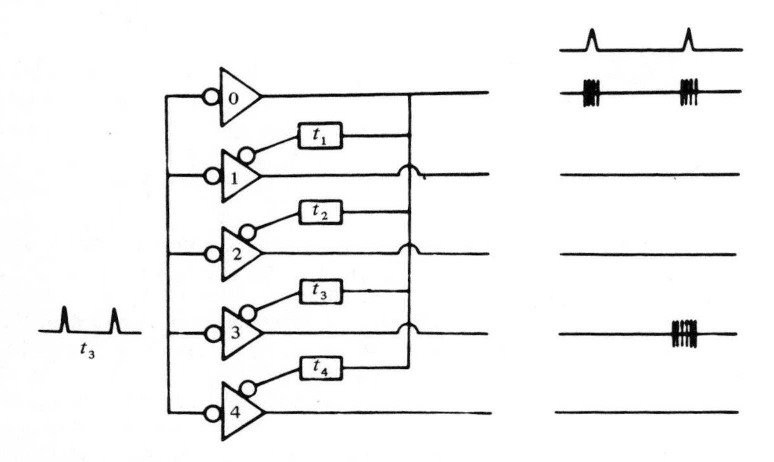

FIG. 3 Neuronal net for interstimulus time detection. Selectivity with respect to a definite time interval is determined by a system of different delays T_1, T_2, T_3, and T_4. The only neuron that responds to signals presented at a fixed interval is that one whose positive feedback delay fits this interval. The neuron without a feed-back mechanism serves for transmission of the incoming excitations to all neurons of the given net. The neurons respond only when the direct and delayed excitations coincide. From left to right: signals presented with a T_3 interval; neuronal net with different positive feed-back delays; neuronal responses. Above are shown the moments of stimuli presentations in order to illustrate the reaction of the "detector of intervals." Designations are the same as in Fig. 1.

indices of the neurons to the mechanism of time counting, we may assume that there exist "time-receptive fields" characterizing the neurons which selectively react to stimuli applied with definite intervals between them. Such a neuronal net may be presented by a scheme with a system of delays. The generation of spikes in the output neuron of the net takes place as a result of coincidence of excitation by an acting stimulus and excitation by a previous stimulus which arrives with a delay equalling the interval between the stimuli. In this hypothetic neuronal net, the interval between the signals, as a separate property, is also coded as an index of the neuron (Fig. 3).

An investigation of neuronal responses of the visual cortex to flashes of light, separated by different intervals, revealed a group of neurons which manifest selectivity with respect to definite intervals (L. R. Chelidze, Chapter 5). Thus, a neuron which shows no reaction when the intervals between the flashes equal 1, 2 and 3 sec, responds when the interval equals 5 sec, and again is reactive at intervals of 10, 15 and 20 sec. The group of neurons, which selectively respond with respect to the intervals between flashes, includes only a small part of the investigated neurons of the visual cortex. Other neurons do not possess such selectivity with respect to the intervals, and their responses are not modified when the intervals between the flashes are changed.

Detectors of complexes. These findings suggest that the perception of some complex, as well as simple, stimuli is also performed with the help of special detectors. This supposition is confirmed by the following effect: light and acoustic stimuli, when applied separately, evoke either a weak response of a specific neuron of the visual cortex or evoke none at all. But when these stimuli are presented simultaneously, a strong neuronal response follows. A transition to isolated applications of the stimuli, after presentation of their complex, reveals an absence of conditioned-reflex changes: neither a light stimulus, nor an acoustic one, if presented separately, provoke any significant neuronal response (Polyansky & Sokolov, Chapter 6). Such intensification of the neuronal responses to the complex of light plus sound, without any appreciable effect of formation of a conditioned reflex, is similar to the selective responses of neurons to other complex properties (for example, to the speed of a moving object). At the same time the detectors of complexes substantially differ from multisensory neurons, on which excitation of various modalities converge and which respond to each of the converging signals.

When schematically presenting the hypothetic neuronal net, which can ensure the detection of complexes, it must be borne in mind that a mere increase in physical intensity of the light and acoustic stimuli does not lead to an increase of the responses in the detectors of complexes. One of the probable mechanisms, which determines the operation of such a net, is axo-axonal inhibition. A detector of a complex (DC) does not respond to an isolated presentation of the light stimulus since a tonically discharging neuron (T) blocks the conduction of excitation with the help of the axo-axonal synapses (Fig. 4).

The acoustic stimulus excites the neuron (S) and blocks the tonic inhibitory

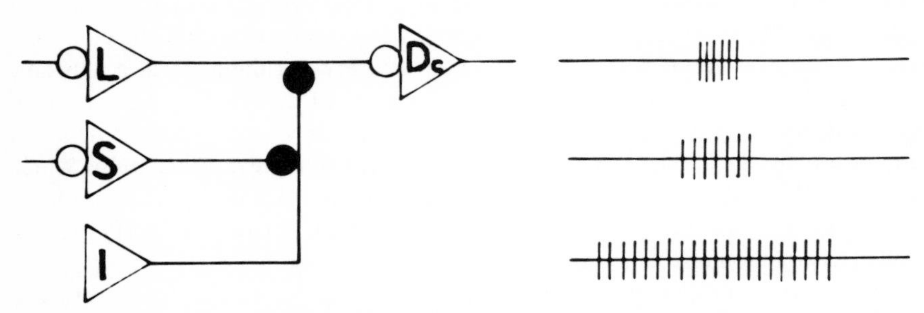

FIG. 4 Neuronal net selectively responding to a complex of "light + sound." The I neuron exerts a continuous tonic inhibitory action on the pathway which conducts excitation from the neuron excited by the light stimulus (L) to the detector of the complex (Dc). The isolated action of the light stimulus, as a result of an axo-axonal inhibition of the pathway which conducts the excitation, proved to be ineffective. The addition of a sound through the neuron which responds to acoustic stimulation (S) in its turn phasically blocks the inhibitory action of the I neuron opening the way for light stimulation. As a result, the detector of the complex (Dc) responds to the light stimulus only at the moment when the acoustic stimulus acts. From left to right: neuronal net for detecting a complex; complex stimulus "light plus sound"; responses of the neurons Dc, S, and I to the action of the complex "light plus sound." Designations are the same as in Fig. 1.

action of the neuron (T) upon it. This, however, does not lead to a response of the DC, if the light component is absent. A simultaneous action of the light and acoustic stimuli eliminates the inhibitory tonic influence and thus ensures the response of the neuron (DC) to a complex stimulus consisting of light plus sound.

Consequently, the available data indicate that the signal properties are detected by special neuronal nets. An output neuron in such a net serves as a code designation of the given property. Then the frequency of the spikes, which are generated by the neuron, characterizes the "degree of expressiveness" or "weight" of this property.

3. Modification of the Property "Weight"

Thus, if the property of the signal is coded as an index of the output neuron in the special neuronal net, then the density of the spikes in the response characterizes the "weight" of the property, i.e., the contribution which is made by the given property to the integral response of the organism. As already stated above, the extinction of the orienting response to a signalling stimulus is accompanied by an increase of the effectiveness of the repeating conditioned signal action with respect to the conditioned response evoked by it.

It may be assumed that with the repetition of the stimulus, the response of neurons, which are specific for the given property, increases, while the response of neurons not involved in coding this property declines.

It proved that such an effect manifests itself in the specific afferent neurons of the superior and inferior colliculi, in the lateral geniculate body, as well as in a part of neurons of the visual cortex. Thus, when the flashes were repeated for a long time, the neurons of the lateral geniculate body did not manifest any signs of extinction but underwent some regular changes. In some of these, activation (potentiation) was observed (a shortening of the latency and an increase in the frequency of generated spikes in the discharge). In other neurons, inactivation developed (an increase of the latency and a decrease in the frequency of spikes in the discharge). In both cases the variability of the discharge diminished with the repetition of the signal. At the same time discharge latency to a flash of light became stabilized (T. G. Beteleva, Chapter 13).

Similar phenomena were observed in some neurons of the visual cortex, where two types of dynamics were found among the great variety of responses. In one type of these neurons, the effect of activation was in evidence, while the other, showed the effect of inactivation. In both cases, the pattern of the spike distribution in the discharge and the periods of excitation and inhibition became more pronounced, i.e., stable. Such changes may be divided in two categories. Some of them are determined only by the repetition of the specific signal and are not eliminated under the action of extra stimuli (lateral geniculate body), while others become deranged as a result of the application of such stimuli. In the latter case, the stabilization of the pattern of the discharge was due to the extinction of the orienting reflex to the extraneous stimuli, which temporarily deranged the established pattern. In one such neuron of the visual cortex, after the omisssion only of one flash or after addition of clicks to the light stimulus, the established pattern of the response was superseded by random spikes, and gradually recovered with further stimulation (A. Bagdonas, Chapter 10).

The effect of limiting the variability of the responses of specific neurons as a result of repeated stimulus presentations may be interpreted as stabilization of connections within the neuronal net which ensures detection of the signal. In other words, the connections within the neuronal ensemble become more stable and are less subject to accidental influences.

The mechanism of intensification of responses in certain neurons, when the stimulus is repeated, and simultaneous weakening of the responses in other neurons, may be explained by the postactivation potentiation of excitatory and inhibitory synapses.

A neuronal net with lateral inhibition may serve as an elementary scheme illustrating this phenomenon. A repetition of a stimulus at one of the inputs of the net, through the mechanism of postactivation potentiation of the synapses, leads to intensification of the excitatory process in the given neuron and to a stronger inhibitory influence on the neighbouring neurons. As a result, the responses of some neurons to a repeated signal increase and decrease in others. The potentiation of the synapses leads to a stabilization of both the excitatory and inhibitory responses, with a simultaneous decline of their dependence on the neighboring influences (Fig. 5).

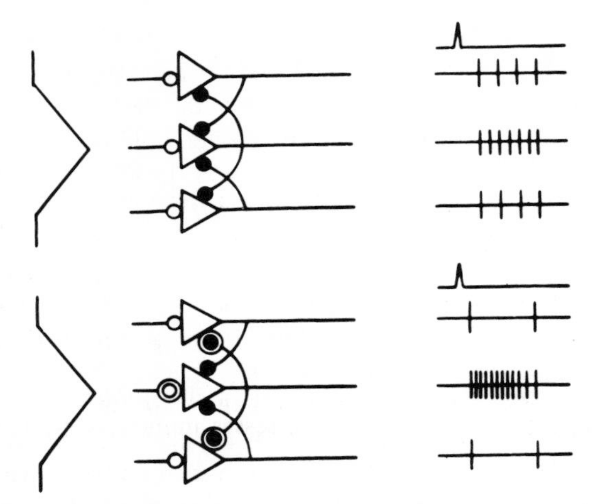

FIG. 5 Neuronal net displaying lateral inhibition. Upper set: a net with lateral inhibitory connections before the onset of stimulus. Lower set: the same net after the application of a series of stimuli which resulted in a potentiation of the excitatory and inhibitory synapses. From left to right: schematic presentation of a signal in different points of the receptive field (the maximum of excitation falls on the neuron in the centre of the receptive field); scheme of the neuronal net; responses of individual neurons at the output of the net (⊙, potentiated synapses).

Before the potentiation of the synapses the difference in the excitation level of various neurons is smaller than after the application of a series of stimuli. The responses of the most excited neuron increase as a result of a potentiated action of the excitatory synapses. Intensification of the lateral inhibition on the border-line neurons leads to the suppression of their responses. Outlined circles–potentiated synapses. Other designations are the same as in Fig. 1.

The neuronal net becomes selectively tuned to the repeated signal. If a new stimulus appears, its action becomes weakened. Thus, a repeated stimulus increases the contrast within the neuronal net. The orienting reaction, deranging the response to the repeated stimuli, influences the neuronal net in such a way that the selective conduction elaborated within it becomes suppressed and is superseded by a general rise of excitation, as a result of which the action of the new signals is facilitated. This effect of deranged stabilization under the influence of the orienting reflex may be hypothetically explained by the mechanism of presynaptic inhibition of the inhibitory lateral connections, due to which the selectivity of the neuronal net becomes reduced.

4. Mechanism of Extinction at the Level of a Single Neuron

"Novelty" detectors. Neurons participating in the act of attention were first discovered by Hubel *et al.* (1959) in the cat's auditory cortex and were called "attention units." The habituation to light signals was observed by O. S. Vinogradova and D. E. Lindsley in 5% of the neurons of the rabbit's visual cortex. But the process of extinction of the response exhibited a particularly

distinct character in the neurons of the hippocampus which react to highly diverse stimuli (O. S. Vinogradova, Chapter 11). The extinction process was observed in 40% of the neurons and consisted of a gradual decrease of the tonic response duration; after 10–15 applications of the stimulus the response completely disappeared but recovered as a result of any change in the stimulus. The extinction of the responses of all hippocampal neurons was characterized by a selectivity similar to that observed at the level of macrocomponents of the orienting reflex.

The fact that hippocampal neurons react to highly diverse stimuli and develop a selective extinction of the responses indicates that the axonal terminals of the neuron, which detect a definite property of the signal, contact with a number of hippocampal neurons, repeatedly duplicating this property. Thus, a single neuron of the hippocampus apparently possesses numerous synaptic contacts, each of which represents a certain property.

The neuronal model of the stimulus as a matrix of potentiated synapses. When explaining the general mechanisms underlying the extinction of the responses of the attention units, it is necessary to consider the effect of intensification of the hyperpolarization wave observed in neurons of the deep layers of the visual cortex as a result of repeated stimulation which gradually inhibit the neuronal spike discharge (V. N. Polyansky & E. N. Sokolov, Chapter 6).

It may be assumed that two basic types of inhibition participate in the generation of the hyperpolarization waves, namely, recurrent and afferent collateral inhibition.

A specific feature characterizing the extinction of the responses of attention units is that the extinction process continues to increase after full disappearance of the response. This indicates that the inhibitory effect does not result from the discharge of a given "attention" neuron, and is therefore not determined by the mechanism of recurrent inhibition. In the case of afferent collateral inhibition, the inhibitory effect, which develops in the interneurons parallel with excitation, does not depend on the degree of excitation of the neuron which summates these two kinds of influences. The gradual development of inhibition is explained in this case by a predominant postactivation potentiation of the synapses of the inhibitory interneurons. As a result of summation of these two influences on the output neuron, the effect of excitation becomes suppressed and the response disappears.

Thus, the formation of the "neuronal model of the stimulus" begins with the detection of separate properties of the signal. This is effected by nets of specific afferent neurons. Individual properties of the signal are coded as an index of the output neuron in the neuronal net. Under the action of a repeatedly applied stimulus, the afferent neuronal nets undergo a transformation and change their parameters in such a way that the repeating properties of the signal are "emphasized," i.e., acquire in some neurons a certain advantage as regards the rate of conduction and the degree of the contribution of the afferent neurons to the specific reflexes. The facilitation of some elements of the neuronal net and the inactivation of others makes it possible to perceive familiar objects during a shorter period of time and more reliably than unfamiliar ones. Numerous elements of the external environment converge on the attention units represented by the pyramidal neurons of the neocortex and hippocampus. It may be admitted that some definite synaptic contact corresponds to each gradation of the signal property. The time parameters of the stimulus and the interval between individual stimuli are, from this point of view, also coded by the index of the synapse with the participation of the detectors of time.

The properties of the signal reach the "attention" neuron in a direct way [as an excitatory postsynaptic potential (EPSP)] and in parallel—through the system of interneurons which contact the "attention" neuron through inhibitory synapses [as an inhibitory postsynaptic potential (IPSP)]. In the course of repeated stimulus presentations, there takes place a potentiation of the synaptic contacts which were involved in the reaction (Fig. 6).

The matrix of the potentiated synapses which code the definite properties of the stimulus, retains the "pattern" of the "signal" and presents a "neuronal model of the stimulus."

The extinction of the orienting reflex includes the potentiation of the synapses of interneurons. The afferent collateral inhibition becomes intensified. The potentiated system of synapses corresponds to the matrix of properties and ensures the selectivity of the extinction of the orienting reflex. When the stimulus is changed, new inhibitory synapses become involved. There is still no mechanism of afferent collateral inhibition for the new stimuli, and their action evokes an orienting response. The response to the new stimulus proves to be stronger, the greater the number of new synapses involved, i.e., the greater the number of properties by which the new stimulus differs from the old one.

Horn (1967) also concluded that the signal is coded as an index of the channel. However he explains the phenomenon of habituation by a depolarization of the axonal terminals of the afferent neurons which converge on the "novelty" neuron.

From this point of view, the effect of disinhibition, which develops after the change of the signal, consists of a repolarization of the endings with the participation of the inhibitory axoaxonal synapses of the neurons of the nonspecific system.

Detectors of novelty as a source of extinguishing components of the spike discharge of the specific neurons. An investigation of vascular reactions (Vinogradova, Sokolov) showed that signals of different biological significance may reach one and the same effector system. Thus, under the action of cold, the reaction of constriction of the vessels of the hand includes two components: constriction connected with the orienting response and constriction connected with the specific thermoregulatory action of the cold. The first effect becomes extinguished, while the second exhibits a tendency to summation and stabilization in the course of the action of the stimulus.

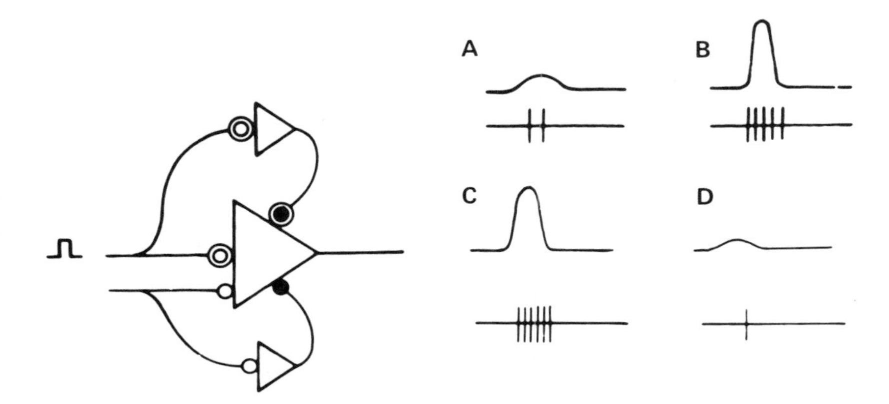

FIG. 6 Neuronal net providing a selective habituation of neuronal reaction to the stimulus. With the repeated application of a stimulus acting through one of the inputs, there takes place a predominant potentiation of the inhibitory synapses in the chain of afferent collateral inhibitory links. As a result of a predominant potentiation of such parallel inhibition, the sum of the EPSP and IPSP diminishes, and the spike reaction at the output of the net is extinguished. A new signal presented through another input does not affect the system of potentiated inhibitory synapses, but evokes an EPSP and an ensuing spike discharge. From left to right: signal at one of the inputs of the net; neuronal net after the extinction of the responses; (A) responses of the interneuron before the onset of extinction; (B) responses of the interneuron after the development of a selective potentiation (with respect to the output signal) of the inhibitory synapses; (C) responses of the main neuron before the development of extinction; (D) responses of the main neuron after the development of extinction: (A, C) First presentation; (B, D) last presentation. In each case the EPSP or the sum EPSP + IPSP is shown above, and the spike activity below. Designations are the same as in Figures 1 and 5.

The neurons of the lower layers of the rabbit's superior colliculus manifested such a convergence of spike discharges of different significance (N. V. Dubrovinskaya, Chapter 9).

Thus, the first application of a flash evokes a sustained suppression of the spike discharges. With its repetition, the late phase became extinct. A sound, as an extra stimulus, does not by itself exert an influence on the background activity of the neuron; however, after the action of an extra stimulus the late phase of inhibition to light recovers, and several repetitions of the flash are again required to reduce the duration of the inhibition to a minimum level.

It may be supposed that two influences are summated on this neuron: a specific signal (the stable component) and a signal coming from the detector of novelty (the extinguishable component).

Similar phenomena were observed by Vinogradova and Sokolov (Chapter 12) in neurons of the caudate nucleus.

5. Effect of Synchronization and Desynchronization of the EEG at the Level of a Single Neuron

Investigation of the responses and background activity of single neurons during repeated applications of a stimulus made it possible to study the correlation between the orienting reflex and the functional state of the cortex. Simultaneous recording of spike activity and the evoked potential at the same point of the visual cortex allowed us to detect neurons on which both the direct influence of the flash, (in the form of a spike discharge and a focal potential corresponding to it), and the indirect influence of the state of the cortex (in the form of a series of slow waves correlating with spike bursts), converged. During the action of a single flash of light, it is possible to record in the visual cortex: (*a*) an evoked potential and a spike discharge not complicated by a series of slow waves; (*b*) a spontaneous series of slow waves, where spike discharges corresponded to the negative phases, and a suppression of the spikes to the positive phases; and (*c*) a combination of an evoked response with a series of slow waves, which either followed the primary response as an aftereffect or preceded it.

If the flash coincided with a series of slow waves that had already started, the discharge evoked by it was lost in the spontaneous alternation of spike bursts and inhibitory intervals. It must be emphasized that the appearance of groups of slow waves in the background activity, with a simultaneous appearance of spike bursts at the negative phase and their inhibition at the positive phase, developed in the course of the experiment during repeated applications of the stimulus.

During rhythmical light stimulations, an intensification of slow waves was observed in the background activity; the alternation of the pauses and of the neuronal spike discharges also occured with long intervals.

The development of bursts in spike activity, coinciding with the slow oscillations of the focal potential under the influence of repeated stimulations, may be schematically presented by a system of neurons which are interconnected according to the principle of recurrent inhibition, so that the neuron is subjected not only to direct excitatory influence, but also to inhibitory influence of the interneurons forming a negative feedback loop. The interneuron itself may also become inhibited under the action of signals produced by "detectors of novelty" (Fig. 7).

As long as the experimental conditions or the applied stimulus are new enough, the interneuron remains inhibited and the negative feedback is switched off. Accidental impulses in the input evoke, with a certain synaptic delay, occasional spikes in the output of the neuron. With the gradual habituation to the experimental conditions, the detectors of novelty cease to produce signals and the interneurons are no longer inhibited. As a result, the negative feedback proves to be switched on. Accidental impulses in the input evoke a long-lasting IPSP during which the activity of the neuron under investigation becomes inhibited for a certain period of time. After the end of the IPSP, spike discharge

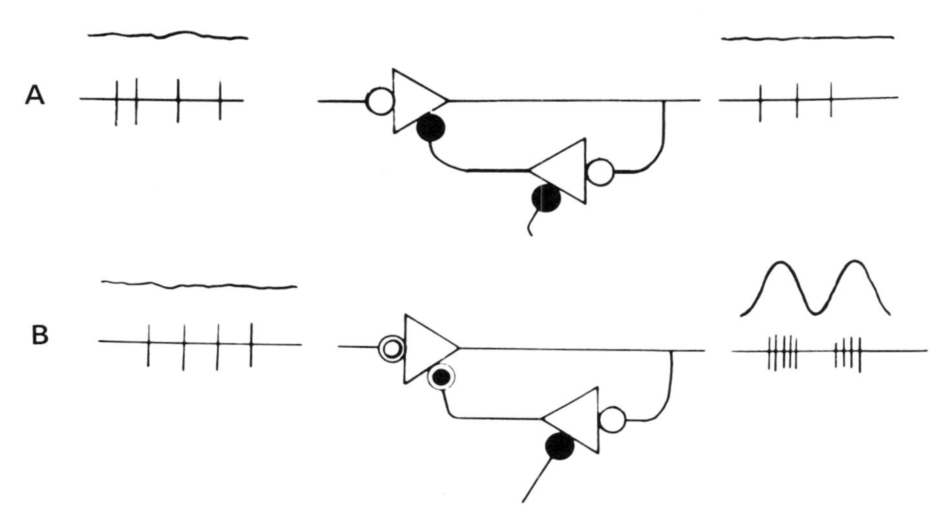

FIG. 7 Scheme for the generation of slow waves and spike bursts. The development of slow waves is based on the predominant potentiation of the inhibitory synapses in the neuronal net displaying recurrent inhibition. (A) Inhibitory connection through an interneuron is ineffective, since the inhibitory synapses have not been potentiated. The main neuron responds with a random sequence of spikes which reproduces the input sequence. (B) As a result of potentiation of the inhibitory synapses, recurrent inhibition comes into action. The neuron responds to a random sequence of spikes with spike bursts separated by silent periods. Spike bursts correspond to EPSP IPSP alternation. The inhibitory synapse on the interneuron indicates the possible arrival of impulses which block the negative feedback; as a result, a desynchronization of the slow waves arises and the spike bursts are replaced by random spikes. From left to right: the sum EPSP + IPSP and the spikes at the input (on the scheme the latter is represented only by an excitatory synapse); neuronal net of recurrent inhibition; synaptic potentials and spikes at the output. Designations are the same as in Fig. 1 and 5.

appears as a result of "postanodic exaltation." The neurons display a "burstlike" activity. The alternation of the IPSP and EPSP forms a series of slow "spontaneous" oscillations of the focal potential. The intensification of inhibition in the chain of the negative feedback during repeated stimulation can be explained by potentiation of the synapses. A considerable contribution to the development of slow waves in the cortex is made by the unspecific system of thalamus. As a result of repeated applications of the stimulus, the thalamic neurons, which respond with single random spikes, develop a burstlike spike activity. The emergence of a series of bursts in the neuronal activity often coincides with a series of slow waves in the neocortex (N. N. Danilova, Chapter 14). A new stimulus simultaneously calls forth a transition of the thalamic neuron from the burst activity to a generation of random spikes and desynchronization of the slow waves in the cortex.

Consequently, a replacement of the burst activity by a series of random spikes in the thalamic neurons serves as an equivalent of the desynchronization reaction in the EEG. These effects can be explained just as in the case of cortical neurons: the detectors of novelty act upon the interneurons which participate in the generation of slow waves, temporarily inhibiting them. Owing to this, a new stimulus evokes an arousal reaction, while the habitual stimulus does not influence the state of sleep or drowsiness.

Besides thalamic neurons, which respond with a transition from burst activity to a generation of random spikes, it is also necessary to take into account the neurons which increase the frequency of spike discharges during the action of a new stimulus. Such neurons constitute the activating system proper.

It may be assumed that there exist separate neuronal mechanisms of two systems of reactions: synchronization–inactivation on the one hand, and activation–desynchronization, on the other.

The development of slow waves, as an expression of a decline in the functional state of the cortex, may be interpreted in informational terms. Thus, as a result of the establishment of an internal cycle (the slow rhythmic waves), the detection of an external signal becomes impeded, and this in its turn leads to a derangement in the realization of reflex acts.

6. Structure of the Orienting Reflex Arc

The analysis of the data relating to the neuronal activity of the different brain structures allows us to revert to a description of the arc of the orienting reflex at the macrolevel.

We shall consider here its structure, basing our analysis on the visual system as one of the afferent mechanisms involved in the orienting reflex. However, similar reasoning is applicable to other analysers, too, since the laws by which the orienting reflex is governed are the same for different modalities. The absence of novelty detectors at the level of the optic nerve, superior and inferior colliculi, lateral geniculate body, and most neurons of the visual cortex, leads to the conclusion that at these levels there takes place a detection of properties, which subsequently effects a selective extinction of the responses to the signal.

The fact that the majority of the hippocampal neurons exhibit a high degree of selectivity of extinction, including such properties as stimulus complexes and time intervals, indicates that the messages which come to the hippocampus were processed at the highest level of the cerebral cortex, where detectors of intervals and complexes have been found. A selective blocking of the signals, which are repeatedly applied during an experiment, develops in the hippocampus with the participation of the interneurons. The inhibitory neurons of the hippocampus, which discharge more and more vigorously with the repeated presentations of the stimulus, apparently may intensify the work of the synchronizing system of the thalamus, and this leads to the development of slow waves in the cortex. The hippocampal neurons, activated by the new stimulus, probably are connected with the activating reticular system. As a result of repeated applications of the

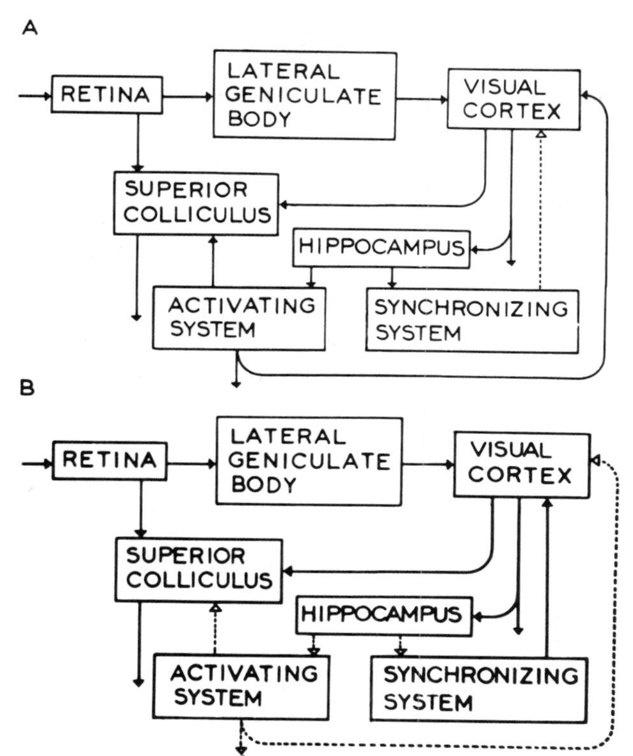

FIG. 8 Scheme for the participation of different brain structures in the responses to the novel (A) and habitual (B) stimuli. The signal from the retina comes through the lateral geniculate body to the visual cortex, where signals are analysed according to their separate properties. Then the detected properties, which are coded as indices of the channels, come to the novelty detectors of the hippocampus, bringing into action the activatory neurons and suppressing the background activity of the inhibitory neurons. The novelty detectors, excited by the stimulus, switch on the activating mechanism of the reticular system, which in its turn influences the visual cortex, modifying activity of the latter. The inhibitory neurons of the hippocampus apparently maintain the activity of the synchronizing system in a tonic way. During the action of a novel stimulus; their spontaneous activity is inhibited; this suppresses the synchronizing system, whose influence on the cortex diminishes. Thus, the novel stimulus evokes the orienting reflex and enhances the functional state of the cerebral cortex due both to the increase in activity of the activating system and to the decrease in operation of the synchronizing mechanism. In addition, the activating influences, which depend on the action of the detectors of novelty, may intensify some special reflexes, e.g., realized through the neurons of the superior colliculus (Fig. 8A). When the stimulus becomes habitual, the process of the detecting signal properties in the specific system does not undergo any essential changes. However the novelty detectors of the A type are no longer excited, while neurons of the I type display sustained activity. As a result, the activating influences disappear, whereas the synchronizing influences not only persist, but become even more intensive, which leads to the development of slow waves in the cortex (Fig. 8B).

Solid lines are connections acting in the given situation. Dashed lines are connections not involved in the response. (A) Action of the novel stimulus; (B) action of the habitual stimulus.

stimulus, when the responses of these neurons become extinguished, the activating system is thereby switched off.

Thus, the decline of the functional state of the cortex is explained by the fact that the activating system is blocked, while the work of the synchronizing system becomes intensified. This decline of the functional state is selective with respect to the signal. A new stimulus evokes the arousal reaction due to the suppression of the synchronizing system and involvement of activating influences. Besides a change of the functional state, there is observed an extinction of some efferent responses which are, apparently, triggered by the novelty signals generated by the A neurons of the hippocampus. Thus, under the action of a new light stimulus, the signal becomes processed in the retina, lateral geniculate body, and visual cortex. Information about different properties of the signal reaches the hippocampus where it excites the A neurons; the latter switch on the activating system and inhibit the interneurons in the synchronizing system.

Therefore, the slow activity becomes suppressed, and the background activity of a number of neurons increases.

Simultaneously, the activating system enhances the excitability of the receptor (apparently with the participation of the sympathetic nervous system} and, as a result of generating "novelty signals," modifies the specific influence of the stimulus on the efferent mechanisms of the superior colliculus and cerebral cortex.

During the action of a habitual stimulus the specific reflexes are preserved, but the influence exerted on them by the activating system, which is controlled by the A-neurons of the hippocampus, disappears. At the same time neurons of the I type, which are no longer inhibited, intensify the activity of synchronizing system, and this leads to the generation of slow waves in the cortex (Fig. 8A, B).

Thus, up to a certain period (development of sleep) the blocking of the activating system, due to the extinction of the responses of the A neurons, creates a favorable background for the realization of stereotyped reflexes.

The orienting reflex, which arises as a result of a change in the stereotype, deranges the established system of reflexes and ensures a rapid and generalized increase of excitability which makes possible a transition to a new level and type of activity.

CONCLUSIONS

1. Repeated applications of stimuli revealed that stable responses characterize the specific afferent neurons that detect the properties of a signal.
2. The properties of a signal may be coded by the index of the neuron: The number of the spikes in the discharge characterizes the "weight" of the detected property.
3. Under the influence of repeated stimuli, the responses in some specific neurons increase, while in others they decrease. In both cases, the variability of

the responses decreases as a result of stimulus repetition. The opposite direction of the changes in the "weights" of the habitual and novel properties may be explained by the potentiation of the synapses in neuronal systems with lateral inhibition.

4. The extinction of responses of "attention" neurons is presumably due to potentiation of the synaptic mechanism of afferent collateral inhibition.

5. A neuronal model of the stimulus may be conceived as a matrix of potentiated synapses that code the system of properties of the repeatedly applied stimulus.

6. Monotonous stimulations result in a decline of the functional state of the cortex which manifests itself in a lowering of EEG frequencies.

7. The phenomenon of synchronization apparently depends on the participation of the inhibitory interneurons of the cortex and nonspecific thalamus in the mechanism of recurrent inhibition.

8. The structural scheme of the orienting reflex includes a connection between the neocortex as the basic mechanism of the analysis of signals, and the hippocampus, as a system of novelty detectors.

9. It is supposed that the hippocampal neurons of the activatory type are predominantly connected with the activating system of the brain, while the neurons of the inhibitory type are predominantly connected with the synchronizing system.

SECTION III

EVOKED POTENTIAL AND ELECTROENCEPHALOGRAPHIC STUDIES

There are four chapters in this section, all of which deal with gross electrophysiological phenomena, but generally from different approaches. Floru (Chapter 17) investigates the effects of arousal upon visual evoked potentials in both cat and man. The effects of arousal appear to be similar, whether induced by stimulation of the reticular formation in cat or increased proprioceptive stimulation in man. Voronin, Bonfitto, and Vasileva (Chapter 18) are concerned with the orienting reflex as it pertains to expectancy as well as novelty. The differential dynamics of the EEG (blocking of the alpha rhythm) and the GSR components of the OR are of particular interest. A new dimension to the EEG aspect of the orienting reflex is set forth by Artemieva and Homskaya (Chapter 19). They have found that even during a state of constant alpha rhythm in man, the EEG still reveals changes which satisfy the criteria for the orienting reflex; this EEG parameter is the degree of asymmetry in the rising and falling phases of each alpha wave. Thus, careful analysis of man's EEG reveals that even when alpha blocking has habituated (extinguished), there are in fact dynamic cerebral processes which continue to reflect the effects of a stimulus for some time. One may view the asymmetry analysis as extending the range of sensitivity of the human EEG, and with it, the time course for cortical habituation. In the final chapter, Simernitskaya reports relationships between human scalp evoked potentials and attention. Although this research does not benefit from computer-assisted extraction of the EP, scrutiny of the records presented reveals that the same stimulus presented under conditions of two sets of verbal instructions results in differential changes in the EP simultaneously at different hemispheric loci. This type of result would seem to have internal controls for general effects, such as arousal level, and is consistent with an "attention" interpretation.

17

PSYCHOPHYSIOLOGICAL INVESTIGATIONS ON ATTENTION

R. Floru

Institute of Psychology, Bucharest

INTRODUCTION

During the past two decades, considerable interest has been focused on the problem of attention and accordingly, there has been a great increase in research devoted to this subject. This interest is accounted for on the one hand by the necessity of solving some practical problems such as maintenance of vigilance under conditions of monotony or of competitive stimulations, and on the other hand, by modern advances in neurophysiology which have contributed to the understanding of the neural substrate of attention.

One of the main results of these investigations was to establish the role of the ascending reticular activating system (ARAS) in the orienting reflex and diffuse attention.

The aims of our studies were: (*a*) to determine the role of cerebral tonus, as an independent variable, in the integration of sensory messages; and (*b*) to establish the relationships between the *orienting reflex, set,* and *attention.*

Investigations concerning the effects of attention on intermittent photic potentials were carried out in numerous experiments in animals (Floru, Staerescu-Volanschi, & Bittman, 1964) and human beings (Floru, Ciofu, Gulian, & Cicata, 1966).

METHODS

1. Thirty adult cats with chronically implanted electrodes were studied. Stainless steel electrodes were placed stereotaxically in the mesencephalic reticular formation (RF), posterior hypothalamus (H), lateral geniculate body (CGL), visual (O), and auditory (A) and motor (F) cortex. The position of the electrodes was histologically checked.

Recordings were performed with a 20-channel electroencephalograph using bipolar and monopolar leads. Intermittent light stimulation (ILS) and acoustic stimulation (an uninterrupted pure tone) (T) of 600 Hz/sec at steady intensity (60 dB) were obtained from a photophonostimulator (electronic generator included). The RF or H were stimulated with square wave pulses of 0.1 and 1 msec at a rate of 200–300 per sec and voltage of 2–7 V. The general experimental conditions were quiet and semidarkness. Each session involved recording of the resting EEG pattern followed by ILS at different rates (1–35/sec); electrical stimulation of RF, H, or acoustical stimulation were added.

2. Forty adult human subjects were also studied in a sound proof chamber under conditions of rest and dim illumination, according to the following general schedule:

A. A period of accomodation, in which the resting EEG, electrodermogram (EDG), and responses evoked by ILS at different frequencies were recorded.

B. The experiment itself in which changes in the photic driving capacity were studied in the following types of experimental conditions:

(a) The application of an unexpected (auditory) stimulus, eliciting an orienting reaction.

(b) The attention of the subject was requested by verbal instruction according to which, at a certain auditory signal, he must: (i) perform a certain movement (flexion of the hand, Morse signal etc.); (ii) describe the order of appearance of certain colored photic extra-stimuli, a change in the rhythm of the photic stimulation (a different sound etc.); (iii) perform an intellectual task (e.g., solve a problem, compute, memorize certain words with or without any sense, build up sentences with given words, answer a question, etc.).

In the same series of experiments, the type of the test remained unchanged but the elements varied; each series of experiments was preceded by a new verbal instruction. The subject's verbal response was noted within approximately one minute after the end of the test.

(c) Conditioned reflexes were elaborated as follows. The subject was informed that during the ILS, he will hear a pair of auditory stimuli separated by a 3–4 sec interval (T_2). At the second stimulus of the pair, the subject had to perform an intellectual or a motor task. This paired stimulation was differentiated from another one (T_1) applied under the same conditions, but at which the subject had nothing to do. The first tone of each pair of stimuli preoriented the subject as to what he had to do at the second signal.

Thus, it was possible to compare the electrical alterations to irrelevant and relevant stimuli, both during their action and in the period of set. Similarly, we followed up the evolution of the EEG pattern, during performance of the intellectual tasks or motor reactions.

RESULTS

1. In cats, against a background of spindle bursts (11–13 Hz/sec) or occasionally slow waves, typical for drowsiness or sleep, ILS slightly changed this

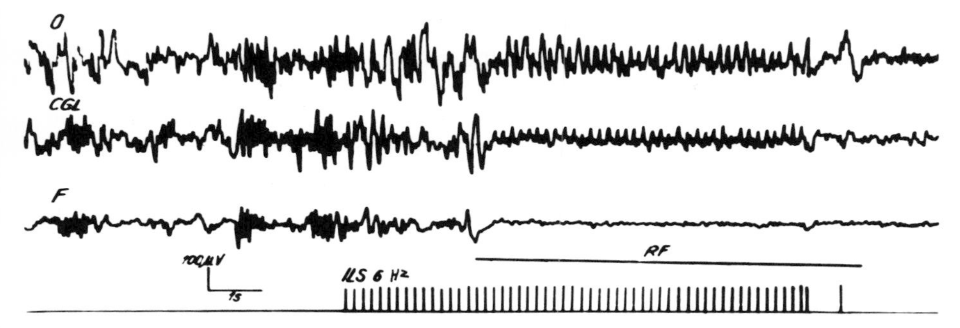

FIG. 1 Reticular facilitation of rhythmic photic potentials in the cat. Before reticular stimulation, the EEG pattern is characterized by spindles accompanying behavioral drowsiness. RF stimulation induced arousal and improvement of the photic driving. ILS, intermittent light stimulation. Electrical RF stimulation (300 Hz/sec, 0.1 msec) is marked by horizontal bar. O, visual cortex; CGL, lateral geniculate body; F, motor cortex, as in Figs. 2–6. Cats with chronically implanted electrodes are shown in this and following Figs. 1–6.

pattern. In some instances occasional photic potentials occurred, either masked or intermingled with the background rhythms; in other cases photic potentials appeared only in the CGL and visual cortex.

Under conditions suggesting a lower level of cerebral tonus in which ILS eventually induced the driving of the intrinsic rhythms, high-rate electric stimulation of the RF or H induced an EEG arousal, and a clear-cut enhancement of the photic responses in the visual system (Fig. 1). The same effect is also conspicuous during auditory stimulation which substantiates the concept that the mechanism lying at its basis is very similar to that of direct stimulation of the ARAS.

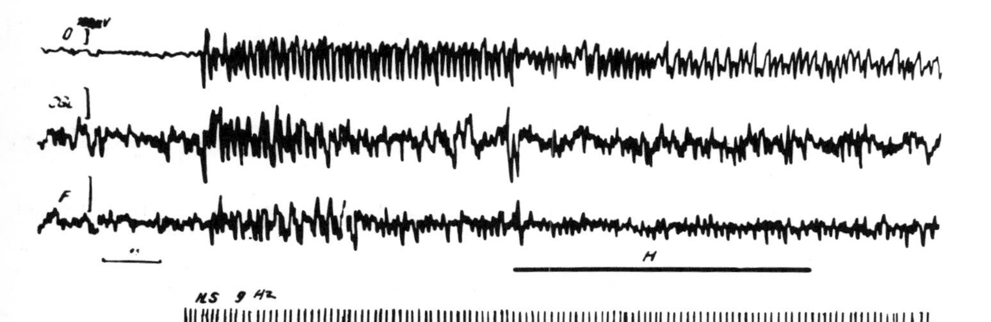

FIG. 2 Depressant effect exerted by hypothalamic stimulation on rhythmic photic responses. Before hypothalamic (H) stimulation, the EEG pattern is characterized by low voltage, fast activity accompanying a behavioral alertness. H stimulation induced reduction of the photic potentials. Electrical stimulation of the posterior hypothalamus with 300 Hz/sec, 0.1 msec.

Against a background of low voltage and fast rhythms, associated with vigilant behavior, ILS induced an obvious photic driving. Under these conditions, electrical stimulation of the RF and H, as well as auditory stimulation, elicited *reduction* in the amplitude of photic responses (Fig. 2). Occasionally, the depressant effect was conspicuous at the same intensity of electrical stimulation which induced an enhancing effect on rhythmic responses during synchronous background; sometimes it was necessary to slightly increase the voltage of electrical stimulation to obtain a depressant effect.

The first series of results suggest definite interrelations between the functional state of the central nervous system and the ability of the cerebral structures to follow the ILS rhythm under reticular stimulation, namely:

(a) During relaxed wakefulness, ILS evoke photic potentials in the specific visual centers as well as in some unspecific structures. A lowering of the functional level, as in the transition from wakefulness to sleep, reduces the cerebral ability to drive the rhythm of afferent impulses; the optimal functional level may be recovered by direct or indirect stimulation of the ARAS.

(b) If stimulation of the ARAS is added to an already existing optimal functional level, a state of an increased *alertness* results, with a decrease in photic driving.

In other words, the effect of reticular stimulation on the photic potentials is actually a cumulative result of the given stimulation and resting activity. In order to check the hypothesis of an occlusive (see also Bremer, 1961) summation, both RF (or H) and auditory receptors were stimulated together (Fig. 3). The facilitatory effect of auditory stimulation was replaced by a clear-cut occlusive effect. In other instances this occlusive effect is not obvious, but after removal

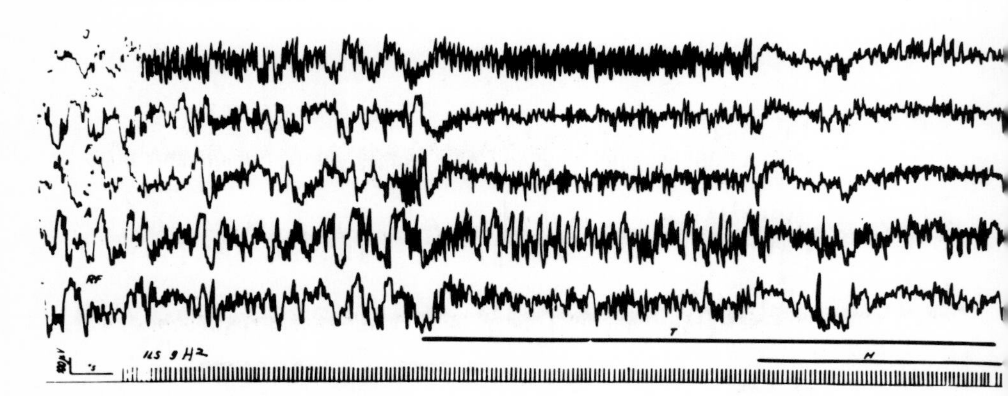

FIG. 3 Occlusive effect exerted by summation of the hypothalamic and auditory stimulations on the rhythmic photic potentials.

Auditory stimulation (T) induces facilitation of the photic potentials; hypothalamic stimulation (H) added to the acoustic stimulus reduces the photic driving. A, auditory cortex; RF, mesencephalic reticular formation.

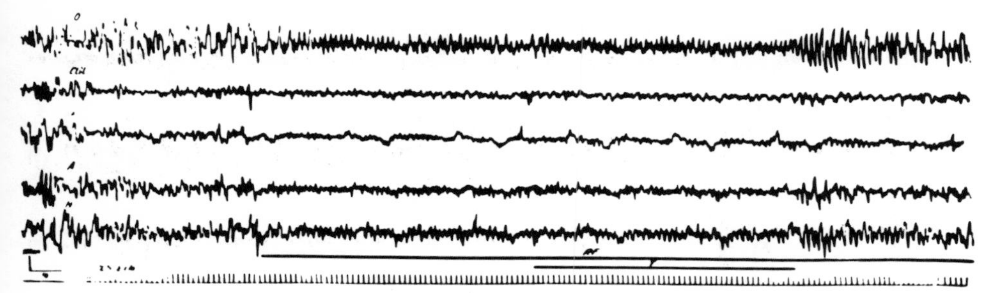

FIG. 4 Effect of summation of reticular and acoustic stimulations on the rhythmic photic potentials.

Reticular stimulation (RF) induces facilitation of the photic potentials; auditory stimulation (T) added to the reticular one does not alter the EEG pattern. After cessation of the auditory stimulation photic potentials are enhanced (rebound). H, posterior hypothalamus.

of one of the two stimulations, a rebound may be observed (Fig. 4). In some instances, electrical stimulation of the RF or H induced an increase of the photic potentials in some structures with a simultaneous decrease in the others. Therefore, there are limits of the functional capacity of neural structures which integrate visual messages; within these limits, stimulation of the ARAS facilitates photic responses. By overpassing these limits, an occlusive effect may take place.

The second investigation concerned the specific functional state of the visual system.

Stimulation of the ARAS causes "dishabituation" of the "habituated" photic potentials such that their amplitude is actually greater than the control values.

The facilitatory effect of the ARAS stimulation could be conditioned. By pairing the ILS (as conditioned stimulus–CS) with acoustical or electrical (RF or H) stimulation (as unconditioned ones–US) on the background of drowsiness, a conditioned facilitation of the photic potentials occurred (Fig. 5). Thus, rhythmic photic stimulation was able to elicit conditioning of the increase in the cerebral tonus induced by RF or auditory unconditioned stimulation. This effect may be extinguished after 5–7 presentations of the CS (ILS) without any reinforcement; it was associated with generalized slow waves on the EEG (Fig. 6).

During elaboration of the conditioned reflex, rhythmic discharges having the same frequency as the CS appeared in the intervals between trials.

2. In man, an unexpected stimulus during ILS produced facilitation of the photic potentials. The facilitatory effect is much more marked when the resting activity indicated a low or medium level of cerebral tonus (drowsiness or relaxed wakefulness).

A similar effect also appeared during proprioceptive stimulation (that arises in the course of motor responses). Alterations of photic driving in the cerebral structures took place not only as a result of an orienting reaction or of a

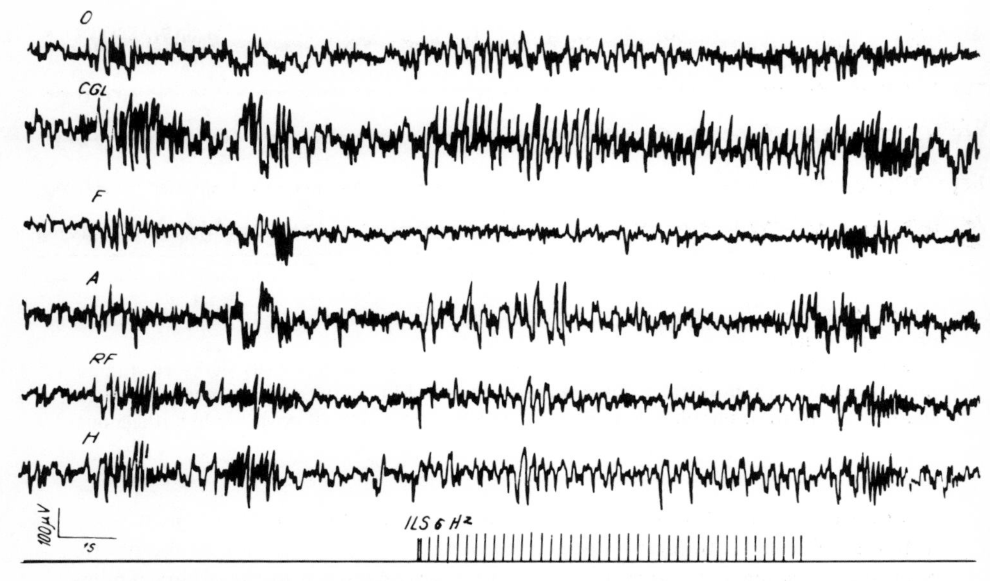

FIG. 5 Conditioned self-facilitation. After the 11th application of ILS paired with acoustic stimulation on a background of drowsiness, ILS presented alone elicits an arousal reaction with self-facilitation of photic potentials.

proprioceptive stimulation but also in the course of a perceptive, motor, or intellectual set.

The preparatory attitude (of waiting for a signal, or a motor or verbal response) expressed itself by an enhancement of photic responses (Fig. 7).

The focusing of attention in the course of the intellectual tasks facilitated the photic potentials (Fig. 8).

If the intellectual task was preceded by a signal (differentiated from a nonsignal stimulus), a preparatory attitude was elaborated (Fig. 9).

By repeating the task the period of facilitation was shortened (Fig. 10).

DISCUSSION

There are many important theoretical and experimental works on the orienting reflex (see Vinogradova, 1961), set, or attention. A neurophysiological approach to attention belongs to Lindsley (1960).

Attention, perhaps more than other psychic processes, should be situated as a bordering concept between neurophysiology and psychology, in order to understand not only its different behavioral expressions but also its intimate mechanism. We consider attention to be a unitary psychophysiological conative–cognitive process. It evolves from an unselective orienting reaction up to diffuse set and focused effective attention (Floru, 1967).

Several mechanisms are involved in the attentional process; our aim was to follow one of these, namely diffuse and selective activation that provides our cognitive responses with readiness, clarity, and vividness.

Whether stimulation of the ARAS is mediated by the collaterals of the specific pathways as in any sensory stimulation (Moruzzi & Magoun, 1949), or by a cortico-reticular loop (Bremer, 1956), the dynamogenic effect is expressed by desynchronization and, within certain limits, by facilitation of the photic rhythmic potentials (Steriade & Demetrescu, 1960). In our experiments, photic potentials are indicators of neural responsiveness to a given sensory testing stimulation and of the indirect effects of attention.

The contradictory results reported by different authors (see Floru, 1967) concerning the effects of direct and indirect stimulation of the ARAS by rhythmic photic potentials may be explained by relationships between "activating" stimuli and resting electrical pattern.

Referring to the orienting reaction, we must recall the fact that Pavlov (1928) emphasized its bivalent role: (a) interruption of the present activity (see external inhibition negative induction); (b) switching the excitability of the sensory system in another direction, becomes the starting point of another series of adaptive reactions. From this point of view, we agree with Sokolov who classifies the orienting reflex as a category dissimilar to that of the adaptive or defensive

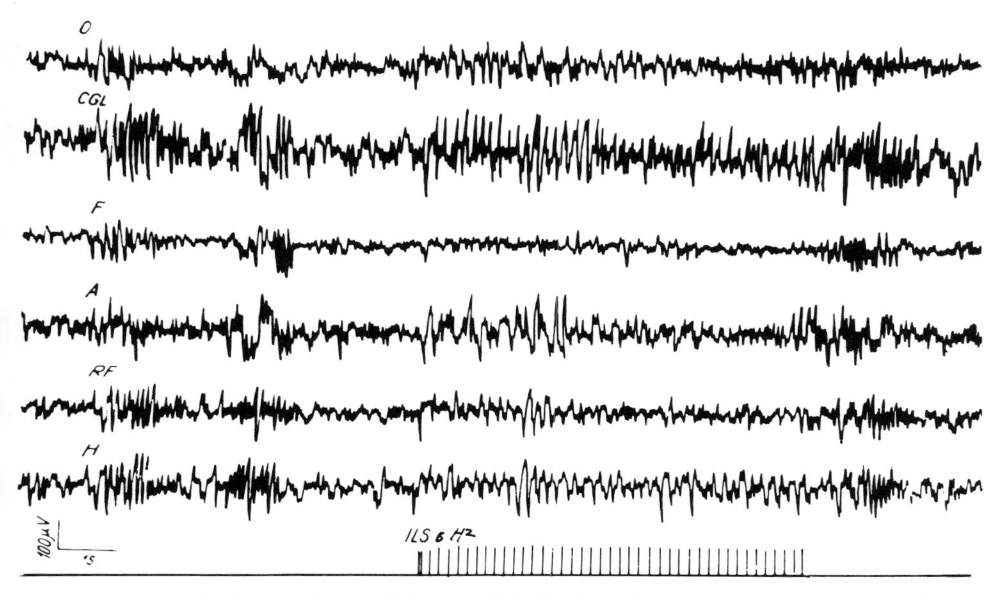

FIG. 6 Extinction of the conditioned self-facilitation. The 6th presentation of the ILS without tonigenic (acoustic stimulus) reinforcement. The conditioned reflex is extinguished and associated with generalized slow waves.

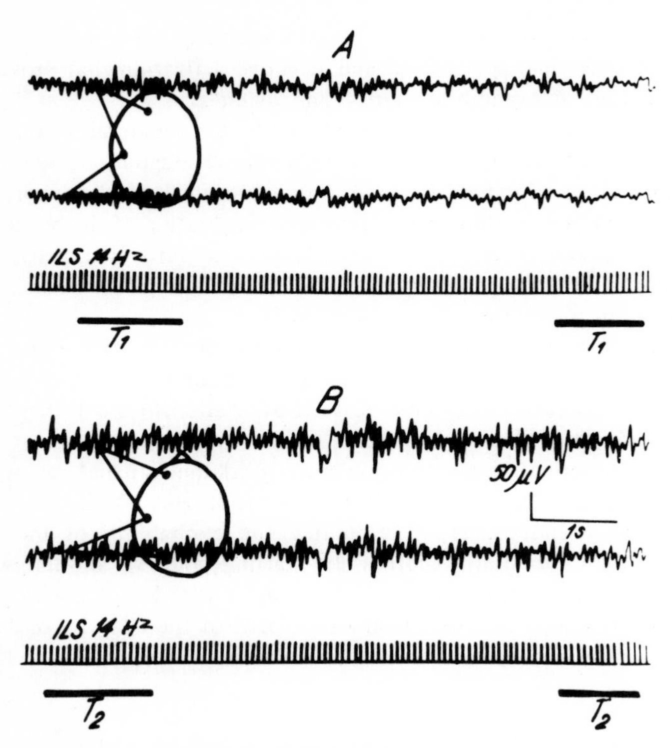

FIG. 7 Facilitation of the photic potentials during set. (A) Control experiment. (B) The subject had to report whether the second T_2 differs in intensity or in pitch from the first one. A higher level of activation corresponds to the preparatory attitude; simultaneously photic potentials are enhanced. T_1, negative acoustic stimulus (100 Hz/sec); T_2, positive acoustic stimulus (1000 Hz/sec).

reflexes (Sokolov, 1960). It is suggested that orienting reactions (together with exploratory and investigatory ones) be classified as preadaptative reactions.

However the orienting reflex in itself does not as yet constitute attention, not only because a physiological phenomenon cannot be identified with a psychological one, but also because the former is not an integral substratum of the latter. Nevertheless, the orienting reaction may be regarded as the turning point to attention as well as to inattention.

Bearing in mind that the first effect of the orienting reaction on a new and unexpected stimulus is interruption of the activity in progress (a disorganizing effect) and also initiation of other activities (a reorganizing activity) its further evolution depends directly upon subsequent significance of the stimulus.

The loss of the orienting significance of a stimulus leads to inattention; the acquisition of a new significance leads to attention. Both are learned and

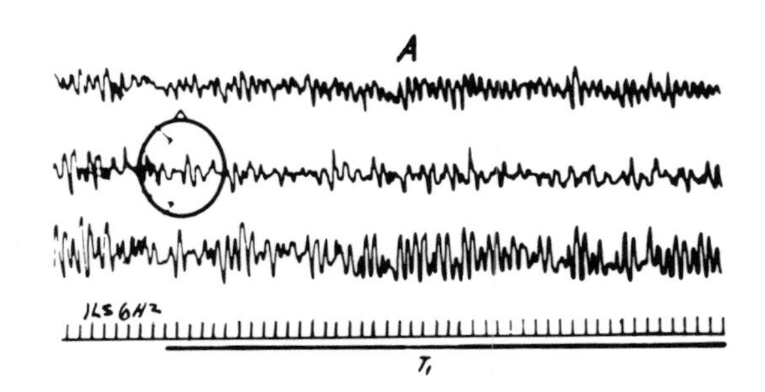

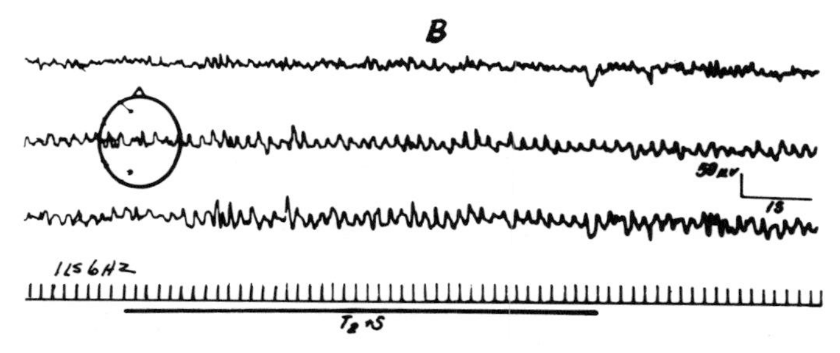

FIG. 8 Facilitation of the photic potentials during effective attention. (A) Control experiment. (B) The subject solves an arithmetical task presented at the beginning of the acoustic stimulus (T_2 + S). In (A) relaxed EEG pattern, undisturbed by the irrelevant signal (T_1). In (B) the EEG pattern is activated and photic potentials are facilitated. T_1, acoustic stimulus of 100 Hz/sec; T_2, acoustic stimulus of 1000 Hz/sec. In previous experiments the subject was habituated to both stimuli.

selective processes. Their common starting point is an unconditioned, rather unselective orienting reaction.

The extinction of this reaction is a strictly selective unresponsiveness (Sokolov, 1966) to the previous orienting stimulus (see Sokolov's "neuronal model of the stimulus"); psychologically, it is a selective, learned inattention.*

The second direction in which the orienting reaction may develop is selective attention, i.e., the selective orientation of the cognitive activity. Its psychological mechanism is the selective activation of cerebral structures which are implied in the given activity.

In order to explain this mechanism we must refer to some known experiments. Segundo and co-workers conditioned the behavioral effects of reticular stimulation (Segundo, 1959). Conditioned self-facilitation, by pairing a visual with a

*This is not the only mechanism of inattention; peripheral and central filtering mechanism (commutation, induction, reciprocal inhibition and others) intervene during shift or focussing of attention (Hernandez-Peon *et al.*, 1956).

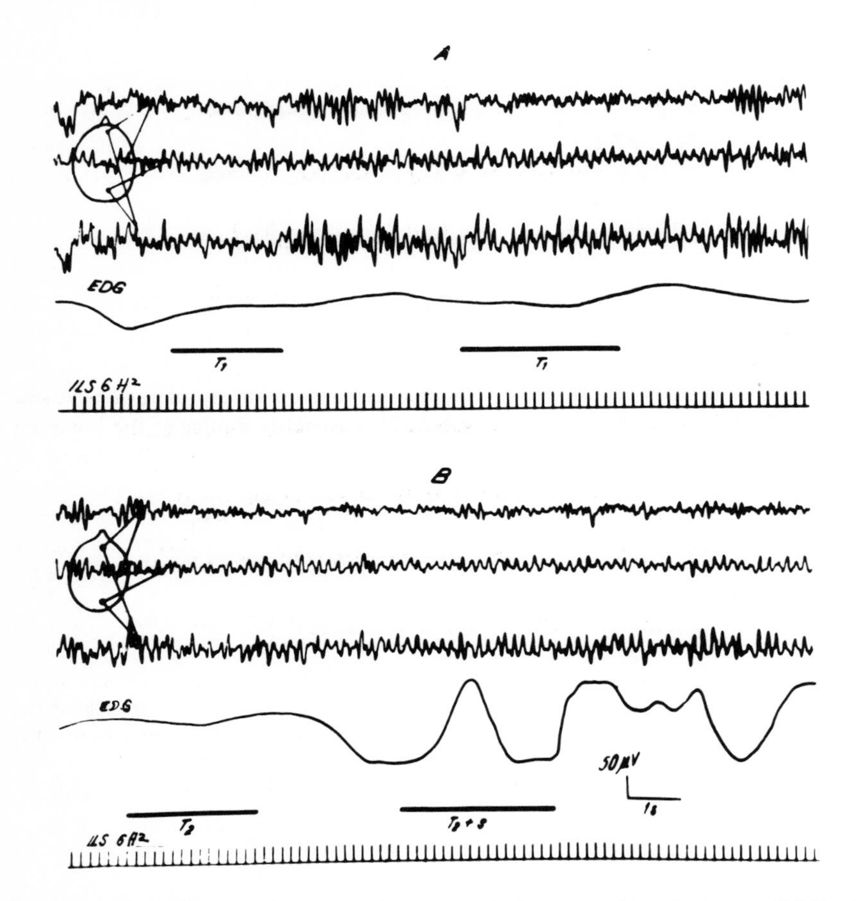

FIG. 9 Facilitation of the photic potentials during set and performance. (A) Control experiment. (B) The subject is requested to give a concise and exact definition in the sphere of his profession (T_2 + S). In previous experiments, following instruction, the second presentations of T_2 were associated with intellectual tasks. In (A) the EEG pattern is slightly activated only during acoustic stimuli; alpha rhythm occurs in intervals. In (B) photic potential are facilitated both during acoustic stimulation, preparatory attitude and performance of the task. The higher degree of activation in (B) than in (A) is detected also in the electrodermogram (EDG) curve. T_1, negative acoustic stimulus (100 Hz/sec); T_2, positive acoustic stimulus (1000 Hz/sec).

reticular stimulation, was also demonstrated (Floru, Staerescu-Volanschi, Bittman, & Nestianu, 1963; Sterescu-Volanschi & Floru, 1964).

The main facts are: the conditioned stimulus signals not only the motor or secretory effect, but also the tonigenic effect generated by the unconditioned stimulus, so that at the end of elaboration of a conditioned reflex, the signal actually is connected with a specific reaction (sensorisensory, sensorivegetative, or sensorimotor) and an unspecific one such as, sensoritonigenic. This fact is based upon the double action of any stimulus–specific and unspecific–i.e., eliciting both the proper effect (auditory, photic, motor, etc.) and the unspecific effect of diffuse activation. From the view point of temporary connections, it has been demonstrated that by the association of rhythmic photic stimuli against a background of drowsiness, with direct or indirect stimulation of the reticular formation, a tonigenic reflex is elaborated in which the photic stimuli produce both their own effect (evoking photic potentials) and a tonigenic effect (facilitation).

Due to summation of the specific and unspecific excitation there is, besides general activation, a dynamogenic excess in the given structures. By repeated reinforcement, this excess is fixed; the stimulus acquires a "cue function." It signals a consumatory reaction but it elicits also a preparatory one. The evoked potentials are facilitated, i.e., self-activated. This aspect is similar to the lowering of the threshold of the "indifferent" stimulus after conditioning. On the other hand, it is known that the unconditioned reflex is also modified (intensity, speed, etc.) after conditioning (Kupalov, 1956).

The learning mechanisms include not only motor, automatic, and verbal responses but also the dynamogenic support to these responses. The convenient adjustment of a dynamic capacity supplying a given activity is a result of learning. We learn not only to react, to perform some activity but also "to tune" the necessary (not more, not less) dynamogenic support for each fragment of behavior. For instance, in our experiments, after instructing the subject that he had to distinguish one pair of stimuli from another, the EEG-pattern desynchronized even *before* the stimuli had been applied, and when the recording was taken against a background of ILS to which he had already been accustomed, the photic potentials were facilitated.

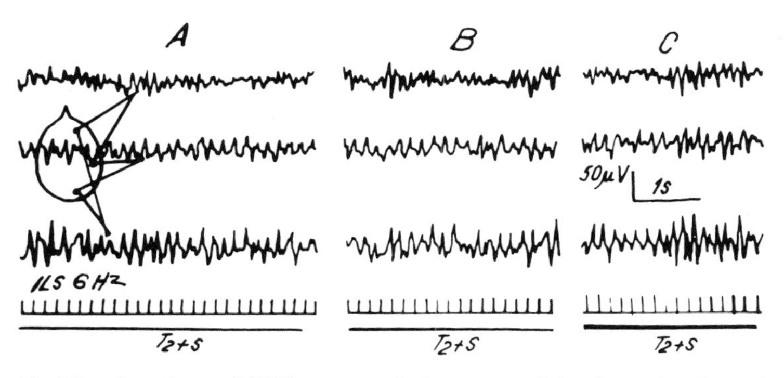

FIG. 10 The alterations of EEG pattern during repeated intellectual tasks. (A) after 7 solutions; (B) after 13 solutions; (C) after 19 solutions. In each experiment, the second presentation of T_2 was associated with a similar intellectual task (the concrete data are different). Cerebral activation lasts shorter and shorter; Alpha rhythm occurs earlier in (C) than in (A), indicating a reduction in redundant activation in the course of training.

Psychologically the subject awaits, is prepared, and "sharpens his feelings" (especially hearing); neurophysiologically, tonic stimulation of the activating system occurs. Moreover, in this first phase, a desynchronized EEG pattern appears both during the positive and the negative stimuli, although, from the perceptive and behavioral point of view, the subject differentiates between the two sounds (for instance a high pitch when a motor reaction ensues, and a low pitch when he has nothing to do).

Study of EEG dynamics and psychophysiological correlation shows that, subsequently, differentiation becomes more accentuated and the EEG pattern reflects the different reactivity of the nervous system to each of the stimuli; facilitation of the photic potentials only appears in the period that separates the pair of positive stimuli. The first stimulus of the pair signals the onset of the task and its type (intellectual, motor, etc.), triggering a selective set, a preparatory attitude to the second stimulus that generally goes with the instruction to perform the task (or question, problem, etc.).

Hence, there is some evidence that the preparatory attitude, expectancy, or set, represents a conditioned orienting reaction whose substrate is the tonic stimulation of the activating system (see also Haider, Spong, & Lindsley, 1964). General tonigenic stimulation, with a low degree of selectivity, is based upon previously elaborated temporary connections and takes place following instruction or an order to pay attention.

Delimitation of the significance narrows down the sphere of tonigenic activation until it is reduced to the short period that separates the signal from the stimulus triggering performance of the task.

With repetition of the intellectual task an "adjustment" is elaborated at the functional level necessary for reception and solving of the task. Effector "adjustment," is therefore an anticipated organization of the tonigenic support of the functional structures which will be requested. Thus, attention brings about a set effect at subsequent repetitions; after a great number of repetitions, preparation itself diminishes and activation may be made evident during isolated application of a positive stimulus, in contrast to the negative stimulus which is not "charged" with the significance of an intellectual activity.

The result of the elaboration of attention is the establishment of a correspondence between the given activity and the tonigenic support that insures the optimal level of the functional structures of the receptor and effector. Although this was not systematically observed, evidence was found of a more intense or reduced facilitation in terms of the difficulty of the task; however, it may be assumed that the "tuning" of the functional level can either raise or lower tonus according to the degree of attention necessary for a given activity.

Worthy of note in the evolution of attention is the parallelism between the systematization of selective facilitation of the functional structures involved in a perceptive, motor, or mental activity and the gradual disappearance of the signs of generalized activation.

It is likely that each new problem, each new activity, requires the functioning of two types of systems: (1) those which analyze and solve problems and (2) another which maintains the background tonus of the analytical and problem solving systems. During the course of training, the system which maintains background tonus becomes activated in a progressively selective manner so that finally only the strictly necessary elements are activated. Selective attention is actually selective activation, selective facilitation of a perceptual, motor, or intellectual act. The reduction (or enhancement) of the extent and intensity of activation in the course of learning is a self-regulating cerebral mechanism that is necessary for the appropriate interaction between an organism and its environment.

18

THE INTERRELATION OF THE ORIENTING REACTION AND CONDITIONED REFLEX TO TIME IN MAN

L. G. Voronin
M. Bonfitto
V. M. Vasilieva

Moscow State University

The problem of the relationship between the innate orienting reaction and the acquired conditioned reflex cannot be regarded as fully solved. As is known, the orienting-investigatory reaction, raising the tone of the cerebral cortex, contributes to the formation of a temporary connection; but when such a connection is elaborated, it may be inhibited by the orienting reflex (Pavlov, 1949).

Investigations carried out by the Department of Physiology of the Higher Nervous Activity at the Moscow State University showed that there are two forms of the orienting reflex. The orienting reflex to novelty, which was called by I. P. Pavlov a "what is it?" reflex, becomes more or less rapidly extinguished, as soon as a temporary connection is formed. However, there is another form of the orienting reaction; it accompanies a stable conditioned reflex which may be called a "when will it be?" or "will it be?" reflex. In application to man, this reaction may be designated as an "expectancy" reflex. It was observed by us also in animals, being expressed in the fact that a synchronized 8–10 per sec rhythm appeared stable during the action of a conditioned signal and apparently reflected the state of "awaiting" food (Voronin & Kotlyar, 1962).

The present chapter presents material which was obtained from investigations carried out on human subjects and which concerns relationships between the orienting reaction and the conditioned reflex to time. It seems to us that such a conditioned reflex allows the revelation of the relationships of the two aforementioned reflexes (orienting and conditioned), to a greater degree than any other conditioned reflex, because regular stimulations create highly favorable conditions for the formation of a "neuronal model of the stimulus" (Sokolov, 1963) and, consequently, for the extinction of the orienting reaction to "novelty." At the same time the "expectancy" reflex manifests itself with particular clarity in such experimental conditions.

METHODS

The recording of EEGs of both occipital and rolandic areas, of GSR (according to the method of Tarkhanov), as well as of EMG of the right forearm and of the oculo-motor reaction (OMR) was accomplished on an "Alvar" electroencephalograph. More than 160 experiments were performed on 50 normal subjects, 18–30 years of age (students and workers of the Moscow State University) who exhibited a well pronounced alpha rhythm. Modifications of the EEG, GSR, EMG, and OMR were recorded during light stimulations (indifferent and signalling). The stimuli were applied at short (2–5 sec), medium (12–15 sec), and long (30 sec) intervals. The duration of the stimulus in the case of short intervals equaled 1 sec, and in the cases of medium and long intervals–2 and 3 sec, respectively. The indifferent or signaling property of the stimulation was ascertained with the help of instructions given before the experiment.

RESULTS

Orienting Response to the Presentation of a Rhythmic Stimulus

a. Indifferent stimulation. A light stimulus with a duration of 1 sec, presented at a constant interval of 2 or 3 sec, at first evokes a protracted blockade of the alpha rhythm. After 5–10 applications of the light, blockade of the alpha rhythm is observed only during stimulus presentation. Thereafter, depression of the alpha rhythm in response to the light decreases and subsequently becomes extinguished. At the beginning of the experiment the GSR to the stimuli is strong. It then becomes extinguished at different rates. Omission of a regular stimulus at any phase of the experiment is not accompanied by modifications of the EEG or by the emergence of a GSR. A state of drowsiness often develops at the end of the experiment; at this period the response to the light recovers in the shape of an alpha-rhythm burst.

In the case of medium intervals between the stimuli (12–15 sec), the character of the dynamics of the GSR is similar to that described above, but the EEG reaction is not reduced or altered, probably because of the longer stimulus duration. When one of the regular stimulations is omitted, a depression of the alpha-rhythm arises at the scheduled time of the stimulation, and the GSR becomes disinhibited.

In the case of long intervals (30 sec and over), the EEG modifications and the GSR arise only in response to the first applications of the stimuli. A state of drowsiness rapidly develops in the subjects. Omission of one regular stimulation does not lead to any changes in the recorded processes.

b. Signalling stimulation. If a signalling meaning is imparted to the light stimulus with the help of a preliminary instruction of the following kind: "When the light is switched on, press the button as rapidly as possible, when it is

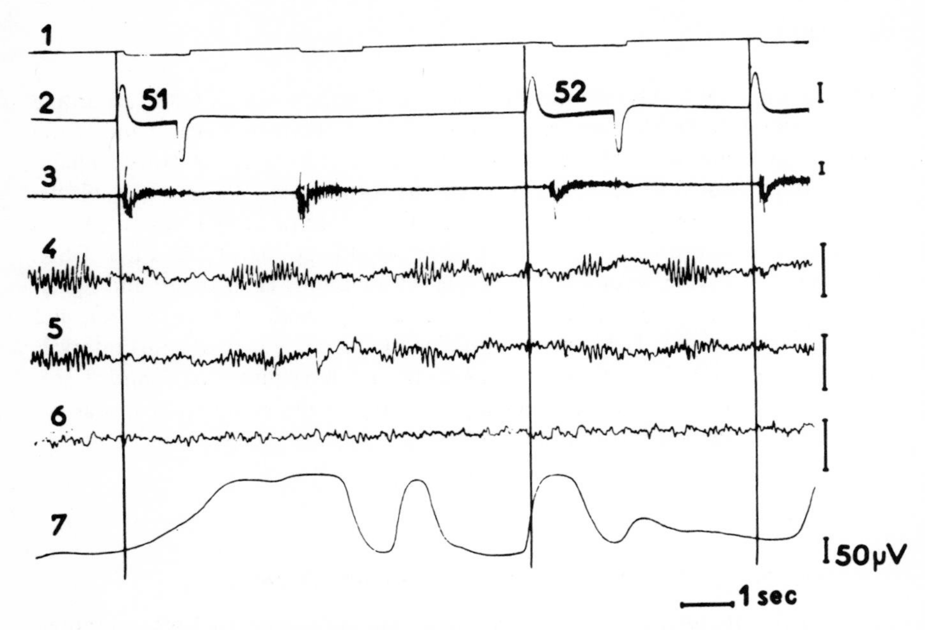

FIG. 1 Omission of one regular stimulus, at a 3 sec interval: 1, subject's response; 2, stimulus; 3, EMG; 4, EEG right occip; 5, EEG left occip.; 6, EEG sensorimotor; 7, GSR. Note the double response and alpha blocking during the interval between stimuli 51 and 52.

switched off, release the button," or "when the light is switched on, rapidly open the eyes, when it is switched off, close the eyes," the dynamics of the recorded processes show a definite change. In the case of an interval of 2–3 sec, a depression of the alpha rhythm to the signal is well pronounced, especially at the beginning, while the GSR becomes gradually extinguished; when the response consists of pressing, this extinction develops slower than when it consists of opening the eyes. The character of the EEG reaction differs depending on the kinds of reinforcement: when the subject opens his eyes in response to the stimulus, a blockade of the alpha rhythm is observed in 100% of the cases and during the whole period of the stimulus presentation, while in the case of a button-pressing response, it is predominantly observed only at the beginning of the stimulation. No state of drowsiness was observed in these experiments. Omission of one regular flash after 15–20 stimulus presentations led to the emergence of a depression of the alpha rhythm and the motor response (EMG or OMR) (Fig. 1).

In the case of a medium interval (12–15 sec), the modifications of the EEG and GSR were of a well pronounced and stable character; there was less extinction than under the action of indifferent stimuli presented in the same conditions. Single omissions of a regular signal in this series of experiments resulted in an intensification of the GSR, as well as in a definite increase of the

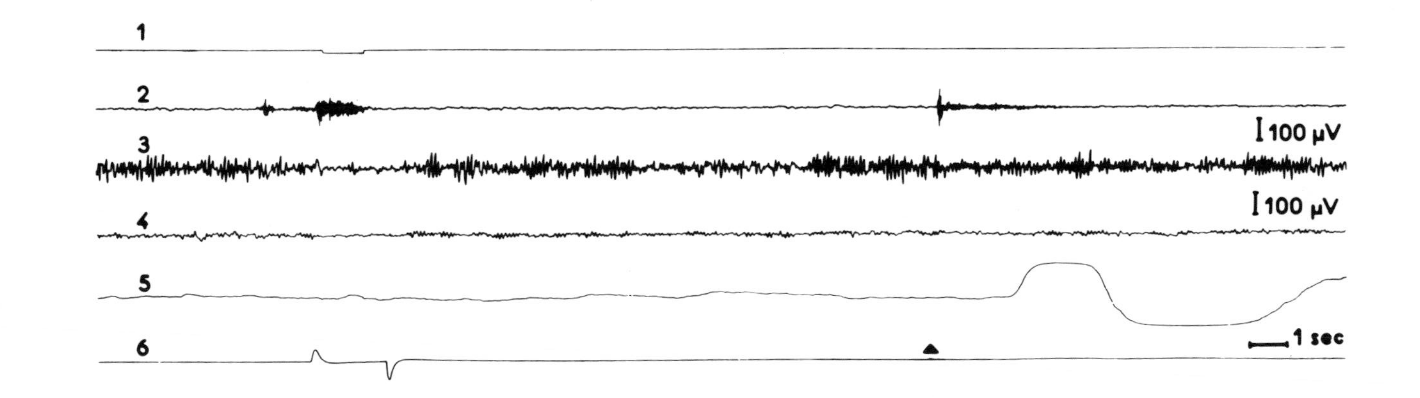

FIG. 2 Omission of one regular stimulus at a 15-sec interval: 1, response; 2, EMG; 3, EEG right occip.; 4, EEG left occip.; 5, GSR; 6, stimulus. The arrow at the bottom of the figure indicates stimulus omission. Note the absence of a motor response, but the occurence of an EMG response, depression of alpha, and a pronounced GSR immediately following.

muscular tone and the blocking of the alpha rhythm at the scheduled time of the omitted stimulation (Fig. 2). Actual pressure of the button or opening of the eyes was not observed in all cases, in contrast to the preceding series of experiments with intervals of 2–3 sec, where they were observed during stimulus omission after only 10–15 stimulations.

When the signals were presented at intervals of 30 sec, the waking state persisted throughout the experiments; modifications of the EEG and GSR in response to the signals were similar to those observed in experiments with intervals of 12–15 sec. If a regular stimulus was omitted, a GSR and a depression of the alpha rhythm appeared in several seconds, but no indications of effector processes (EMG or motor response) were observed.

Anticipatory Orienting Reaction ("Expectancy" Reflex)

In the course of regular stimulations, modifications of the recorded processes began to appear during the intervals between stimuli. In order to distinguish the elaborated EEG changes from the background ones, each cycle, from the onset of one signal to the onset of the next one, was divided into an equal number of periods. In each, the presence or absence of a blockade of the alpha rhythm was marked. Data relating to the corresponding periods of 15–20 successive cycles (or of the whole experiment) were then summed, and, on the basis of the results obtained, graphs were formed showing the frequency of emergence of the alpha-rhythm depression in each period of the interval between the stimuli. During statistical processing of the GSR, the intervals were divided into a smaller number of periods. When forming graphs illustrating the frequency of emergence of the GSR, we took into account the latency of this reaction (1.5 sec).

a. Indifferent stimulation. A significant emergence of an alpha-rhythm blockade before indifferent stimulation was observed in only 4 subjects in the first half of the experiment at intervals of medium (15 sec) duration (Fig. 3A). At the same time, there was observed a GSR which began to weaken approximately by the 30th stimulation and almost fully disappeared by the end of the experiment (Fig. 3B). Thus, modifications of the EEG and GSR preceding the stimuli appeared regularly after a definite number of stimulations and then became extinguished. In the case of short (2–3 sec) and long (30 sec and over) intervals between the indifferent stimulations, no such modifications of the EEG and GSR were in evidence.

b. Signaling stimulation. When a signaling meaning is imparted to the stimulus with the aid of a preliminary instruction, definite electrographic changes begin to manifest themselves before the onset of the next stimulation. Such changes of electrographic reactions depend on the duration of the intervals between the stimuli. Thus, in the case of short intervals (3 sec), there is observed an anticipatory blockade of the alpha rhythm but no anticipatory GSR (Fig. 1; Fig. 4A and B) while for medium intervals (12–15 sec.) there is an anticipatory blockade of the alpha rhythm and a weak anticipatory GSR (Fig. 4C and D). In the case of long intervals (30 sec and over) the frequency of emergence of an alpha-rhythm depression before the next stimulation is not distinguished from that in the background, but the frequency of the anticipatory GSR is very high (Fig. 4E and F; Fig. 5).

After each experiment, the subjects were questioned to find out their subjective evaluation of their participation. Such evaluation differed depending on the experimental conditions. In the case of short intervals, the subjects noticed that the stimulus was switched on rhythmically at definite intervals, but did not

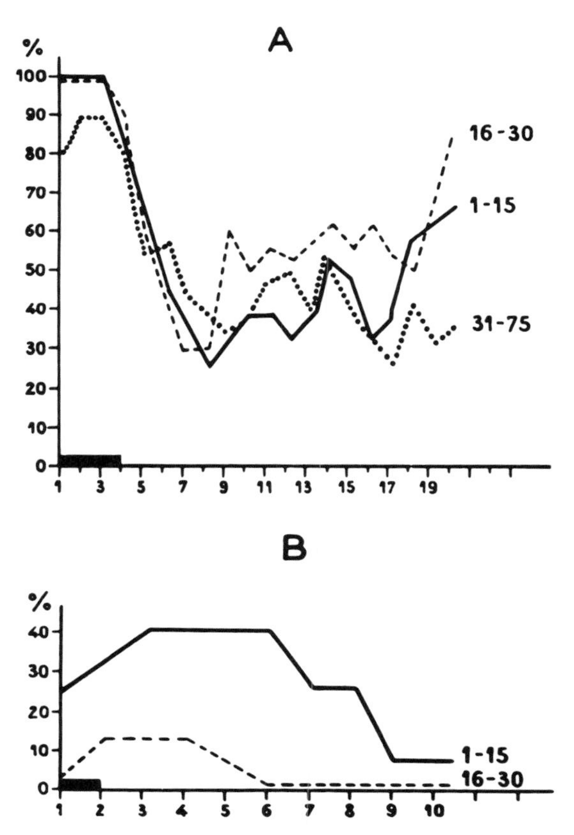

FIG. 3 Averaged data relating to one experiment. Modifications of the EEG (A) and GSR (B) throughout the interval between regular stimuli. Indifferent stimuli; interval, 15 sec. Abscissa, number of sections in each interval; ordinate, frequency of emergence of an alpha-rhythm blockade or of a GSR. The figures indicate the number of serial numbers of the stimuli on the basis of which each curve was formed. The dash along the abscissa indicates the duration of each stimulus. Note increases in anticipatory alpha blockade (3A) during the first half of the experiment (1–15 and 16–30) and their absence later (31–75). Anticipatory GSR also extinguished during the course of the experiment (3B).

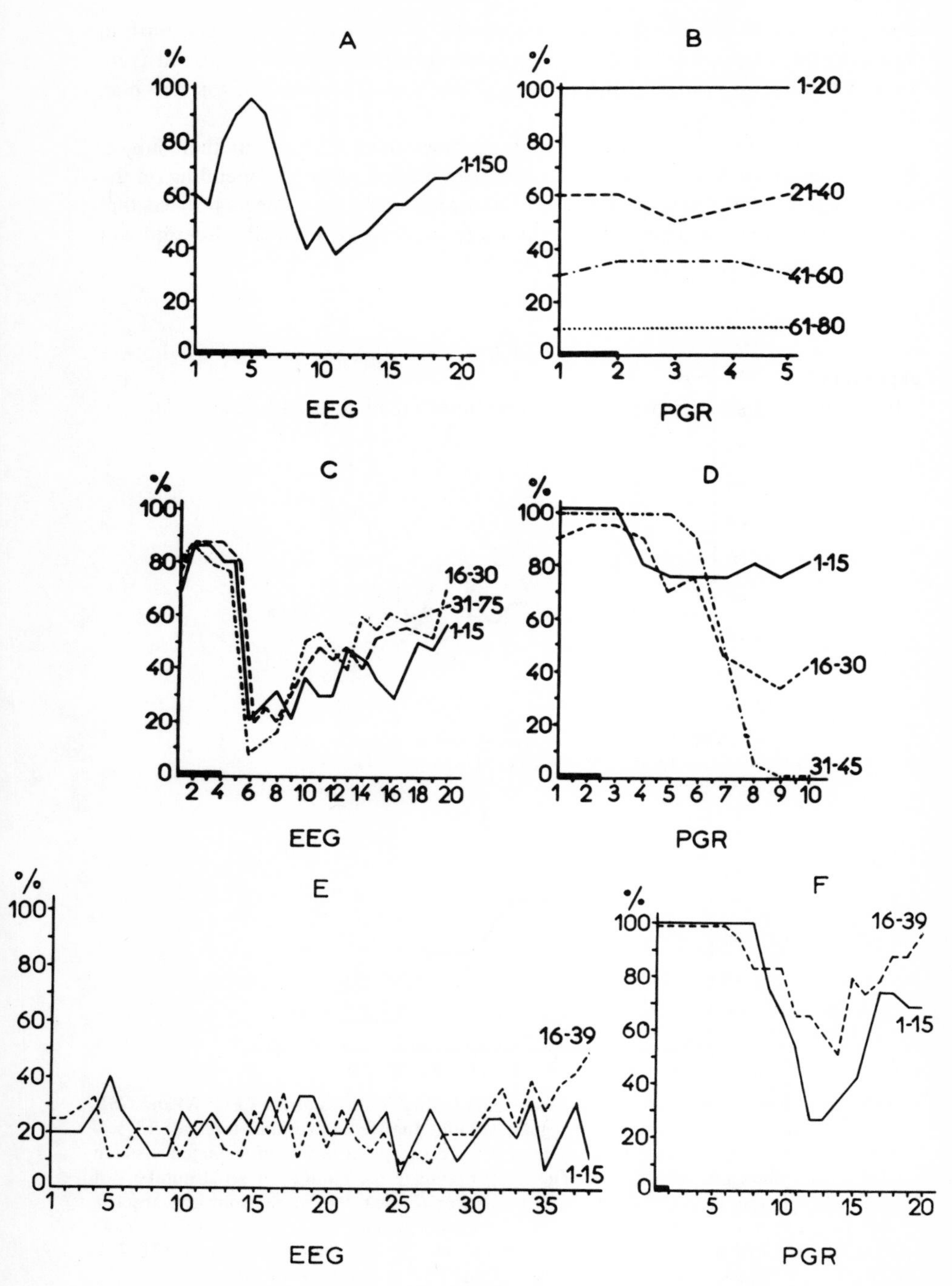

A
%
100
80
60
40
20
0
1
5
10
15
20
1-150
EEG
B
%
1-20
21-40
41-60
61-80
2
3
4
PGR
C
16-30
31-75
1-15
6
8
12
14
16
18
D
1-15
16-30
31-45
7
9
E
16-39
25
30
35
F

FIG. 4

attempt to determine the duration of the interval by means of counting. In the medium interval condition (12–15 sec), some subjects sometimes tried to determine the duration of the interval by counting, and thereby to consciously foresee the time when the next, regular stimulus would be switched on. In the case of indifferent stimulation, the subjects' interest in the course of the experiment usually proved to be of a short duration, while in the case of signalling stimulation, their participation in the experiment differed: some of them determined the moment of the subsequent stimulation by way of counting, and prepared themselves for the required response; others showed no interest whatever in the course of the experiment and merely accomplished the required action in response to each stimulation, in accordance with the preliminary instruction. For long intervals, all subjects exhibited little interest in the experiment.

In these particular experiments we sometimes failed to establish a correlation between the degree of expressiveness of the changes preceding the stimulus and the evaluation by the subjects of their attitude towards the experiment.

In order to ascertain the dependence of these changes on verbal processes, experiments were carried out with the application of the following instruction: "During the course of the experiment, watch for the moment when the regular stimulus is switched on, and respond to it as quickly as possible." In this case the anticipatory changes assumed a more pronounced and constant character in all subjects. In experiments with long intervals (30 sec and over), the frequency of emergence of a GSR before stimulation increased almost to 100% but no increase of the significant anticipatory depression of the alpha rhythm was observed (Fig. 6A, B).

DISCUSSION

Our data show that the parameters of the processes (EEG, GSR, and others), which arise during the presentation of regular stimulation and during the intervals between them, are determined by the duration of these intervals, by the indifferent or signalling property of the stimulus, and by the character of the signalling property and the subject's attitude to the course of the experiment (set, attention, automatism) which depend on many factors, sometimes not yielding to any control.

First, the following question arises: With the help of which physiological processes does the time factor determine the (1) character of the response of the

---

FIG. 4 Averaged data relating to one experiment: Signaling stimuli. (A) frequency of emergence of an alpha-rhythm blockade at a short interval; (B) frequency of emergence of a GSR ("PGR") at a short interval; (C) frequency of emergence of an alpha-rhythm blockade at a medium interval; (D) frequency of emergence of a GSR at a medium interval; (E) frequency of emergence of an alpha-rhythm blockade at a long interval; (F) frequency of emergence of a GSR at a long interval. Other designations are as in Fig. 3.

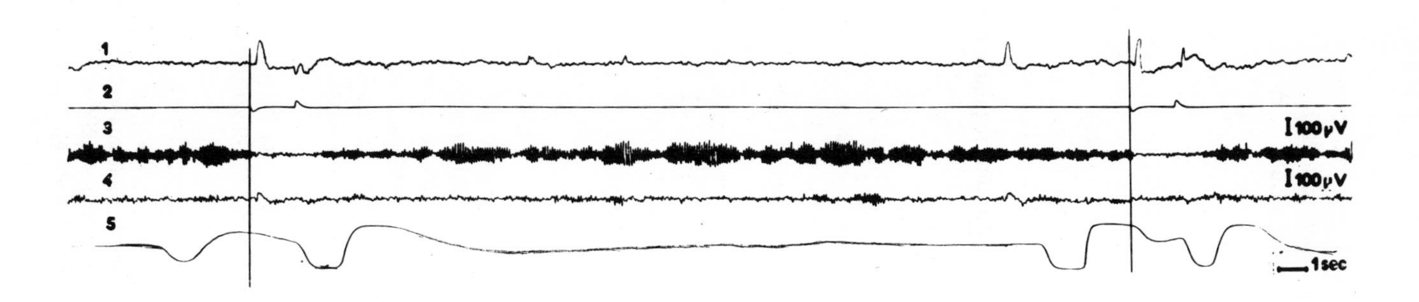

FIG. 5 A long interval. An example of an anticipatory GSR in the absence of an anticipatory blockade of the alpharhythm. 1, oculomotor reaction; 2, stimulus; 3, EEG right occipital; 4, EEG sensorimotor; 5, GSR.

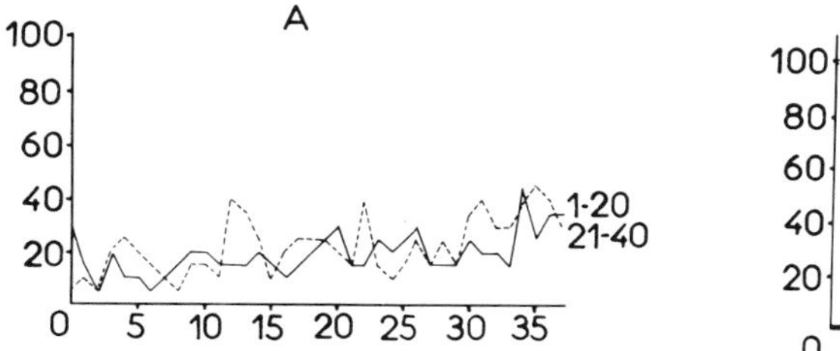

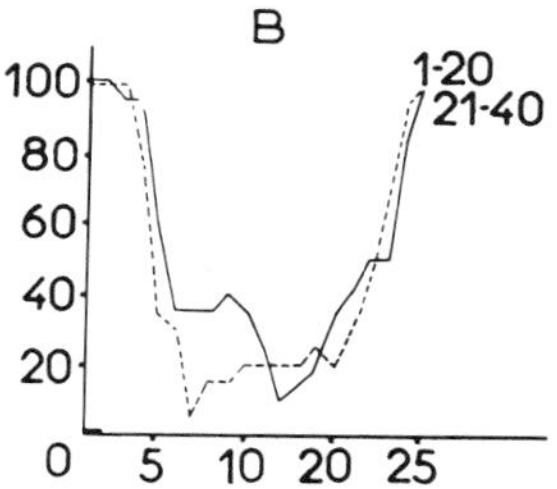

FIG. 6 Frequency of emergence of an alpha-rhythm blockade (A) and a GSR (B) at a long interval after an activating instruction. The graphs present the data obtained from processing only the last 10 sec of the interval. Other designations are as in Fig. 3.

nervous system to the direct action of a stimulus regularly presented, and (2) the establishment of connections between the stimuli (which are revealed by the development of anticipatory reactions)?

According to classical concepts, the orienting reflex is a reaction to "novelty," when a stimulus is first presented or when the parameters of an already habitual stimulus are changed (its quality, appearance at an unusual moment, or absence in the usual place).

In recent years the concepts of the orienting reflex have been further developed (Sokolov, 1958, 1964) by a number of investigators. Investigations have shown that factors other than novelty influence the orienting reflex, specifically the significance of the stimulus, and the probability of emergence of a signalling or unknown stimulus within a given space of time (Sokolov, 1964).

According to the results of our investigation, the electrographic changes (depression of the alpha rhythm, GSR) may reflect different processes; therefore, a study of the specific features of their dynamics makes it possible to reveal the participation of different mechanisms in the orienting reflex, depending on the form and stage of the experiment.

When indifferent or signaling stimuli are regularly presented at short intervals, an orienting reaction to their "novelty" is observed only in response to the first stimulations; however, a depression of the alpha rhythm to the light persists for a long time (apparently due to specific sensory activity) while the GSR is already extinguished. A rapid extinction of the orienting reflex to regular stimulation at short intervals is, apparently, due to the fact that soon after the first applications, the probability of emergence of a stimulus equals 100% (Sokolov, 1964). In the case of an indifferent stimulus, the orienting reflex to particular stimuli and to the entire course of the experiment is, as a rule, of a short duration; because of this, the formation of a connection between the regular stimuli is also very unstable. In the case of a signalling stimulus, the orienting reflex proves to be of a more pronounced and protracted character. A stable connection is formed between the regular stimuli and the corresponding motor reactions on the basis of trace processes from the preceding stimuli; evidence for this connection is given by the emergence of more or less stable anticipatory reactions before each stimulus. As a result of this process, a stereotype of excitations is formed in the nervous system (a "neuronal model of the stimulus," Sokolov, 1963a, b, c), and due to this, the orienting reflex becomes extinguished. This assumption is based on the following facts.

At short intervals, the GSR becomes quite rapidly extinguished both to indifferent and signaling stimuli. Nor does it appear before the next regular stimulus. But in the course of the experiment, the EEG reactions during the interval show a definite tendency to gradually increase and are most pronounced by the end of the interval (Fig. 4A). From this it follows that the EEG reaction during the interval reflects the establishment of a connection and the appearance of conditioned changes at a definite phase of the trace process. The absence of any appreciable increase in muscular tone before the stimulus and the emergence of movement in the usual place, when one of the regular stimuli is omitted, also indicate that after the formation of a connection at short intervals the orienting reflex to each stimulus, as well as to the course of the experiment, is no longer of essential significance. This is apparently due to the fact that at short intervals a stable motor stereotype establishes itself quite rapidly; in other words, the motor conditioned reflex becomes automatized, as a result of which the "expectancy" orienting reaction disappears. The formation of such a stereotype of excitation at short intervals between the stimuli is determined by trace processes, and in the case of signalling stimuli, mainly by traces of kinesthetic stimulations.

At a longer interval between the stimuli, when the trace processes no longer play an essential part in the formation of a connection due to their weakening with the lapse of time, some other mechanisms come into action. The strongly pronounced character of the changes, both in response to stimuli and during the interval, and their stability in the course of the whole experiment indicate that a modification of only the temporal relationships between the stimuli (an increase of the interval from 2–3 to 12–15 sec)—all other conditions being equal—markedly enhances the role of the orienting processes in the formation of a response both to a directly acting signaling or indifferent stimulus and to the preparation for its accomplishment. In the case of these temporal relationships between stimulations, the conditions for the automatization of the elaborated connections are less favorable than in the case of short intervals; this is reflected in the stability of the EEG and GSR responses to the direct action of the stimulation, as well as in the stability of the elaborated GSR and EEG modifications that precede the stimulation. The orienting reflexes to each stimulus and to the course of the experiment as a whole are of a quite stable character. This stability is indicated by changes taking place before the subsequent stimulation, by a rise of muscular tone during the last seconds before the onset of a signalling stimulus, and by the absence of a complete motor reaction when one of the regular stimuli is omitted. Consequently, not only the signaling property of the stimulus but also the time factor are of great importance for maintaining the orienting process on a high level.

It is obvious that the orienting reaction (expectancy reaction) may sometimes disappear in conditions of longer intervals, when the motor conditioned reflexes become automatized as a result of long training. Thus, the two kinds of orienting reflex mentioned before may be absent in conditions of short intervals between the stimuli, as well as in the case of long training of the motor conditioned reflex.

The fact that, after a certain training of the temporary connections, the orienting reflex arises only by the end of the interval gives rise to the question of the mechanisms which are responsible for this phenomenon. Our data, as well as data available in the literature, allow us to state with confidence that an essential role in this process is played by the ability of the nervous system to register time periods, a phenomenon which is called "biological clock," "reflex to time," etc. in the literature. Although the essence of this mechanism still remains uncertain, we can make use of its general concept for elaborating a working scheme of the processes which determine the emergence of responses to regular, indifferent, and signalling stimuli at intervals with such a duration when the trace processes no longer play an essential role in the formation of connections. When the interval is increased, the changes arising in response to the stimulations and the anticipatory reactions developing in the course of the experiment assume a stronger and more constant character. It may therefore be supposed that an increase in the interval makes the time of stimulation more uncertain, since a precise fixation of the moment when the expected stimulus is switched on proves to be impeded. Apparently in those conditions, when due to an increase of the interval above a definite level, the nervous system has difficulty in determining the moment of onset of the expected stimulation, the orienting mechanisms become activated; this indicates that the time of the onset of the next stimulus is no longer fixed by the nervous system with a high degree of probability. This is testified to by the fact that when the interval between the signalling stimuli is increased to 30 sec, there is, in almost 100% of all cases, an anticipatory GSR which reflects best the uncertainty and the orienting reflex.

Reinforcement of the signalling meaning of the stimuli with the help of a corresponding instruction makes it possible to increase the interval between them. This leads to the assumption that a special instruction allows the admissible interval to increase to a considerable degree, as a result of an additional activation of the orienting reflex through the second signalling system (heightened attention to the experiment) and utilization of such mechanisms, such as oral counting, which replaces the biological "clock."

19

FORM OF THE EEG WAVE AS AN INDICATOR OF ACTIVATION PROCESSES

E. Yu. Artemieva
E. D. Homskaya

Moscow State University

A number of experiments have been carried out in recent years which suggest that the study of the form of single EEG waves yields significant data. These studies provide evidence that the characteristics of the wave form may be a useful addition to usual spectral frequency parameters of the EEG when investigating various functional states (Artemieva & Meshalkin, 1965; Artemieva, Meshalkin, & Homskaya, 1965; Genkin, 1963; Saltzberg & Burch, 1957, 1959).

Average Level of Asymmetry

One of the characteristics of the EEG wave form—an average level of asymmetry of ascending and descending fronts (rise and fall times) of alpha waves—was investigated by Genkin (1963). A special parameter, Δ_T, was introduced for measuring the average asymmetry of the duration of the fronts during a specified period time, *T*. Genkin found that in normal adult subjects the average level of asymmetry in the stable state of rest remains constant; but changes (drowsiness, fatigue, intellectual work) are accompanied by changes in the level of asymmetry (Genkin, 1963). A definite correlation was obtained between the level of asymmetry and the performance of tasks which require fixed attention (Genkin, 1964). It was also established that changes in the level of asymmetry depend on the initial state: For example, the higher the initial level of asymmetry in the background resting state, the higher it is during the active state.

Thus, according to the results obtained by Genkin, there exists a correlation between the average level of asymmetry of the alpha waves and the functional state of the subject.

The existence of such a correlation suggests that the alpha-wave asymmetry indicator is related to the nonspecific forms of cerebral activity which occur during orienting reactions. This assumption could be proved by finding a correlation between the asymmetry level and the laws of the orienting reflex. The goal of this investigation was to verify this assumption.

The "behavior" of the Δ_T parameter was investigated in normal subjects (20–30 years of age) who had a stable alpha-rhythm in different experimental conditions. The complete program of an experiment was carried out on 8 subjects. The relatively small number of subjects is accounted for by the difficulties of data processing, and also by the necessity to specially select subjects having a stable alpha rhythm.

One or two experiments were performed on each subject. We achieved a stable extinction of the alpha-rhythm depression to intermittent acoustic stimuli (60 dB, 0.5 sec, 6–12 sound pulses in each trial) before starting the actual experiments in order to have alpha-rhythm for calculating Δ_T. Recording was done on an "Alvar" encephalograph. The EEG was recorded from the parietooccipital regions of the left and right hemispheres (by the bipolar method). The subject sat with his eyes closed in a dark screened chamber. The absence of any

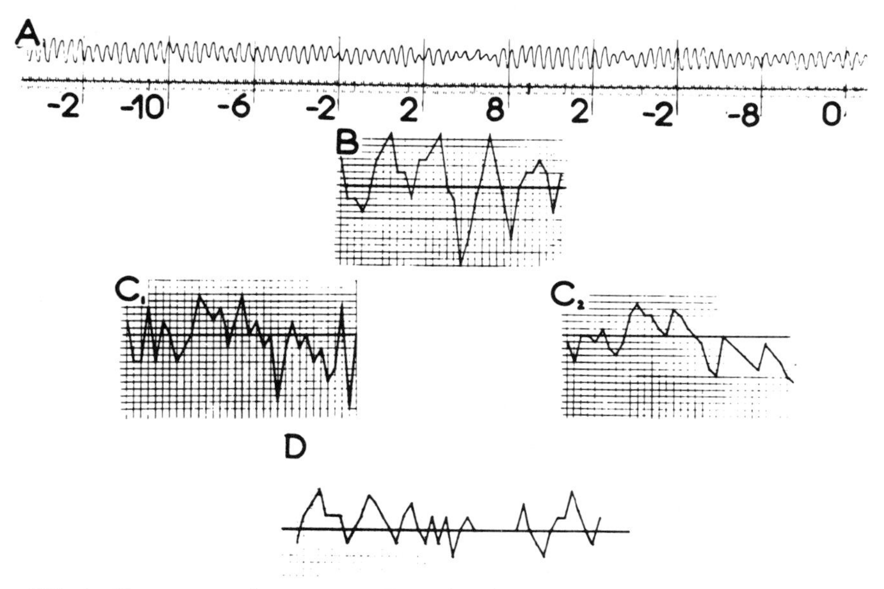

FIG. 1 Time course of asymmetry. Examples of the oscillatory form of the EEG waves (Subject L; Recording from the occipito-central region of the left hemisphere). (A) real EEG record with marks of the automatic analyser for asymmetry. When thoroughly examining the record, one can see the full period within which one transformation of the form of the wave took place. Figures below: values of asymmetry of the Δ_1 for the corresponding sections of the record. (B) regular G waves; the third wave from the left corresponds to the record shown in (A). (C_1 and C_2) "unorganized" G waves: C_1, oscillations of Δ_1; C_2, oscillations of Δ_2 estimated for overlapping intervals with a shift of 1 sec. (D) example of disorganization of G waves which preceded the spontaneous depression of the alpha rhythm. In this and other figures, each square represents one sec.

alpha-rhythm depression in response to 10 or more successive stimulus presentations was accepted as the criterion of extinction of alpha blockade. Three series of experiments were then carried out: (*1*) presentation of uniform intermittent acoustic stimuli, (*2*) introduction of a verbal instruction ("Count the sounds in the trains"), and (*3*) rescinding the instruction ("Don't count any more"). Twenty to twenty-five trials were presented in each series. During the presentation of sounds in all three series, alpha-rhythm depression was observed once or twice in only three subjects. In the rest of the cases, the EEG during the acoustic stimulus was visually indistinguishable from the background EEG and did not manifest any signs of alpha-rhythm depression. An example of such a record is presented in Fig. 1A.

Equal periods of the EEG records (2–5 sec) were subjected to analysis; they related to the presentation of a train of acoustic signals and to the prestimulus background activity. To estimate Δ_T for each period, the following procedure was used. The EEG was automatically evaluated every 25 msec (Artemieva & Meshalkin, 1965). The analyzer determined whether the phase of the alpha rhythm was ascending or descending, at each sample point. An ascending phase was denoted by a vertical line, and a descending phase by a dot, which were automatically printed on the write-out, below the EEG. The following formula was used to calculate Δ_T for each period:

$$\Delta_T = \frac{f_a - f_d}{n},$$

where f_a is the number of times a sample revealed an ascending front, f_d is the number of times a sample revealed a descending front, and n is the number of waves during the period which was analyzed.

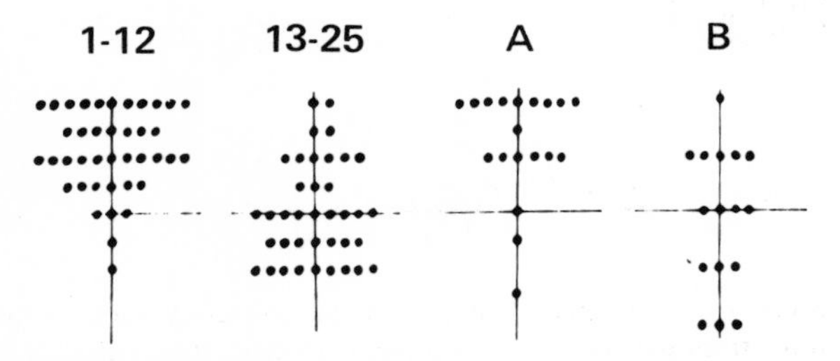

GRAPH 1. Relationships between the levels of asymmetry for background activity preceding a stimulus and asymmetry during action of the acoustic stimulus. The numbers denote trials of stimulus action; (A) when the signals were counted; (B) when the signals were no longer counted. Dots above the horizontal line indicate more asymmetry in the background than during the acoustic signal; dots below the line, vice versa. [*Ed. Note:* The text failed to provide a clear indication of what a single dot represented. However, it is clear that asymmetry was reduced, relative to the prestimulus background period, during trials 1–12 and also during counting of the acoustic stimulus (A).]

A typical record is shown in Fig. 1A. This covers a span of 10 sec during which time Δ_T was calculated for each 1-sec period. For example, $\Delta_T = -10$ (the second period in Fig. 1A) by (15–25)/10.* A $\Delta_T = 0$ would indicate no asymmetry during a period (last period of Fig. 1A).

Experiments showed that the average value of asymmetry of the ascending and descending fronts of the alpha waves changes in differing modes during the first and subsequent presentations of sounds. To the 1st–12th sound (in the first series of experiments) all subjects exhibited definite reductions of asymmetry in comparison with spontaneous activity (Graph 1, "1–12"). During subsequent presentations of acoustic stimuli in the same series of experiments (13th–25th presentations), the asymmetry of the alpha wave did not change significantly; the relation between the values of asymmetry for spontaneous activity and those observed during an action of the acoustic stimuli was random (Graph 1, "13–25").

The statistical significance of the above-described results was established according to the *F* test and its simplest form for analysis of variance.

When the stimulus assumed a signal meaning (by means of the verbal instruction "Count the sounds"), the reduction in asymmetry of the alpha waves reappeared (Graph 1A). But when there was no longer a requirement to count the stimuli, the asymmetry did not differ between the spontaneous activity and activity during the action of the acoustic stimulus (Graph 1B).

It must be specially pointed out that the changes of asymmetry described above could in no way be merely a reflection of frequency changes in the EEG: first, in almost all cases, the basic frequency of the rhythm—as far as it may be judged on the basis of a visual analysis—did not change during the stimulus presentations (in one case this frequency constancy was verified by special analysis); second, as already stated previously, Genkin could not find a correlation between the Δ_T and the EEG frequency.

Time Course of the Asymmetry Level

It might be assumed that not only the asymmetry level itself, but also its temporal and spatial dynamics, are closely connected with the functional state of subject.

Indeed sometimes periodic changes of the form of the alpha-wave may be observed even visually in the EEG, when recorded at a sufficiently high paper speed. One such case is presented in Fig. 1A. In less distinct cases such changes may be detected by means of special methods. For example, when investigating the "behavior" in time of the parameter Δ_1 (mean value of asymmetry for each second), we divided the EEG record into 1-sec periods and evaluated the Δ_1 for every one. The method of this evaluation was explained previously. The only difference is that, in the given case, $T = 1$ rather than 2–5 sec.

**Ed. Note:* Δ_T values apparently were multiplied by 10 to remove the decimal point.

These graphs show that in normal subjects in the state of rest, there are regular, rhythmic oscillations of asymmetry (Δ_1) with a period of 6–8 sec (Fig. 1B). We call these oscillations "G waves." The duration of the period does not depend upon the average level of asymmetry. Such oscillations in asymmetry are recorded all over the surface of the scalp and have the same period in all channels. It must, however, be pointed out that in some observations in the state of rest no periodic oscillations of Δ_1 could be detected. Apparently, the so-called "state of rest" includes a number of different states. Particularly distinct G waves, (i.e., oscillations of the degree of asymmetry having a period of 6–8 per sec), visible without a special analysis, are recorded only within a narrow range of states which may be conventionally called "active rest." In the case of slight changes in the subject's state which are due to fatigue, stress conditions, etc., detection of the G waves becomes complicated. For example, as a rule, it is impossible to record G waves in students during the academic examination period, although at other times distinct and regular G waves are in evidence. But even in these states, the G waves can be detected in almost all cases with the help of special methods. We show two of them here. First, in the original record one picks out the sections with well expressed rhythmic oscillations, and Δ_1 is estimated only for those sections. The points are plotted on a graph. Then, if we know the period of the G waves of this person, we can compare the empirical estimate of Δ_1 with the predicted one. Sometimes one manages to obtain very strong confirmation of the existence of G waves. The second method is to use Δ_2 or Δ_3 (i.e., perform the Δ_T calculations for periods of 2 or 3 sec).

During presentation of sensory stimuli as well as during intellectual activity (for example, when imagining a familiar picture, clenching a fist, or computing mentally) the G waves become disorganized (Artemieva, Meshalkin, & Homskaya, 1965). In this case the special statistical methods described above no longer help to detect regular G waves (Fig. $1C_1$, $1C_2$).

It is noteworthy that such a destruction of the G waves was often observed by us immediately before spontaneous alpha-rhythm depression (see Fig. 1D). This is another proof that the dynamics of asymmetry depend upon the level of alertness and that the changes of asymmetry are finer indicators of the functional state than the previously known measures (e.g., EEG frequency).

This suggests that the irregular, chaotic changes of the form of the EEG wave correspond to an enhancement of alertness as is in the case of a depression of the alpha-wave, when a reaction of desynchronization appears.

In order to establish the orienting nature of the oscillations Δ_1, we carried out experiments on the same 8 subjects according to the same program described previously. We investigated the change of the average level of asymmetry during the presentation of intermittent acoustic stimuli–indifferent and signaling. Disruption of the G waves obviously depends on the number of the presented sound. It is observed in response to the 1st–7th presentations and disappears during the subsequent presentations. When the instruction "Count the sounds" is given, the derangement of the periodicity of the Δ_1 oscillations reappears.

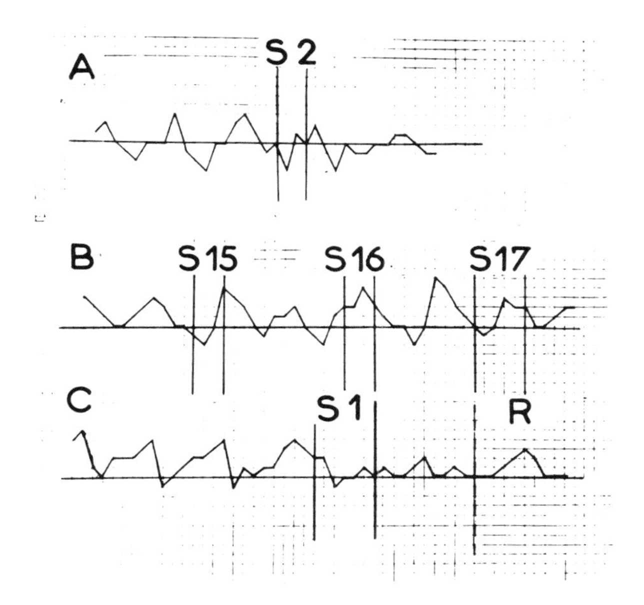

FIG. 2 Periodic oscillations of the alpha-waves' asymmetry (G waves) during presentation of the intermittent acoustic stimulus in subject G. (A) the 2nd sound presentation; the periodicity of the oscillations of asymmetry is deranged. (B) the 15th, 16th, and 17th sound presentations; the periodicity of the oscillations of asymmetry is unaffected. (C) after the instruction "Count the sounds"; the first presentation of the signal evokes chaotic changes in the alpha waves' asymmetry. R, response.

This can be seen from the graphs in Fig. 2 which present a time-course of Δ_1 in one of the subjects during the three series of experiments. In response to the 2nd sound, the periodicity of oscillations of the alpha-waves' asymmetry is greatly reduced (Fig. 2A). However, during presentation of the 15th, 16th, and 17th acoustic stimuli, the G waves are of a relatively regular character (Fig. 2B). When the subject counts the sounds in a train (Fig. 2C) the periodic oscillations of asymmetry are again abolished.

Thus, the "behavior" of the average value of the asymmetry and the dynamics of Δ_1 (G waves) in the background activity and during the action of nonsignaling and signaling stimuli are similar to the modifications of different components of the orienting reflex within the system of unconditioned and conditioned reflexes. They disappear as a result of numerous repetitions of the stimulus (like extinction of the orienting responses to indifferent stimuli) and recover when a signal meaning is imparted to the stimulus (cf. recovery of the orienting reactions during the formation of a conditioned connection).

Spatial–Temporal Dynamics of the Asymmetry Level

The problem of the spatial changes of the oscillations of asymmetry of the EEG waves is much less clear. When recording the EEG from the parietooccipi-

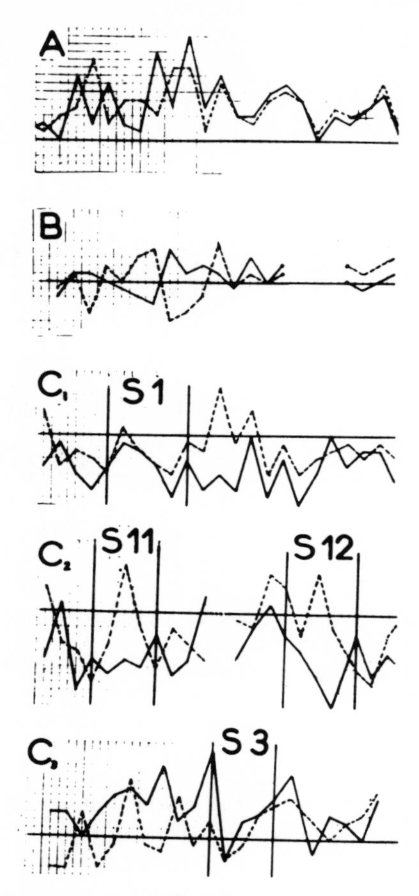

FIG. 3 Spatial dynamics of the oscillations of asymmetry (Subject L; occipitocentral recording; solid line right, dotted line, left hemisphere). (A) following a period characterized by random ratio between the oscillations' phases, a period of longlasting synchronization arises. (B) synchronization of the oscillations before the spontaneous depression of the alpha rhythm, and immediately after it. (C_1) spatial synchronization of the oscillations during the first presentation of the intermittent sound; (C_2) desynchronization due to numerous presentations of the sound; (C_3) resynchronization of the oscillations of asymmetry after the instruction "Count the sounds." Details in the text.

tal, central, and posterofrontal regions of the left and right hemispheres (monopolar recording) it was found that distinct G waves are moving (display successive phasic changes) along the surface of the scalp from the frontal toward the occipital region with a velocity of 3–6 cm per sec.

It proves particularly difficult to analyse records with "unorganized" asymmetry oscillations, from which we failed to detect the G waves by means of all the known methods.. Such oscillations seem to bear a random character. However, during simultaneous EEG recording from similar regions of the left and right hemispheres, a coincidence of the phases of the Δ_2 changes is sometimes observed within a period of 20–40 sec. An example of such a coincidence lasting 25 sec can be seen in Fig. 3A, which presents oscillations of the Δ_2 in a 50-sec section of the EEG, recorded by a bipolar method from the occipito-central parts of the left and right hemispheres in Subject L.

Periods of such coincidences appear almost invariably before spontaneous depression of the alpha-rhythm and immediately after it (Fig. 3B); then oscillations with random phasic correlations are recorded again. A coincidence or non-coincidence of the phases of irregular Δ_1 and Δ_2 oscillations do not depend upon the average asymmetry level. These facts, which were observed in all of the investigated subjects without exception, make us doubt whether the irregular asymmetry oscillations bear a random character. As to their spatial organization, it may be assumed that the spatial characteristics of asymmetry likewise reflect the subject's functional state.

The fact that the spatial correlation of the asymmetry oscillations is connected with the level of alertness is also confirmed by the coincidence of the form of the irregular asymmetry oscillations during intellectual activity (mental computation, etc.).

Such coincidences of the form of the Δ_2 irregular oscillations were also observed in experiments with the application of indifferent and signaling sounds. During the first applications of sounds, nonsynchronous oscillations of the Δ_2 in similar parts of both hemispheres became synchronized (Fig. $3C_1$), whereas subsequent sound presentations did not evoke such an effect (Fig. $3C_2$). However, the command "Count the sounds" resulted in the reappearance of synchronization (Fig. $3C_3$).

DISCUSSION

The utilization of the form of the EEG wave as an indicator of functional state proves to be fully justified in the light of the previously mentioned facts. The existence of such a correlation with the functional state might be assumed on the basis of a number of studies (Callaway, 1962; Callaway & Lansing, 1957; Yeager, 1960), which demonstrate the functional difference of various EEG phases. An analysis of the form of the EEG waves revealed at least three indicators of the functional state: the average level of the waves' asymmetry, the periodic oscillations of this level, and the spatial characteristics of the asymmetry changes. All three parameters possess some common features and apparently reflect the activity of a slow-acting system which regulates the functional states. The temporal parameters of this system coincide with the temporal characteristics of the system responsible for the slow bioelectrical activity which was investigated by Aladjalova (1964). It must be assumed that besides quick-acting mechanisms regulating the state of activity, there exists a slow-acting system whose operation has not yet been sufficiently studied.

As to the spatiotemporal dynamics of asymmetry, it should be pointed out that the correlation between spatial synchronization of the bioelectrical activity and a definite functional state was repeatedly discussed in the literature. Thus, in a series of experiments concerned with elaboration of a conditioned reflex, Livanov demonstrated that, at a definite stage of the coupling of the conditioned connection, there emerges a synchronization of the bioelectrical activities from some areas of the cerebral cortex which participate in the process of such coupling (Livanov, 1960).

The experiments described above showed that all three characteristics of the wave (the asymmetry level, its temporal, and temporo-spatial dynamics) are correlated with the orienting reflex. The characteristics of the form of the wave show a change during the first presentations of indifferent stimuli; they disappear with the repetition of the stimulus and reappear when a signal meaning is imparted to the stimulus. The investigation showed that the analysis of the form of the EEG wave may provide valuable information concerning the functional state of the subject.

20

REFLECTION OF THE PROCESSES OF LOCAL ACTIVATION IN THE DYNAMICS OF EVOKED POTENTIALS

E. G. Simernitskaya

Moscow State University

The problem of correlation between the EEG and mental functions has been investigated by many authors. The investigation of evoked potentials (EP), which reflect local synchronous excitation of a group of neurons in response to afferent stimulation, has been of particular significance.

From the literature it follows that EP, especially their secondary and later components, can be regarded as sensitive indicators of the functional state of the central nervous system. They are closely associated with the processes of attention (Dawson, 1958; Garcia-Austt, Bogacz, & Vanzulli, 1959), and depend on the signaling function or meaning of a stimulus (Dumont & Dell, 1958; Bremer & Stoupel, 1959; Cobb & Dawson, 1960; Shumilina, 1965; and others). Even the color and the structure of the object perceived (Chapman & Bragdon, 1964; Spehlmann, 1965) exert marked influences on parameters (e.g., amplitude or configuration) of the EP.

Some investigators consider that when attention is drawn to a stimulus by means of a verbal instruction, this correlates with an increase of EP amplitude (Davis, 1964; Dawson, 1958; Garcia-Austt *et al.*, 1959). Others believe that in the same conditions the amplitude decreases (Samson, Samson-Dollfus, & Pinchan, 1959). Apparently, these contradictory data reflect the fact that the attention reaction is accompanied, on the one hand, by an increase of the excitability of the sensory system, and on the other, by selective inhibition of all accessory forms of activity (Sokolov, 1959). From this point of view, the specific character of the verbal instructions, with the help of which attention is drawn to the stimulus, is a highly important factor influencing the pattern of EP modifications.

Questions relating to the influence of verbal instructions on the EP in man have not yet been sufficiently studied. Their elucidation is of importance for ascertaining the type of interaction between the sensory systems and mental activity.

The purpose of the work presented here was to investigate the correlation between the EP and the specific character of the subject's activity, organized with the help of verbal instructions. We proceeded from the hypothesis that different forms of activity demand the participation of various cortical areas. Therefore, it should be expected that different verbal instructions, used to modify the signalling function or meaning of one and the same stimulus, will be accompanied by diverse phenomena of local cortical activation which can be revealed with the help of EP.

METHODS

In the course of the investigation we compared EP to flashes (50 msec) under two conditions: when the flashes served as indifferent and as signalling stimuli. The investigation consisted of 7 series of flashes, each series including 10 flashes.

In the first series, the light stimuli were preceded only by a general instruction of the following type: "Sit quietly; you mustn't do anything." Then, a signaling meaning was imparted to the flashes by one of the following three verbal instructions:

1. "The 8th–10th flashes will be followed by an electric shock to the left hand." (However a painful reinforcement was never applied).
2. "Watch the duration of the flashes, and after each second flash state which of them was shorter." (In the course of the entire investigation the duration of the flashes was constant).
3. "When you see a flash, press the button with the right hand as quickly as possible."

The choice of the instructions mentioned above was determined by the fact that each of them was aimed at activating a definite cortical zone constituting the central representation of a corresponding analyser. Thus, an expected electrocutaneous stimulation was regarded as involving the sensorimotor area of the right hemisphere. An expected differentiation of the flash duration would involve the visual areas of the cortex. A motor response to flashes would involve the visual areas and the motor area of the left hemisphere. In accordance with this, the EEG was recorded from the parietooccipital and central–frontal regions of both hemispheres.

After each series involving one of the three instructions, the task was cancelled and the following series was given under "quiet" conditions. Comparison of the EP from these control conditions with those from the preceding "instructed" condition permitted determination of the extent to which modifications of EP were really caused by the instruction.

Evoked potentials were recorded on an "Alvar" 10-channel electroencephalograph. Sections of the record, having a duration of 500 msec (100 msec prior to the action of the stimulus and 400 msec after it), were subsequently magnified.

Single records obtained in this way were measured graphically with a (digitizing) interval of 10 msec separately for each of the four analysed regions (parietooccipital and central–frontal regions of the right and left hemisphere) in all 7 series of the investigation.

Ten normal subjects were used. Data from individuals were pooled for purposes of statistical analysis. Differences between instructional and control series were evaluated by the *t* test. Statistical processing of the EP was accomplished on an electronic computor of the M-20 type according to a special program.

RESULTS AND DISCUSSION

EP Dynamics during Expectation of an Electrical Shock

As stated above (see "Methods"), in the case of the instruction, "The 8th–10th flashes will be followed by an electric shock in the left hand," no painful

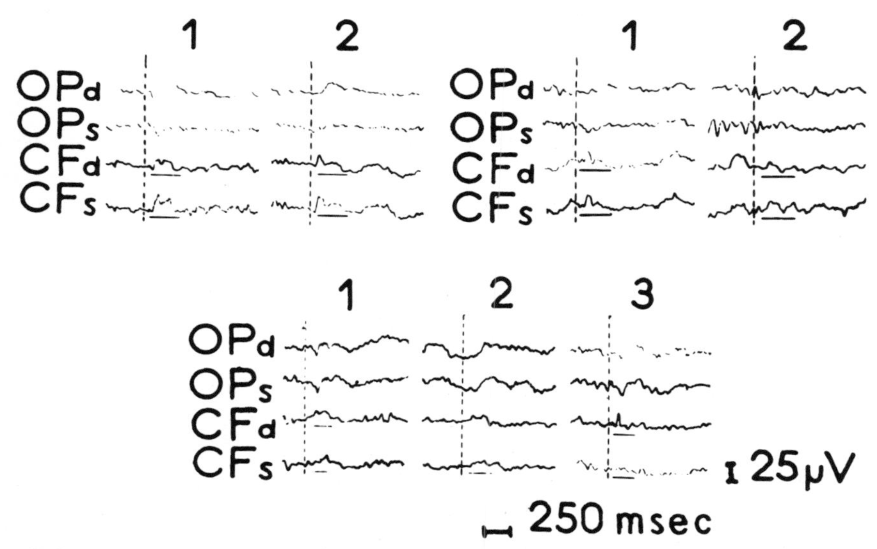

FIG. 1 Dynamics of EP during expectation of an electrical shock. (Top) Records of Subject E. (Left) General effects–1, presentation of an indifferent stimulus; 2, 2nd flash in series in which shock is expected, note changes in all regions. (Right) Specific effects–1, 8th (signaling) flash is accompanied by specific depression of second negative wave in central–frontal regions; 2, 2nd flash in subsequent control series, EP modifications are gone. (Bottom) Subject B. Responses to 7th flash in series during presentation of an indifferent stimulus (1) and expectation of shock (2). Response to 10th flash, which shock would presumably follow, is characterized by specific increase in the amplitude of the first negative wave in the central–frontal region of the right hemisphere only. In this and following figures, vertical dashed line indicates flash presentation. The horizontal line denotes EP from the central–frontal region. Polarity–negative is up. Abbreviations–OP, occipitoparietal; CF, central–frontal; d, right; s, left.

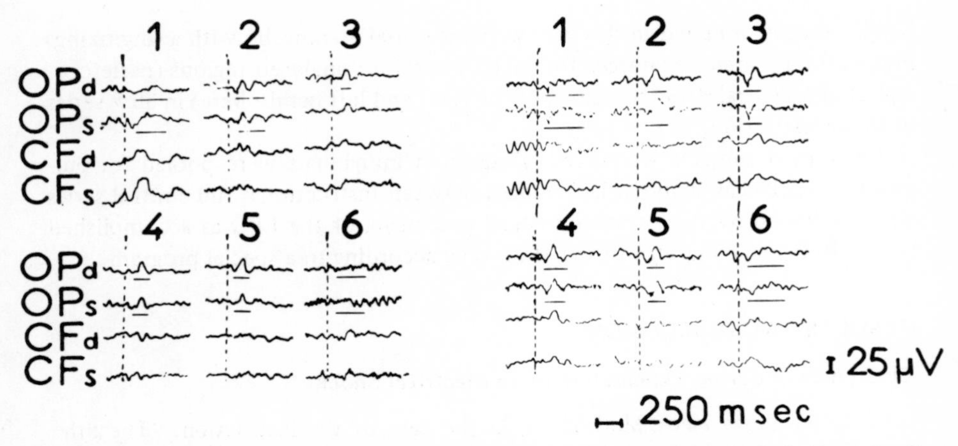

FIG. 2 Dynamics of the EP during discrimination of flash duration. (Left) Records of subject E. 1–4, first four flashes of the series, respectively; 4, 10th flash; 6, flash during subsequent control series. (Right) Records of subject B. Numbers correspond to the same conditions as for subject E.

reinforcement was really applied. The task consisted of investigating the EP modifications which arise in normal subjects when they expect a painful stimulation.

An analysis of the results obtained shows that this instruction was accompanied by local EP modifications in the central–frontal regions. Even in those cases when the instruction at first evoked generalized EP modifications in various cortical areas, these modifications gradually became selective as the 8th–10th (signaling) flashes approached, and manifested themselves only the anterior areas of the hemisphere.

Figure 1 (top left) presents EP to flashes in subject E during anticipation of electrical shock and in the control series after cancellation of the given instruction. An EP to the indifferent stimulus (No. 1) in the parietooccipital regions has the form of a slow positive–negative wave. The positive wave has a latency of 50 msec, while the secondary response, in the shape of a slow negative wave, appears with a latency of 110 msec and reaches its maximum in 300 msec. In the anterior regions, the EP is a two-humped negative wave, where the first peak is observed in 100 msec and the second in 250 msec.

Figure 1 (top left), No. 2 and Fig. 2 (top left), No. 1 demonstrate two types of EP modification recorded in this subject when electrical shock was expected.

In Fig. 1 (top left), No. 2 the modifications of the EP are of a generalized character and manifest themselves in the EP both in the anterior and posterior regions of the cerebral hemispheres. The positive wave of the primary response is abolished in the parieto-occipital regions; the peak latency of the negative wave is reduced to 150 msec from 300 msec, and the total duration of the EP is reduced. In the central–frontal regions the first negative wave greatly increases and the time during which the maximum is reached diminishes.

With the gradual approach of the 8th–10th signalling flashes, (Fig. 1, top right, No. 1), the EP in the parietooccipital regions return to the general form seen during the indifferent stimulus (Fig. 1, top left, No. 1). However, the central–frontal regions exhibit a depression of the second peak of the negative wave. After the instruction is countermanded (subsequent control series), the modification of the central–frontal EP disappears by the time the second flash is presented (Fig. 1, top right, No. 2).

In other subjects, EP modifications (resulting from expectation of an electrical shock to the left hand) are exactly timed to the moment when the signalling stimuli appear and are recorded only in the central–frontal region of the right hemisphere. Data relating to subject B may serve as an illustration of this type of response. Figure 1 (Bottom Nos. 1 and 2) presents responses to the seventh flash in a series of indifferent stimuli, and in series with the expectation of a painful stimulus respectively. We can see that the character of the EP in the central–frontal regions is essentially similar for both conditions. However, in response to the tenth flash, which must be "followed by an electrical shock to the left hand," a sharp increase in the amplitude of the first negative wave of the initial response is observed in the central–frontal region of the right hemisphere, and the total duration of the EP decreases due to a reduction of the subsequent negative wave (Fig. 1 bottom, No. 3).

Thus, the data obtained show that expectation of a painful stimulus leads to selective modifications of the EP in the central-frontal regions of the cortex of one or both hemispheres. These modifications consist of an increase in amplitude of the negative component having a peak at 70–100 msec. Since these EP modifications were not obtained in response to a real painful stimulus, but only under the action of a verbal instruction, it may be assumed that the neurophysiological processes, which are reflected in the characteristics of the EP, bear a direct relation to the intimate mechanisms regulating specifically human mental activity.

EP Dynamics during the Discrimination of Flash Duration

The subjects were given the following instruction: "Watch the duration of the flashes and after each second flash state which of them was shorter."

As stated above, the duration of the flashes remained constant during the entire investigation; however, their short duration (50 msec) did not allow the subjects to realize it. Every time after the subject's answer, a verbal reinforcement was introduced ("correct" or "wrong"). The purpose of this series was to study the parameters and distribution of the EP in the cortical areas in conditions of concentrated visual attention.

Figure 1A (Top No. 1) presents the EP to single flashes having no signalling meaning, in subject E. The parameters of these responses were described previously.

The character of the EP during the performance by this subject on the flash duration task to the 1st, 2nd, 3rd, 4th, and 10th flashes is shown in Fig. 2 (left).

From the figure it follows that the EP arising in response to the first flash does not differ from the "indifferent" one; however, beginning with the second stimulus, considerable EP modifications take place in the parietooccipital regions of the brain; the latency to the peak of the negative wave is reduced markedly (300–150 msec), and the duration of this wave increases.

An analysis of the configuration of the EP presented in Figs. 1 (top, No. 1) and 2 (left) shows that the EP modifications observed in this case in the parietooccipital regions of the cortex consist of a split of the slow negative wave into two components and formation of an additional peak. The appearance of a peak with a subsequent increase of its amplitude is accompanied by a gradual decline in the amplitude of the late component of the negative wave up to its full depression (Fig. 2 (left), Nos. 3, 4, 5). As a result, the total duration of the EP is shortened, indicating, according to published data, a rise in the level of cortical excitability.

In the central-frontal regions, EP modifications appear only by the 4th flash (Fig. 2 (left), No. 4) and are expressed not in an increase of their amplitude (as was observed when an electrical shock was expected), but, on the contrary, in a depression of the EP which persists up to the end of this series.

Thus, the increase of the electrical activity in the occipito-parietal areas of the cortex, which is characteristic of the mobilization of visual attention, takes place against a background of a reciprocal decline of excitability in the central-frontal parts of the brain. Cancellation of the instruction (control series) eliminates the effect of activation and brings the parameters of the EP back to their initial values (Fig. 2A (left), No. 6).

Similar data obtained in subject B are presented in Figure 2 (right). During the discrimination of flash duration, EP with high amplitude, negative, alphalike waves are recorded in the parietooccipital regions of both hemispheres. Just as in the previous observation, such a form of response does not arise at once, but only with the second stimulus presentation. As a rule, the negative wave is preceded by a positive wave with a latency of 50–60 msec. The total duration of the EP is reduced to approximately half that of EP to "indifferent" flashes (Fig. 1B (bottom), No. 1).

In the central–frontal regions the EP are represented by a slow negative wave which in most cases has two peaks. As distinct from the previous observation for subject E, where visual attention led to a depression of the EP, the EP is not greatly depressed. However, the character of the EP is changed between the defensive and flash duration tasks. A comparison of the EP of the central–frontal region of the right hemisphere with the "indifferent" negative wave (Fig. 1 (bottom), No. 1) shows that the defensive situation is accompanied by an intensification of the first component (Fig. 1 (bottom), No. 3), and the performance of the visual duration task by an intensification of the second component of the negative wave (Fig. 2 (bottom), Nos. 3, 4).

Thus, in certain conditions a split of the seemingly single negative EP wave into two components may be observed; the functional significance of each of these components is apparently different. These data show that one and the same central–frontal cortical zone participates differently in various forms of mental activity.

From these data it follows that, depending on the character of the verbal instruction which makes the subject expect an electrical shock or discriminate the duration of flashes, there is a *selective* increase of excitability in the central–frontal or parietooccipital regions of the cerebral cortex, respectively.

Dynamics of EP during the Performance of Voluntary Motor Responses

The investigation of EP dynamics during the performance of a motor task included the following series of experiments. The subjects were given the instruction: "When you see a flash, press the button with the right hand as quickly as possible."

Figure 3 presents EP to flashes which are signals of voluntary motor reactions in subject E. The first application of the stimulus (Fig. 3, No. 1) does not evoke any appreciable modifications of EP whose form and parameters do not differ essentially from the "indifferent" ones (Fig. 1 top left, No. 1). After the second flash (Fig. 3, No. 2), a split of the slow negative wave into two components appears in the posterior regions of the hemispheres, as was also observed during the discrimination of flash duration (Fig. 2A (left), No. 2). However, in this case, the split of the negative wave is not accompanied by a depression of its late component, whose amplitude can not only remain within the previous limits (Fig. 3, No. 2), but exceed the "early" component of the EP (Fig. 3, No. 4).

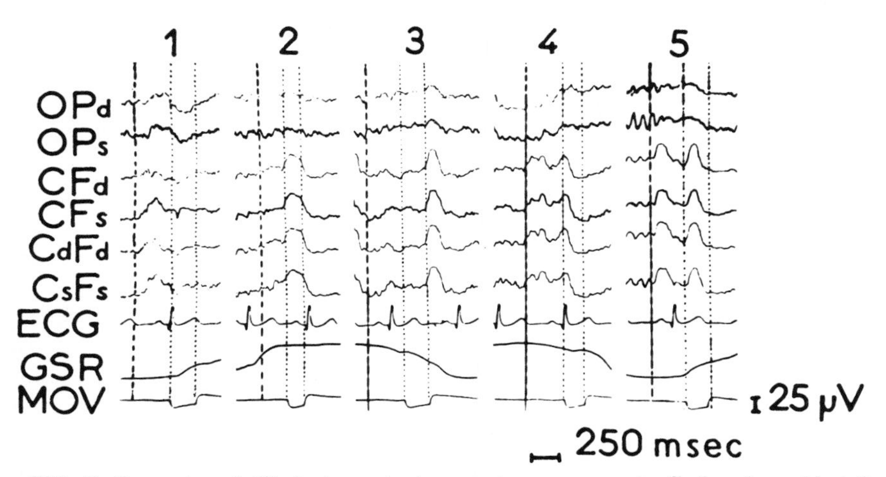

FIG. 3 Dynamics of EP during voluntary motor responses to flashes for subject E. Figures indicate the first five flashes in the series, respectively. The vertical dotted lines denote pressing the button. In addition to cortical records, the electrocardiogram (ECG), galvanic skin response (GSR), and moment of pressing the button (MOV) are also given.

This fact confirms the aforesaid assumption that the shortening of the latency between the flash and the peak of the negative wave observed during the performance of a motor task was determined, not by a change of the wave elicited by "indifferent" flashes, but by the appearance of an additional component of the EP. The different character of the EP modifications in the occipitoparietal cortex during attempted discrimination of flash duration and the performance of voluntary motor reactions indicates the specific character of the participation of this region of the brain in the actualization of the two types of responses.

In the central–frontal regions, the EP also undergo substantial changes in comparison with the "indifferent" EP. A comparison of the temporal parameters of the EP recorded in these experimental conditions with the time of the voluntary motor reactions shows that the EP may enter into different relationships with the response, preceding it (Fig. 3, No. 1), coinciding with it (Fig. 3, No. 2), or even appearing after its termination (Fig. 3, No. 3). With automatization of the motor skills, two high-amplitude waves are recorded in the central–frontal regions; one of these waves precedes the motor response and the other accompanies it (Fig. 3, No. 5).

The EP modifications described above in the parietooccipital and central–frontal regions, i.e., in the posterior and anterior areas of the hemisphere, indicate a selective participation of these cortical structures in the formation of different functional systems.

The results of the visual analysis of the data were confirmed by statistical analyses. It was found that different stimulus meanings led to significant differences of the mean values of the EP in some sections of the evoked response. In particular, both the expectation of an electrical shock and discrimination of flash duration were accompanied by statistically significant modifications in the EP of the central–frontal regions of the right hemisphere. However, in the first case they appear during the 70th msec and in the second case during the 240th msec, i.e., they reflected the dynamics of different components of the response.

A comparison of EP modifications in the central-frontal region of the left hemisphere during differentiation of visual stimuli and during the performance of voluntary motor responses shows that, although the EP have the same latency, in the latter case the EP duration exceeds that of the first one (the discrimination of flashes).

In the posterior regions of the hemisphere the character of the EP modifications also differed depending on the verbal instructions. In the case of an instruction which demanded a quick response to the flashes, statistically significant EP modifications appeared in the parietooccipito region of the right hemisphere 20 msec earlier than in the case of differentiation of flash duration which required a retention of the their traces rather than their rapid perception. Also, during differentiation of flash duration, a second period of significant differences was observed, but it proved to be absent during the performance of the motor task.

Thus, an investigation of the bioelectrical reactions of the cerebral cortex in normal subjects during different forms of activity revealed a dependence of the zone of local EP modifications on the specific character of verbal instructions. The facts obtained by us show that when the stimulus is included in different functional systems, the differences of the EP in comparison with the EP to "indifferent" stimuli may not manifest themselves in all cortical areas. The absence of such differences in some regions of the brain may be accompanied by modifications of the EP in others; this indicates that certain forms of mental activity do not require a corresponding analysis of the stimuli and consequently do not involve the responsible cortical area in the reaction.

It may be assumed that the contradictory data (published in the literature) concerning the influence of different verbal instructions on human EP are the result of not taking into account the dynamics of varied EP modifications at different cortical loci, in various experimental conditions.

However, the selective participation of one and the same cortical zone in various forms of mental activity is not confined to the presence or absence of a reaction in certain parts of the brain; it includes diverse and statistically significant changes of responses which become modified depending upon the specific features of the subject's activity.

On the basis of the data obtained, it may be concluded that the application of the method of evoked potentials for studying the cerebral organization of mental processes is fully justified. The analysis of evoked potentials, manifesting substantial changes within fractions of a second, brings us closer to the comprehension of the physiological mechanisms which lie at the base of the rapidly-acting system of regulation of activation processes. The operation of this system, which ensures a rapid change of the functional states in various parts of the cerebral cortex, is of great interest from the point of view of research on the physiological foundation of mental activity.

REFERENCES

Akimoto, H., & Creutzfeldt, O. Reaktionen von Neuronen des optischen Cortex nach elektrischer Reizung unspezifischer Thalamuskerne. *Arch. Psychiat. Nervenkr.*, 1957, **196**, 494–519.

Akimoto, H., Saito, Y., & Nakamura, Y. Effects of arousal stimuli on evoked neuronal activities in cat's visual cortex. In R. Jung (Ed.), *Neurophysiologie und Psychophysik des Visuellen Systems.* Symposium, Berlin, 1960. Berlin: Springer-Verlag, 1961. Pp. 363–371.

Aladjalova, N. A. Slow electrical processes in the brain. *Progress in Brain Research,* 1964, 7, 1–243.

Albe-Fessard, D., Rocha-Miranda, C., & Oswaldo-Cruz, E. Activités évoquées dans le noyau caudé du chat en réponse à des types divers d'afférences. II. Étude microphysiologique. *Electroencephal clin. Neurophysiology,* 1960, **12**, 649–661.

Aleksandrov, V. J. Regulatory processes in cells. In *Manual of cytology,* Vol. 2. Moscow, Leningrad: Science Press, 1966. Pp. 590–620. (In Russian)

Altman, Y. A. Electrophysiological investigation of different parts of the auditory system of the cat during extended rhythmic stimulation. *Physiological Journal of the USSR,* 1960, **46**(5), 526–536. (In Russian)

Altman, Y. A. Electrical responses of different parts of the auditory system during prolonged rhythmic auditory stimulation. Dissertation abstract, Leningrad, 1961. (In Russian)

Alving, B. O. Spontaneous activity in isolated somata of *Aplysia* pacemaker neurons. *Journal of General Physiology,* 1968, **51**, 29–45.

Amassian, V. S. Microelectrode studies of cerebral cortex. *International Revues of Neurology,* 1961, **3**, 67–136.

Andersen, P., & Andersson, S. A. *Physiological basis of the alpha rhythm.* New York: Appleton-Century-Crofts, 1968.

Andersen, P., Andersson, S. A., & Lomo, T. Nature of thalamocortical relations during spontaneous barbiturate spindle activity. *Journal of Physiology,* 1967, **192**, 283–307.

Andersen, P., Brooks, C. M., Eccles, J. C., & Sears, T. A. The ventro-basal nucleus of the thalamus: Potential fields, synaptic transmission and excitability of both presynaptic and postsynaptic component. *Journal of Physiology,* 1964, **14**, 348–369.

Andersen, P., & Eccles, J. C. Inhibitory phasing of neuronal discharge. *Nature,* 1962, **196**, 645–647.

Andersen, P., Eccles, J. C., & Loyning, Y. Recurrent inhibition in the hippocampus with identification of the inhibitory cells and its synapses. *Nature,* 1963, **198**, 541–542.

Andersen, P., Eccles, J. C., & Loyning, Y. Location of postsynaptic inhibitory synapses on hippocampal pyramids. *Journal of Neurophysiology,* 1964, **27**, 592–607. (a)

Andersen, P., Eccles, J. C., & Loyning, Y. Pathway of postsynaptic inhibition in the hippocampus. *Journal of Neurophysiology,* 1964, **27**, 608–619. (b)

Andersen, P., Eccles, J. C. & Sears, T. A. The ventrobasal complex of the thalamus: Types of cells, their responses and their functional organiation. *Journal of Physiology,* 1964, **174**, 370–390.

Andersen, P., & Sears, T. A. The role of inhibition in the phasing of spontaneous thalamo-cortical discharge. *Journal of Physiology,* 1964, **173**, 459–480.

Anokhin, P. K. Electrophysiological analysis of the cortico-subcortical relations during positive and negative conditioned reflexes. Pavlovian Conference on Higher Nervous Activity. *Annals of the New York Academy of Science,* 1963, **92**, 899–938.

Arden, G. B. Complex receptive fields and responses to moving objects in cells of the rabbit's lateral geniculate body. *Journal of Physiology,* 1963, **166**, 468–488.

Arduini, A., & Whitlock, A. Spike discharges in pyramidal system during recruitment waves. *Journal of Neurophysiology,* 1953, **16**, 430–436.

Artemieva, E. Yu., & Meshalkin, L. D. Analysis of Δ parameter in asymmetry of the length of the ascending and descending fronts of alpha-rhythm. *Nov. Issled. Pedagog. Nauk,* 1965, **4**, 222–227. (in Russian)

Artemieva, E. Yu., Meshalkin, L. D., & Homskaya, E. D. The periodical oscillations of the asymmetry of ascending and descending fronts of alpha-waves. *Nov. Issled. Pedagog. Nauk.,* 1965, **4**, 249–251. (in Russian)

Barlow, H. B., Hill, R. M., & Levick, W. R. Retinal ganglion cells responding selectively to direction and speed of image motion in the rabbit. *Journal of Physiology,* (Lond.) 1964, **173**, 377–407.

Bekhtereva, N. P. Dynamics of biological potentials of subcortical areas in humans. In *Problems in modern neurophysiology.* Moscow, Leningrad, 1965. Pp. 100–133. (In Russian)

Bell, C., Buendia, W., Sierra, G., & Segundo, J. P. Mesencephalic reticular responses to natural and to repeated sensory stimuli. *Experientia,* 1963, **19**, 308–312.

Bell, C., Sierra, G., Buendia, N., & Segundo, J. P. Sensory properties of neurons in the mesencephalic reticular formation. *Journal of Neurophysiology,* 1964, **27**, 961–987.

Bennett, M. V. L., Nakajima, Y., & Pappas, G. D. Physiology and ultrastructure of electronic junctions. III. Giant electromotor neurons of *Malapterurus electricus. Journal of Neurophysiology,* 1967, **30**, 209–235. (a)

Bennett, M. V. L., Nakajima, Y., & Pappas, G. D. Physiology and ultrastructure of electronic junctions. I. Supramedullary neurons. *Journal of Neurophysiology,* 1967, **30**, 161–179. (b)

Bennett, M. V. L., Pappas, G. D., Aljure, E., & Nakajima, Y. Physiology and ultrastructure of electronic junctions. II. Spinal and medullary electromotor nuclei in mormyrid fish. *Journal of Neurophysiology,* 1967, **30**, 180–208.

Bennett, M. V. L., Pappas, G. D., Giméniz, M., & Nakajima, Y. Physiology and ultrastructure of electronic junctions. IV. Medullary electromotor nuclei in gymnotid fish. *Journal of Neurophysiology,* 1967, **30**, 236–300.

Beteleva, T. G. Dependence of the responses of single neurons in the rabbit on the intensity of the light stimulus. *Pavlov Journal of Higher Nervous Activity,* 1964, **14**(6), 1048–1054. (In Russian)

Beteleva, T. G. A change in the responses of neurons of the lateral geniculate to repeated presentations of a light. *Pavlov Journal of Higher Nervous Activity,* 1967, **17**(6), 1052–1057. (In Russian)

Bishop, G. H., & O'Leary, J. Components of the electrical response of the optic cortex of the rabbit. *American Journal of Physiology,* 1936, **117**, 292–308.

Bremer, F. Neurophysiological mechanisms in cerebral arousal. In G. E. W. Wolstenholme &

M. O'Connor (Eds.), *The Nature of Sleep* (Ciba Foundation Symposium, London). Boston: Little, Brown & Co., 1961. Pp. 30–56.

Brodal, A. The hippocampus and the sense of smell. A review. *Brain,* 1947, **70,** 179–222.

Bruner, G. & Tauc, L. Long-lasting phenomena in the molluscan nervous system. *Social and Experimental Biology,* 1965, **20,** 457–475.

Buchwald, N. A., Heuser, G., Wyers, E. J., & Lauprecht, C. W. The "Caudate-Spindle." III. Inhibition by high frequency stimulation of subcortical structures. *Electroencephalography and Clinical Neurophysiology,* 1961, **13,** 525–530.

Buchwald, N. A., Hull, C. D., & Trachtenberg, M. C. Concomittant behavioral and neural inhibition and disinhibition in response to subcortical stimulation. *Experimental Brain Research,* 1967, **4,** 58–72.

Buchwald, N. A., Wyers, L. J., Lauprecht, C. W., & Heuser, G. The "Caudate-Spindle." IV. A Behavioral index of caudate-induced inhibition. *Electroencephalography and Clinical Neurophysiology,* 1961, **13,** 531–537.

Buchwald, N. A., Wyers, E. J., Okuma, T., & Heuser, G. The "Caudate-Spindle." I. Electrophysiological properties. *Electroencephalography and Clinical Neurophysiology,* 1961, **13,** 509–518.

Bullock, T. H., & Horridge, G. A. *Structure and function of the nervous system in invertebrates.* Vol. 2. New York: Freeman & Co., 1965.

Bures, J., & Buresova, O. Plastic changes of unit activity based on reinforcing properties of extracellular stimulation of single neurons. *Journal of Neurophysiology,* 1967, **30,** 43–113.

Buser, P., & Imbert, M. Sensory projections to the motor cortex in cats: A microelectrode study. In W. A. Rosenblith (Ed.), *Sensory Communication.* New York, London: The MIT Press, Wiley, 1961. Pp. 597–626.

Buser, P., Rougeul, A., & Perret, C. Caudate and thalamic influences on conditioned motor responses in the cat. *Boletin del Instituto de Estudios Medicos y Biologicos (Mexico),* 1964, **22,** 293.

Butkhusi, S. M. The influence of caudate stimulation on cortical evoked potentials. *Transactions of the Institute of Physiology, Academy of Science of the Georgian SSR,* 1965, **14,** 89, (In Russian)

Callaway, E. Factors influencing the relationship between alpha activity and visual reaction time. *Electroencephalography and Clinical Neurophysiology,* 1962, **14,** 674–682.

Callaway, E., & Yeager, C. L. Relationship between reaction time and EEG alpha-phase. *Science,* 1960, **132,** 1765–1766.

Carpenter, D. O. Temperature effects on pacemaker generation, membrane potential, and critic firing threshold in *Aplysia* neurons. *Journal of General Physiology,* 1967, **50,** 1469–1484.

Cazard, P., & Buser, P. Réponses sensorielles recueilles au niveau du cortex moteur chez le lapin. *Electroencephalography and Clinical Neurophysiology,* 1963, **15,** 403–412.

Chapman, R. H., & Bragdon, H. R. Evoked responses to numerical and nonnumerical visual stimuli while problem solving. *Nature,* 1964, **203,** 1155–1157.

Cherkes, V. A. The physiology of the basal ganglia of the brain. *Advances in Contemporary Biology,* 1966, **61,** 39. (In Russian)

Cobb, W. A., & Dawson, G. D. The latency and form in man of the occipital potentials evoked by bright flashes. *Journal of Physiology,* 1960, **152,** 108–121.

Creutzfeldt, O. D. Relations between EEG and neuronal activity. *Electroencephalography and Clinical Neurophysiology,* 1961, **13,** 651. (Abstract).

Creutzfeldt, O., Bell, F. R., & Adey, W. R. The activity of neurons in the amygdala of the cat following afferent stimulation. In *Progress in Brain Research,* 1963, **3,** 31–49.

Creutzfeldt, O. D., Fromm, G. H., & Kapp, H. Influence of transcortical d-c currents on cortical neuronal activity. *Experimental Neurology,* 1962, **5,** 436–452.

Creutzfeldt, O. D., & Jung, R. Neuronal discharge in the cat's motor cortex during sleep and arousal. In G. E. W. Wolstenholme & M. O'Connor (Eds.), *The Nature of sleep* (Ciba Foundation Symposium, London). Boston: Little, Brown, 1961. Pp. 131–170.

Cross, B. A., & Green, J. D. Activity of single neurones in the hypothalamus: Effect of osmotic and other stimuli. *Journal of Physiology,* 1959, **148,** 554–569.

Danilova, N. N. Complex effects of intermittent light stimulation. *Science Reports of the Higher School of Biological Sciences,* 1961, **3,** 86–92. (In Russian)

Danilova, N. N. Periodicity of spontaneous activity of the single neuron. *Pavlov Journal of Higher Nervous Activity,* 1966, **16**(4), 678–683. (In Russian)

Darion-Smith, J., & Yokota, T. Corticofugal effects on different neuron types within the cat's brain stem activated by tactile stimulation of the face. *Journal of Neurophysiology,* 1966, **29,** 185–206.

Davis, H. Enhancement of evoked cortical potentials in humans related to a task requiring a decision. *Science,* 1964, **145,** 182–183.

Dawson, G. D. Central control of sensory inflow. *Electroencephalography and Clinical Neurophysiology,* 1958, **10,** 351.

Demetrescu, M., & Demetrescu, M. The inhibitory action of the caudate nucleus in cortical primary receiving areas in the cat. *Electroencephalography and Clinical Neurophysiology,* 1962, **14,** 37–52.

Douglas, R. J. The hippocampus and behavior. *Psychological Bulletin,* 1967, **67,** 416–442.

Douglas, R. J., & Pribram, K. H. Learning and limbic lesions. *Neuropsychologia,* 1966, **4,** 197–220.

Drachman, D. A., & Arbit, J. Memory and the hippocampal complex. *Archives of Neurology,* 1966, **15,** 52–61.

Dubner, R. Single cell analysis of sensory interaction in anterior lateral and suprasylvian gyri of the cat's cerebral cortex. *Experimental Neurology,* 1966, **15,** 255–273.

Dubrovinskaya, N. V. Dynamics of neuronal responses to light stimuli in the cortex of rabbit. Unpublished Dissertation, University of Moscow, 1967. (In Russian)

Duensing, F., & Schaefer, R. Zum Problem der Differenzierung der Neurone in der Formation reticularis. In R. Werner (Ed.), *Jenenser EEG Symposion.* 30 Jahre Elektroenzephlographie, 17. bis 19, 1959. Berlin: Verlag Volk und Gesundheit, 1963.

Dumont, S., & Dell, P. Facilitations spécifiques et non-spécifiques des réponses visuelles corticales. *Journal de Physiologie (Paris).* 1958, **50.** 261–264.

Dumont, S., & Dell, P. Facilitation réticulaire des mécanismes visuels corticaux. *Electroencephalography and Clinical Neurophysiology,* 1960, **12,** 769–796.

Dunlop, C. W., Webster, W. R., & Day, R. H. Amplitude changes of evoked potentials at the inferior colliculus during acoustic habituation. *Journal of Auditory Research,* 1964, **4,** 159–169.

Eccles, J. C. Inhibition in thalamic and cortical neurons and its role in phasing neuronal discharges. *Epilepsia,* 1965, **6,** 89–115. (a)

Eccles, J. C. Possible ways in which synaptic mechanisms participate in learning, remembering and forgetting. In D. Kimble (Ed.), *The anatomy of memory.* Vol. 1. Palo Alto: Science and Behavior Books, 1965, Pp. 12–87. (b)

Evans, E. F., & Whitfield, I. C. Classification of unit responses in the auditory cortex of the unanaesthetized and unrestrained cat. *Journal of Physiology,* 1964, **171,** 476–493.

Evarts, E. Photically evoked responses in visual cortex units during sleep and waking. *Journal of Neurophysiology,* 1963, **26,** 229–248.

Evarts, E. V., Corwin, F., & Huttenlocher, P. R. Recovery cycle of visual cortex of the awake and sleeping cat. *American Journal of Physiology,* 1960, **199,** 373–376.

Fernandez-Guardiola, A., Ayala, F., & Kornhauser, S. EEG, heart-rate and reaction time in humans: Effect of variable vs. fixed interval-repetitive stimuli. *Physiology and Behavior,* 1968, **3,** 231–240.

Fernandez-Guardiola, A., Toro-Donoso, A., Aquino-Cias, J., & Guma-Dias, E. Peripheral and central modulation of visual input during "habituation." In *Progress in Brain Research.* Vol. 22. Amsterdam: Elsevier, 1967. Pp. 388–399.

Fifkova, E., & Marsala, J. Stereotaxic atlas for the rabbit. In J. Bures, M. Petran, & J. Zachar. *Electrophysiological methods in biological research.* (2nd ed.) New York: Academic Press, 1962. Pp. 426–443.

Floru, R. Psihofiziologia atentiei. Bucuresti: Editura Stiintifica, 1967.

Floru, R., Staerescu-Volanschi, M., & Bittman, E. Role of the ascending activating system in the integration of sensorial messages. *Revue Roumaine de Physiologie,* 1964, **1,** 63.

Floru, R., Staerescu-Volanschi, M., Bittman, E., & Nestianu, V. Autofacilitarea reflex conditionata. *St. Cerc. Fiziol.,* 1963, **8,** 21.

Frank, K. Basic mechanisms of synaptic transmission in the central nervous system. *IRE Transactions on Medical Electronics,* 1959, **2,** 85–88.

Friend, J. H., Suga, N., & Suthers, R. A. Neural responses in the inferior colliculus of echolocating bats to artificial orientation sounds and echoes. *Journals of Cellular Physiology,* 1966, **67,** 319–332.

Fuster, J. M. Excitation and inhibition of neuronal firing in visual cortex by reticular stimulation. *Science,* 1961, **133,** 2011–2012.

Fuster, J. M., Creutzfeldt, O. D., & Straschill, M. Intracellular recording of neuronal activity in the visual system. *Zeitschrift für Vergleichende Physiologie,* 1965, **49,** 605–622.

Galambos, R., Sheatz, G., & Vernier, V. G. Electrophysiological correlates of a conditioned response in cats. *Science,* 1955, **123,** 376–377.

Garcia-Austt, E., Bogacz, J., & Vanzulli, A. Changes in photic EEG responses in man due to attention, habituation and conditioning. *Twenty-first International Congress of Physiological Sciences* (Symposia and Special Lectures, Buenos Aires, 1959). Amsterdam: Excerpta Medica Foundation, 1959 (Abstract). Pp. 103–104.

Genkin, A. A. The asymmetry of the phases of EEG in mental activity. *Doklady Akademii Nauk SSSR,* 1963, **149** (No. 6), 1460–1463. (in Russian)

Genkin, A. A. The length of the ascending and descending tract of the EEG as a source of information about neurophysiological processes. Unpublished Ph.D. thesis, Leningrad, 1964 (in Russian).

Gershuni, G. V. The significance of temporal characteristics in the organization of the activity of the auditory system. In *Contemporary problems of the electrophysiology of the central nervous system.* Moscow: Science Press, 1967. Pp. 65–70. (In Russian)

Gleser, V. D. *The mechanism of visual image detection.* Moscow, Leningrad: Science Press, 1966. (In Russian)

Gloor, P. Etudes electrographiques de certaines connexions rhinencéphaliques. In T. Alajouanine (Ed.), *Physiologie et pathologie du rhinencephale.* Vol. II, 1961. Pp. 1.

Grastyan, E. The hippocampus and higher nervous activity. In M. A. B. Brazier (Ed.), *The central nervous system and behavior.* Vol. 1. New York: Josiah Macy, Jr. Foundation, 1959. Pp. 409–430.

Grastyan, N., Lissak, K., Madarasz, J., & Donhoffer, H. Hippocampal electrical activity during the development of conditioned reflexes. *Electroencephalography and Clinical Neurophysiology,* 1959, **11,** 409–430.

Grundfest, H. Some determinants of repetitive electrogenesis and their role in the electrical activity of the central nervous system. Proceedings of the International Symposium on Comparative and Cellular Pathophysiology of Epilepsy. *Excerpta Medica International Congress Series,* 1965, **124,** 19–46.

Haider, M., Spong, P., & Lindsley, D. B. Attention, vigilance, and cortical evoked potentials in human. *Science,* 1964, **145,** 180–182.

Harker, J. Endocrine and nervous factors in insect circadian rhythms. *Cold Spring Harbor Symposium Quantitative Biology,* 1960, **25,** 279–287.

Heath, R. G., & Hodes, R. Induction of sleep by stimulation of the caudate nucleus in

Macacus rhesus and man. *Transactions of the American Neurological Association,* 1952, 77, 204–210.

Hernandez-Peon, R. Central neuro-humoral transmission in sleep and wakefulness. In K. Akert, C. Bally, & J. P. Schade (Eds.), *Sleep mechanisms. Progress in Brain Research.* Vol. 18. Amsterdam: Elsevier, 1956. Pp. 96–117.

Hernandez-Peon, R. Neurophysiological correlates of habituation and other manifestations of plastic inhibition (internal inhibition). In H. H. Jasper & G. D. Smirnov (Eds.), *The Moscow colloquium on encephalography of higher nervous activity. Electroencephalography and Clinical Neurophysiology,* 1960, Suppl. 13, 101–112.

Hernandez-Peon R., Chaves-Ibarra, G., Morgane, F. J., & Timo-Iaria, C. Limbic cholinergic pathways involved in sleep and emotional behavior. *Experimental Neurology,* 1963, 8, 93–111.

Hernandez-Peon, R., Scherrer, H., & Jouvet, M. Modification of electrical activity in cochlear nucleus during "attention" in unanaesthetized cats. *Science,* 1956, **123**, 331–332.

Hernandez-Peon, R., Jouvet, M., & Scherrer, H. Auditory potentials at cochlear nucleus during acoustic habituation. *Acta Neurologica Latinoamerica,* 1957, **3**, 144–156.

Heuser, G., Buchwald, N. A., & Wyers, E. J. The "Caudate-Spindle." II. Facilitatory and inhibitory caudate-cortical pathways. *Electroencephalography and Clinical Neurophysiology,* 1961, **13**, 519–524.

Holmgren, B., & Frenk, S. Inhibiting phenomena and "habituation" at the neuronal level. *Nature,* 1961, **192**, 1294–1295.

Hori, Y., Toyohara, I., & Yoshii, N. Conditioning of unitary activity by intracerebral stimulation in cats. *Physiology and Behavior,* 1967, **2**, 255–259.

Hori, Y., & Yoshii, N. Conditioned change in discharge pattern for single neurons of the medial thalamic nuclei of the cat. *Psychological Reports,* 1965, **16**, 241–242.

Hori, Y., Yoshii, N., Hayashi, Y., & Takeguchi, H. Unit activities of medial thalamic nuclei under hypothermia. *Medical Journal of Osaka University,* 1963, **13**, 425–435.

Horn, G. Some neural correlates of perception. In J. D. Carthy and C. L. Duddington (Eds.), *Viewpoints in biology.* Vol. 1. London: Butterworths, 1962. Pp. 242–285.

Horn, G. Neuronal mechanisms of habituation. *Nature,* 1967, **215**, 707–711.

Horn, G., & Hill, R. M. Habituation of the response to sensory stimuli of neurons in the brain stem of rabbits. *Nature,* 1964, **202**, 296–298.

Horn, G., & Hill, R. M. Responsiveness to sensory stimulation of units in the superior colliculus and subjacent tectotegmental regions of the rabbit. *Experimental Neurology,* 1966, **14**, 199–223.

Horvath, F. E., & Buser, P. Modulation par le noyau caudé de l'excitabilité de neurones du noyau ventral latéral du thalamus chez le chat. *Journal de Physiologie (Paris),* 1965, **57**, 628–629.

Horvath, F. E., Soltysic, S., & Buchwald, N. A. Spindles elicited by stimulation of the caudate nucleus and internal capsule. *Electroencephalography and Clinical Neurophysiology,* 1964, **17**, 670–676.

Hubel, D. H. Single unit activity in striate cortex of unrestrained cats. *Journal of Physiology (London),* 1959, **147**, 226–238.

Hubel, D. H., Henson, C. O., Rupert, A., & Galambos, R. "Attention" units in the auditory cortex. *Science,* 1959, **129**, 1279–1280.

Huttenlocher, P. R. Evoked and spontaneous activity in single units of medial brain stem during natural sleep and waking. *Journal of Neurophysiology,* 1961, **24**, 451–468.

Jarcho, J. W. Excitability of cortical afferent systems during barbiturate anesthesia. *Journal of Neurophysiology,* 1936, **12**, 447–457.

Jasper, H. H. Diffuse projections systems: The integrative action of the thalamic reticular system. *Electroencephalography and Clinical Neurophysiology,* 1949, **1**, 405–420.

Jasper, H. H. Transformation of cortical sensory responses by attention and conditioning. *IBRO Bulletin,* 1964, **3,** 80.

Jasper, H. H., Ricci, G. F., & Doane, B. Patterns of cortical neuronal discharge during conditioned responses in monkeys. In G. E. Wolstenholme & C. M. O'Conner (Eds.), *Neurological basis of behavior.* Boston: Little, Brown, 1958. Pp. 277–294.

Jung, R. Neuronal Integration in the visual cortex and its significance for visual information. In W. A. Rosenblith and J. Wiley (Eds.), *Sensory communication.* Cambridge, Massachusetts: The MIT Press, 1961. Pp. 627–674.

Jung, R. Integration in the neurons of the visual cortex. In *Theory of connection in the sensory systems* (Collected Articles). Moscow, 1964. P. 375.

Jung, R., Creutzfeldt, O., Grusser, O. The microphysiology of cortical neurons. *Germ. Month.,* 1958, **8,** 269–276.

Jung, R., & Hassler, R. The extrapyramidal motor system. In J. Field *et al.* (Eds.), *Handbook of physiology.* Sect. I, Neurophysiology, Vol. II. American Physiological Society, Washington, D.C.: Williams & Wilkins, 1960. Pp. 863–927.

Jung, R., Kornhuber, H. H., & Da Fonseca, J. S. Multisensory convergence on cortical neurons. *Progress in Brain Research.* Vol. 6. Amsterdam: Elsevier, 1963. Pp. 207–240.

Kalinin, P. N., & Sokolova, A. A. The influence of stimulation of the midbrain reticular formation on neurons of the motor cortex in an unanesthetized rabbit. *Pavlov Journal of Higher Nervous Activity,* 1968, **18,** 528–531. (In Russian)

Kamikawa, E., McIlwain, J. T., & Adey, W. R. Response patterns of thalamic neurons during classical conditioning. *Electroencephalography and Clinical Neurophysiology,* 1964, **17,** 485–496.

Kandel, E. R. Cellular studies in learning. In G. C. Quarton *et al.* (Eds.), *The neurosciences.* Vol I. New York: The Rockefeller University Press, 1967. Pp. 666–689.

Katsuki, Y. Neural mechanisms of the auditory senses in the cat. In *Communication theory in sensory systems.* Moscow: World Publishers, 1964. Pp. 288–306. (In Russian)

Kiang, N. Y. *Discharge patterns of single fibers in the cat's auditory nerve.* Cambridge, Massachusetts: MIT Press, 1965.

Kirvlis, D. P., Pozin, N. V., & Sokolov, E. N. Model for detection of signal intensities. *Biophysics,* 1967, **12.** (In Russian)

Kogan, A. B. Concerning the physiological nature of the electrical correlates of attention. In *Biological and physiological problems of psychology* (Abstracts of Communications). *Eighteenth International Congress of Psychology.* Moscow: Science Press, 1966. P. 161. (In Russian)

Kondratjeva, I. N. On the inhibition of neurons in the visual cortex. *Pavlov Journal of Higher Nervous Activity,* 1964 **14**(6), 1069–1078. (In Russian)

Kondratjeva, I. N., & Volodin, V. M. Recovery cycles of neuron responses to flashes in the visual area of the cerebral cortex in rabbits. *Pavlov Journal of Higher Nervous Activity,* 1966, **16**(5), 874–881. (In Russian)

Kopylov, A. G. An electrophysiological investigation of central neurons of mollusks. In *The nerve cell.* Leningrad: Leningrad University Press, 1966. Pp. 131–146. (In Russian)

Kopytova, F. V., & Rabinovich, M. G. Microelectrode study of the conditioned reflex to time. *Pavlov Journal of Higher Nervous Activity,* 1967, **17**(6). (In Russian)

Kostyuk, P. G. *Neuronal reflex arcs.* Moscow: Medical Press, 1959. (In Russian)

Kotlar, B. I., & Shulgovsky, B. B. Bioelectric activity of neurons in the dominant focus. Paper presented at the fifth conference on the Electrophysiology of the Central Nervous System, Tbilisi, 1966, 161–62. (In Russian)

Kratin, Y. G. Change in the primary response of the auditory system in the cerebral cortex to an infrequent rhythmic stimulus. In *Physiology and pathology of higher nervous activity.* Moscow, Leningrad: Science Press, 1965. Pp. 77–83. (In Russian)

Kratin, Y. G. *Electrical responses of the brain to inhibitory signals.* Leningrad: Science Press, 1967. (In Russian)

Krauthamer, G. M., & Albe-Fessard, D. Electrophysiologic studies of the basal ganglia and striopallidal inhibition of non-specific afferent activity. *Neuropsychologia,* 1964, **2**, 73–83.

Krauthamer, G., & Albe-Fessard, D. Inhibition of nonspecific sensory activities following striopallidal and capsular stimulation. *Journal of Neurophysiology,* 1965, **28**, 100–124.

Krauthamer, G., & Feltz, P. Quelques propriétés d'unités du thalamus intralaminaire répondant aux influx des corps striés ou d'origine périphérique. *Journal de Physiologie (Paris),* 1965, **57**, 638–639.

Krnjevic, K., Randic, M., & Straughan, D. W. Cortical inhibition. *Nature,* 1964, **201**, 1294–1296.

Krupp, P., & Monnier, M. Die funktionalle Dualitat des mesencephalen Reticularsystems. II. Aktivirender und hemmender Einfluss auf enzelne Neurone des motorischen Cortex. *Pflüger's Archiv für die Gesamte Physiologie,* 1964, **278**(6), 586–596.

Kupalov, P. S. Some problems of the higher nervous system. Rapport au XX-eme Congres international de physiologie, Bruxelles, 1956, 45.

Lansing, R. Relation of brain and tremor rhythms to visual reaction time. *Electroencephalography and Clinical Neurophysiologie,* 1957, **9**, 497–504.

Laufer, M., & Verzeano, M. Periodic activity in the visual system of the cat. *Vision Research,* 1967, 7, 215–229.

Laursen, A. M. Corpus striatum. *Acta Physiologica Scandinavica,* 1963, Suppl. 211, **59**, 1–106.

Lehmann, D., Koukkou, M., & Spehlmann, R. Neuronal effects of caudate stimulation in the visual cortex. *Proceedings of the International Union of Physiological Sciences* (Twenty-second International Congress, Leiden, 1962). Vol. II. Amsterdam: Excerpta Medica Foundation International Congress Series, No. 48. Abstract No. 1099.

Lewis, E. R. An electronic model of the neuron based on the dynamics of potassium and sodium ion fluxes. In R. F. Reiss (Ed.), *Neural theory and modeling.* Stanford, California: Stanford University Press, 1964. Pp. 154–189.

Li, C. L., & Chou, S. N. Cortical intracellular synaptic potentials and direct cortical stimulation. *Journal of Cellular and Comparative Physiology,* 1962, **60**, 1–16.

Li, C. L., Cullen, C., & Jasper, H. H. Laminar microelectrode studies of specific somatosensory cortical potentials. *Journal of Neurophysiology,* 1956, **9**, 111–130.

Li, C. L., & Jasper, H. H. Microelectrode studies of the electrical activity of the cerebral cortex in the cat. *Journal of Physiology,* 1953, **131**, 117–140.

Lindsley, C. B. Attention, consciousness, sleep and wakefulness. In J. Field, H. W. Magoun, & V. E. Hall (Eds.), *Handbook of Physiology.* Sect. I, Neurophysiology, Vol. 3. American Physiological Society, Washington, D.C.: Williams & Wilkins, 1960, Pp. 1553–1593.

Livanov, M. N. Concerning the establishment of temporary connections. *Electroencephalography and Clinical Neurophysiology,* Supplement 13, ("The Moscow Colloquium on Electroencephalography of Higher Nervous Activity"). 1960, 185–198.

Long, R. G. Modification of sensory mechanisms by subcortical structures. *Journal of Neurophysiology,* 1959, **22**, 412–427.

Luco, J. V. Plasticity of neural function in learning and retention. In M. A. B. Brazier (Ed.), *Brain function.* Vol. II. Berkeley, Los Angeles: University of California Press, 1964. Pp. 135–159.

Machne, X., Calma, J., & Magoun, H. Unit activity of central cephalic brain stem in EEG arousal. *Journal of Neurophysiology,* 1955, **18**, 547–558.

Machne, X., & Segundo, J. P. Unitary responses to afferent volleys in amygdaloid complex. *Journal of Neurophysiology,* 1956, **19**, 232–240.

McLardy, T. Hippocampal formation of brain as detector-coder of temporal patterns of information. *Perspectives in Biology and Medicine,* 1959, **2,** 443–452.

McLean, P. D. The limbic system and its hippocampal formation: Studies in animals and their possible application to man. *Journal of Neurosurgery,* 1954, **11**, 29–44.

Meissner, W. W. Hippocampal functions in learning. *Journal of Psychiatric Research,* 1966, **4**, 235–304.

Meshchersky, R. M. *Techniques for microelectrode investigation.* Moscow: Medical Press, 1960. (In Russian)

Meulders, M. Étude éléctrophysiologique des mechanisms conditionant le passage d'impulses visuale dans le system nerveux central. Verland van de Koninhi Vlaamse Acad. voor Geneesk. Vol. 27. Van Belgie, 1965. P. 199.

Mkrticheva, L. I., & Samsonova, V. G. Functional characteristics of neurons of the visual centers of a frog depending on light intensity. *Pavlov Journal of Higher Nervous Activity,* 1965, **14**(3), 125.

Mollica, A., Moruzzi, G., & Naquet, R. Decharges reticulaires înduites par la polarization du cervolet: Leur rapport avec la tonus postural et la reaction d'eveil. *Electroencephalography and Clinical Neurophysiology,* 1953, **5**, 571–583.

Monnier, M., & Gangloff, H. Atlas for stereotaxic brain research on the conscious rabbit. In *Rabbit brain research.* Vol. I. Amsterdam: Elsevier, 1961. Pp. 1–76.

Monnier, M., Hosli, L., & Krupp, P. Moderating and activating systems in medio-central thalamus and reticular formation. In R. Hernandez-Peon (Ed.), *The physiological basis of mental activity* (Proceedings of a symposium, Mexico City, October, 1961). *Electroencephalography and Clinical Neurophysiology,* 1963, Suppl. 24, **15.**

Morgane, P. J., & McFarland, W. L. The neuroanatomical correlates of functional specializations in the dolphin (*Tursiopa truncatus*). *Proceedings of the Twenty-Third International Congress on Physiological Sciences,* Tokyo, 1965. Amsterdam: Excerpta Medica Foundation, 1965.

Morrell, F. Information storage in nerve cells. In W. S. Fields & W. Abbot (Eds.), *Information storage and neural control* (Tenth Annual Scientific Meeting of Houston Neurological Society). Springfield, Illinois: C. C Thomas, 1963. Pp. 189–229.

Moruzzi, G., & Magoun, H. W. Brain stem reticular formation and activation of the EEG. *Electroencephalography and Clinical Neurophysiology,* 1949, **1**, 455–473.

Mountcastle, V. B., Davis, P., & Berman, A. L. Response properties of neurones of cat's somato-sensory cortex to peripheral stimuli. *Journal of Neurophysiology,* 1957, **20,** 373–407.

Murata, K., Cramer, H., & Bach-y-Rita, P. Neuronal convergence of noxious, acoustic and visual cortex of the cat. *Journal of Neurophysiology,* 1965, **28,** 1223–1239.

Nadvodnyuk, A. I. *An electrophysiological investigation of neurons and interactions among them in the subesophageal ganglion of a snail. (Helix pomatia).* Candidate Thesis, Kisinev, 1967. (In Russian)

Narikashvili, S. P., Arutyunov, V. S., & Moniava, E. S. The influence of reticular impulses on the activity of single neurons in the visual cortex. *Pavlov Journal of Higher Nervous Activity,* 1965, **15,** 1004–1010. (In Russian)

Negishi, K., Bravo, M. C., & Verzeano, M. The action of convulsants on neuronal and gross wave activity in the thalamus and the cortex. In R. Hernandez-Peon (Ed.), *The Physiological basis of mental activity* (Proceedings of a symposium, Mexico City, October, 1961). *Electroencephalography and Clinical Neurophysiology,* 1963, Suppl. 24, **15,** 90–96.

Negishi, K., & Verzeano, M. Recordings with multiple microelectrodes from the lateral geniculate and the visual cortex of the cat. In *The visual system: Neurophysiology and psychophysics* (Freiburg Symposium). Berlin: Springer-Verlag, 1961. Pp. 288–295.

Nelson, P. G., Erulkar, S. D., & Bryan, J. S. Responses of units of the inferior colliculus to time-varying acoustic stimuli. *Journal of Neurophysiology,* 1966, **29,** 834–860.

Okuda, O. On the relationship between the electrocorticogram and pyramidal tract cell activity of the cat. *Journal of the Physiological Society of Japan,* 1963, **5**, 236–253.

Okuda, Y. Subcortical structures controlling lateral geniculate transmission. *Tohoku Journal of Experimental Medicine,* 1962, **76**, 350–364.

Okudzhava, V. M. Discussion of the speech by Fernandez-Guardiola and co-authors. In *Reflexes of the brain.* Moscow: Science Press, 465–466.

Pavlov, I. P. *Lectures on conditioned reflexes* (translated by W. H. Gantt). New York: International Publishers, 1928.

Pavlov, I. P. *Complete Works.* Vol. 4. Moscow and Leningrad: USSR Acad. Sci. Press, 1947–1949. Pp. 144–162. (In Russian)

Pearlman, A. L. Evoked potentials of rabbit visual cortex: Relationship between a slow negative potential and excitability cycle. *Electroencephalography and Clinical Neurophysiology,* 1963, **15**, 426–434.

Penfield, W., & Milner, B. Memory deficit produced by bilateral lesions in the hippocampal zone. *AMA Archives of Neurological Psychiatry,* 1958, **79**, 475–497.

Perkel, D. H., Moore, G. P., & Segundo, J. P. Continuous-time simulation of ganglion nerve cells in *Aplysia. Biomedical Science and Instrumentation,* 1963, **1**, 347–357.

Polyansky, V. B. Excitatory potentials of the visual cortex of the rabbit and their relations to the impulse activity of neurons. Thesis for a candidate's degree. (In Russian)

Polyansky, V. B. Distribution of evoked potentials in rabbit's cerebral cortex elicited by photic stimuli of different brightness. *Pavlov Journal of Higher Nervous Activity,* 1963, **13**(2). (In Russian)

Polyansky, V. B. On the relation between spike discharges and evoked potentials in the visual cortex of the alert rabbit. *Pavlov Journal of Higher Nervous Activity,* 1965, **15**(5), 903–910. (In Russian)

Polyansky, V. B. Cycles of excitability of visual cortical neurons in the alert rabbit in response to double flashes. *Pavlov Journal of Higher Nervous Activity,* 1967, **17**(4) 714–721. (In Russian)

Polyantsev, V. A., & Serbinenko, M. V. Studies of excitability in the cerebral cortex by the method of evoked potentials. *Works of the Institute of Normal and Pathological Physiology of Academy of Medical Sciences, USSR,* 1962, **6**, 8–10. (In Russian)

Purpura, D. P. Nature of electrocortical potentials and synaptic organizations in cerebral and cerebellar cortex. *International Reviews of Neurobiology,* 1959, **1**, 47–163.

Ramon y Cajal, S. Structure of Ammon's Horn and the fascia dentada *Anales de la Sociedad Espanola de Historia Natural,* 1893, **22** (pages not available). (In Spanish)

Redding, F. K. Modification of sensory cortical evoked potentials by hippocampal stimulation. *Electroencephalography and Clinical Neurophysiology,* 1967, **22**, 74–83.

Renshaw, B., Forbes, A., & Morison, B. K. Activity of isocortex and hippocampus: Electrical study with microelectrodes. *Journal of Neurophysiology,* 1940, **3**, 74–105.

Ricci, G., Doane, B., & Jasper, H. H. Microelectrode studies of conditioning: Technique and preliminary results. *Proceedings of the Fourth International Congress of Neuronal Sciences, Brussels, 1957,* 401–415.

Rose, J. E., Brugge, J. F., Anderson, D. J., & Hind, J. E. Phase-locked response to low frequency tones in single auditory nerve fibers of the squirrel monkey. *Journal of Neurophysiology,* 1967, **30**, 769–793.

Rutledge, L. T., & Duncan, J. A. Extracellular recording of converging input on cortical neurons using a flexible microelectrode. *Nature,* 1966, **210**, 737–739.

Saharov, D. A. A functional organization of giant neurons in *Mollusca. Advances in Contemporary Biology,* 1965, **60**, 365–383. (In Russian)

Saito, Y., Maekawao, K. Takenaka, S. & Kasamatzu, A. Single cortical unit activity during EEG arousal. Proceedings of Sixth Annual Meeting of Japan EEG Society. *Electroencephalography and Clinical Neurophysiology,* 1957, Suppl. 9, 95–98.

Saltzberg, B., & Burch, N. R. A new approach to signal analysis in electroencephalography. *I.R.E. Transactions on Medial Electronics,* 1957, 24.

Saltzberg, B., & Borch, N. R. A rapidly convergent orthogonal representation for EEG time series and related methods of automatic analysis. *I.R.E. Wescon Convention Record, Part B,* 1959, 35–43.

Samson, M., Samson-Dollfus, D., & Pinchan, S. Coupled electroretinographic and electroencephalographic recording. *Electroencephalography and Clinical Neurophysiology,* 1959, **11,** 387.

Sawa, M., & Delgado, J. M. R. Amygdala unitary activity in the unrestrained cat. *Electroencephalography and Clinical Neurophysiology,* 1963, **15,** 637–650.

Scheibel, M. E., & Scheibel, A. B. The response of reticular units to repetitive stimuli. *Archivio Italiano di Biologica,* 1965, **103,** 279–299.

Scheibel, M. E., & Scheibel, A. B. Patterns of organization in specific and nonspecific thalamic fields. In D. P. Purpura & M. D. Yahr (Eds.), *Tha thalamus.* New York: Columbia University Press, 1966. Pp. 13–46.

Sedgwick, E. M., & Williams, T. D. The response of single units in the caudate nucleus to peripheral stimulation. *Journal of Physiology (London),* 1967, **189,** 281–298.

Sepp, E. K. *Story of the development of the nervous system in vertebrates.* Moscow: Medical Press, 1959. (In Russian)

Sharpless, S. K. Reorganization of function in the nervous system: Use and disuse. *Annual Review of Physiology,* 1964, **26,** 357–388.

Sharpless, S. K., & Jasper, H. H. Habituation to the arousal reaction. *Brain,* 1956, **79,** 655–680.

Shkolnik-Yarros, E. G. Some interneuronal connections in the cerebral cortex. *Pavlov Journal of Higher Nervous Activity,* 1965, **15,** 1063–1071. (In Russian)

Shulgina, G. I. The investigation of the activity of cortical neurons during initial stages of the elaboration of the conditional reflex. In *Contemporary problems of the electrophysiology of the central nervous system.* Moscow: Science Press, 1967. Pp. 296–308. (In Russian)

Shumilina, A. I. Experimental analysis of systematic changes in unit activity in cortex and subcortical structures. In *Contemporary problems of physiology and pathology of nervous system.* Moscow: Medical Press, 1965. Pp. 240–256.

Simons, L. A., Dunlop, C. W., Webster, W. R., & Aitkin, L. M. Acoustic habituation in cats as a function of stimulus rate and the role of temporal conditioning of the middle ear muscles. *Electroencephalography and Clinical Neurophysiology,* 1966, **20,** 485–493.

Skrebitsky, V. G. Microelectrode investigation of the activity of visual neurons in the waking rabbit during the orienting reaction. *Biological and physiological problems of psychology* (Abstracts of Communications Eighteenth International Congress of Psychology). Moscow: Science Press, 1966. Pp. 169–170. (In Russian)

Skrebitsky, V. G., & Bomstein, O. S. Modulation of the neuronal activity in the visual cortex of an unanesthetized rabbit by means of different nonvisual stimuli. *Physiological Journal,* 1967, **53,** 129–137. (In Russian)

Skrebitsky, V. G., & Gapich, L. I. Microelectrode study of the extinction of the response to sound stimuli in the neurons of the visual cortex of the unanesthetized rabbit. *Physiological Journal,* 1963, **53,** 906–914. (In Russian)

Skrebitsky, V. G., & Shkolnik-Yarros, E. G. Functions and structures of the visual analyzer. Symposium on Visual and auditory analyzers, Moscow, 1967. Pp. 101–105. (In Russian)

Skrebitsky, V. G., & Voronin, L. L. Intracellular records of single unit activity of the visual cortex in non-anaesthetized rabbits. *Pavlov Journal of Higher Nervous Activity,* 1966, **16,** 864–873. (In Russian)

Slipjer, E. J. *Whales.* New York: Basic Books, 1962.

Sokolov, E. N. The modelling properties of the nervous system. In *Cybernetics, thought, and life.* Moscow: Idea Publishers. Pp. 242–279. (In Russian)

Sokolov, E. N. (Ed.) *Orientation reflexes and problems in higher nervous activity, normal and pathological.* Moscow: Acad. Pedag. Nauk RSFSR Inst. Defektol., 1959. (In Russian)

Sokolov, E. N. The neural model of the stimulus and the orienting reflex. *Problems in Psychology,* 1960, **4**, 61–72.

Sokolov, E. N. Higher nervous functions: The orienting reflex. *Annual Review of Physiology,* 1963, **25**, 545–580. (a) (In Russian)

Sokolov, E. N. The modelling processes in the central nervous system of animal and man. *Gagra Symposium,* 1963, **4**, 183–202. (b) (In Russian)

Sokolov, E. N. The orienting response as a cybernetic system. *Pavlov Journal of Higher Nervous Activity,* 1963, **13**(5), 816–830. (c) (In Russian)

Sokolov, E. N. *Perception and the conditioned reflex.* Oxford and New York: Pergamon Press, 1963. (d).

Sokolov, E. N. (Ed.) *The orienting reflex and problems in normal and pathological reception.* Moscow: Prosveshchenic, 1964. (In Russian)

Sokolov, E. N. Inhibitory conditioned reflex at the single unit level. *Proceedings of the Twenty-Third International Congress in Physiology,* 1965, **4**, 340–343.

Sokolov, E. N. Neuronal mechanisms of the orienting reflex. In *Orienting reflex, alertness and attention* (Eighteenth International Congress of Psychological Sciences). Moscow, 1966. Pp. 31–36. (In Russian)

Sokolov, E. N. An investigation of memory at the level of the single neuron. *Pavlov Journal of Higher Nervous Activity,* 1967, **17**(5), 909–924. (In Russian)

Sokolov, E. N., Arakelov, G. G., & Levinson, L. B. Parallel inhibition in spontaneously discharging neurons in ganglia of the mollusc *Limnaea stagnalis.* In E. M. Kreps (Ed.) *Evolutionary neurophysiology and neurochemistry.* Leningrad: Science Press, 1967. Suppl. Pp. 3–11. (In Russian)

Sokolov, E. N., Arakelov, G. G., & Pakula, A. An adaptation of the pacemaker neurons in the visceral ganglion of a snail *Limnaea stagnalis* to the penetrated microelectrodes. In V. V. Parin (Ed.), *Systems organization of the physiological functions.* Moscow: Medical Press, 1969. Pp. 65–74. (In Russian)

Sokolov, E. N., Chelidze, L. R., & Korzh, N. N. Control involving prediction in human motor and electroencephalographic reactions. *Pavlov Journal of Higher Nervous Activity,* 1969, **19**(1), 83–89. (In Russian)

Sokolov, E. N., Pakula, A., & Arakelov, G. G. The after effects due to intracellular electric stimulation of the giant neuron "A" in the left parietal ganglion of the mollusc *Limnaea stagnalis.* In K. H. Pribram & D. E. Broadbent (Eds.), *Biology of memory.* New York, London: Academic Press, 1970. Pp. 175–190.

Sokolov, E. N., Vinogradova, O. S., & Paramonova, N. P. Analysis of physiology of learning in hard-of-hearing children. In E. N. Sokolov (Ed.), *Orienting reflex and sensory processes in normal state and pathology.* Moscow, 1964. Pp. 21. (In Russian)

Sokolova, A. A. Microelectrode investigation of the arousal reaction as a response to various stimuli. *Biological and physiological problems of psychology* (Abstracts of Communications of Eighteenth International Congress of Psychology). Moscow: Science Press. 1966. Pp. 168. (In Russian)

Sokolova, A. A., & Lipenetskaya, T. D. Microelectrode study of the motor cortex in an anesthetized rabbit. *Pavlov Journal of Higher Nervous Activity,* 1966, **16**, 1055–1063. (In Russian)

Spehlmann, R. The averaged electrical responses to diffuse and to patterned light in the human. *Electroencephalography and Clinical Neurophysiology,* 1965, **19**(6), 560–569.

Spehlmann, R., Creutzfeldt, O. D., & Jung, R. Neuronale Hemmung im motorischen Cortex nach elektrischer Reizung des Caudatum. *Archiv. der Psychiatrie der Nervenkrankeit,* 1960, **201**,332–354.

Spencer, W. A., & Kandel, E. R. Hippocampal neuron responses to selective activation of recurrent collaterals of hippocampofugal axons. *Experimental Neurology,* 1965, **4**, 149–161.

Spiegel, E. A., & Szekely, E. G. Prolonged stimulation of the head of the caudate nucleus. *AMA Archives of Neurology,* 1961, **4**, 55–65.

Spinelli, C., Pribram, K., & Weingarten, M. Centrifugal optic nerve responses evoked by auditory and somatic stimulation. *Experimental Neurology,* 1965, **12**, 303–319.

Steriade, M., & Demetrescu, M. Unspecific system of inhibition and facilitation of potentials evoked by intermittent light. *Journal of Neurophysiology,* 1960, **23**, 602–617.

Strumwasser, F. Post-synaptic inhibition and excitation produced by different branches of a single neuron and the common transmitter involved. *Proceedings of the International Union of Physiological Sciences* (Twenty-second International Congress, Leiden, 1962). Vol. II. Amsterdam: Excerpta Medica Foundation International Congress Series, No. 48. Abstract No. 801.

Suga, N. Single unit activity in cochlear nucleus and inferior colliculus of echolocating bats. *Journal of Physiology,* 1964, **172**, 449–474.

Suga, N. Analysis of frequency-modulated sounds by auditory neurons of echolocating bats. *Journal of Physiology,* 1965, **179**, 20–53.

Supin, A. J. Some mechanisms of the rhythmicity of electrical activity of the cerebral cortex. *Pavlov Journal of Higher Nervous Activity,* 1967, **17**(2), 287–294. (In Russian)

Svidersky, L., & Karlov, A. A. Rhythm generator (pacemaker) in motor center of the cicada *Lyristes plebejus.* In E. M. Kreps (Ed.), *Evolutionary neurophysiology and neurochemistry.* Leningrad: Science Press, 1967. Pp. 45–53. (In Russian)

Szentagothai, J. Synaptic architecture of the spinal motoneuron pool. *Electroencephalography and Clinical Neurophysiology,* 1967, Suppl. 25 ("Recent Advances in Clinical Neurophysiology"), 4–29.

Tarlov, E. C., & Moore, R. I. The tecto-thalamic connections in the brain of the rabbit. *Journal of Comparative Neurology,* 1966, **126**, 403–422.

Tauc, L. The activity of the mollusc neuron. *Endeavour,* 1966, **25**, 39–44.

Tauc, L., & Hughes, G. M. Modes of initiation and propagation of spikes in the branching axons of molluscan central neurons. *Journal of General Physiology,* 1963, **46**, 533–549.

Tissot, P., & Monnier, A. M. Dualite du systeme thalamique de projection diffuse. *Electroencephalography and Clinical Neurophysiology,* 1959, Suppl. 11, 675–686.

Tyc-Dumont, S. Contribution àl'étude du controle d'origine réticulaire des intégrations sensori-motrices. Thèse, Paris, 1964.

Vardapetian, G. A. Dynamic classification of single unit responses in the auditory cortex of the cat. *Pavlov Journal of Higher Nervous Activity,* 1967, **17**(1), 95–103. (In Russian)

Vasilevsky, N. N. Correlation between the background activity of neurons in the somatosensory cortex and their functional organization. *Physiological Journal,* 1965, **51**, 711–716. (In Russian)

Verzeano, M. Sequential activity of cerebral neurons. *Archives Internationales de Physiologie et de Biochimie,* 1955, **63**, 458–476.

Verzeano, M. Activity of cerebral neurons in the transition from wakefulness to sleep. *Science,* 1956, **124**, 366–367.

Verzeano, M., & Calma, I. Unit-activity in spindle bursts. *Journal of Neurophysiology,* 1954, **17**, 417–428.

Verzeano, M., & Negishi, K. Neuronal activity in cortical and thalamic networks. *Journal of General Physiology,* **43**(suppl.), 177–195.

Vigouroux, R. P. Physiologie du rhinencephale non-olfactif. Unpublished manuscript, Université d'Aix-Marseille, 1959.

Vinogradova, O. S. Habituation of neuronal reactions in different brain structures with special reference for hippocampus. *Proceedings of the 18th International Psychology Congress,* Moscow, 1966, 5, 55.

Viktorov, I. V. *Neuronal structure and interneuronal connections of the fourth layer of the frontal cortex in certain mammals and man.* Dissertation, Moscow, 1966. (In Russian)

Vinogradova, O. Inhibitory conditioned reflex at single unit level. *Proceedings of the Twenty-third International Congress of Physiology,* 1965. **4**, 340–343.

Vinogradova, O. S., & Lindsley, D. F. The extinction of responses to sensory stimuli in the single units of the visual cortex of the unanaesthetized rabbit. *Pavlov Journal of Higher Nervous Activity,* 1963, **13**(2), 207–217. (In Russian)

Vinogradova, O. S., & Sokolov, E. N. The orienting reflex and stimulus intensity. *Problems in Psychology,* 1955, **2.** (In Russian)

von Euler, C. Excitatory and inhibitory mechanisms in hippocampus. In D. B. Tower & J. P. Schade (Eds.), *Structure and function of the cerebral cortex* (Second International Meeting of Neurobiologists, 1959). Amsterdam: Elsevier, 1960. Pp. 272–277.

von Euler, C., & Green, J. D. Excitation, inhibition and rhythmical activity in hippocampal pyramidal cells in rabbit. *Acta Physiolica Scandinavica,* 1960, **48,** 110–125.

Voronin, L. G. The influence of extracellular polarization of neurons in rabbit's sensorimotor cortex on their evoked activity. *Pavlov Journal of Higher Nervous Activity,* 1966, **16,** 667–677. (In Russian)

Voronin, L. G., & Kotlar, B. I. Bioelectrical activity of some brain structures during elaboration and extinction of the food-getting reflex. *Pavlov Journal of Higher Nervous Activity,* 1962, **12,** 547–555. (In Russian)

Voronin, L. G., & Skrebitsky, V. G. Extra- and intracellular investigation of neuronal responses to acoustic and photic stimuli in the motor cortex of nonanaesthetized rabbits. *Pavlov Journal of Higher Nervous Activity,* 1967, **17,** 523–533. (In Russian)

Waziri, K., Frazier, W. T., and Kandel, E. R. Prolonged alterations in the efficacy of inhibitory synaptic transmission in *Aplysia californica. Proceedings of the Twenty-third International Congress on Physiological Sciences,* Tokyo, 1965. Amsterdam: *Excerpta Medica Found.,* 1965. Abstract No. 910.

Webster, W. R., Dunlop, C. W., Simons, L. A., & Aitkin, L. M. Auditory habituation: A test of a centrifugal and a peripheral theory. *Science,* 1965, **148,** 654-656.

Whitlock, A., Arduini, A., & Moruzzi, G. Microelectrode analysis of pyramidal system during transition from sleep to wakefulness. *Journal of Neurophysiology,* 1953, **16,** 414–429.

Wilson, P., Pecci-Saavedra, J., & Doty, R. Extraoptic influences upon transmission in primate lateral geniculate nucleus. *Fed. Proceed.,* 1965, **24,** 206. (Abstract)

Yarmizina, A. L., Sokolov, E. N., & Arakelov, G. G. Identification of neurons of the left parietal ganglion of *Limnaea stagnalis. Cytology,* 1968, **10,** 1384–1389. (In Russian).

Yarmizina, A., Sokolov, E. N., & Arakelov, G. G. Identification of neurons of the left parietal ganglion of *Limnaea stagnalis. Neurosciences Translations,* 1969–70, #10, 119–127. (Originally appeared in Russian in *Tsitologiya,* 1968, **10,** 1384–1389.)

Subject Index